TRIVENT

PUBLISHING

TRIVENT
Ethics in Science & Technology

The Bioethics of the "Crazy Ape"

Edited by

Oguz KELEMEN, Gergely TARI

Applied Ethics: From Bioethics to Environmental Ethics
Volume II

The Bioethics of the "Crazy Ape"
Ed. Oguz Kelemen, Gergely Tari
ISBN 978-615-6405-01-2
Applied Ethics: From Bioethics to Environmental Ethics, Volume II
Series ed. János I. Tóth
Responsible publisher: Teodora C. Artimon
ISSN 2630-8185
http://trivent-publishing.eu/ethics.html

First published in 2019 by Trivent Publishing

Trivent Publishing
1119 Budapest, Etele út 59–61
Hungary

For more information, visit our website: http://trivent-publishing.eu

TABLE OF CONTENTS

Introduction

Why the "Crazy Ape"?

The Second International Conference on Bioethics in the New Age of Science (BNAS) was held at the University of Szeged (Hungary) on the 10th and 11th of May 2018. The University of Szeged is the alma mater of the Nobel Prize winner and scientist Albert Szent-Györgyi, well known for his pure scientific work and his discovery of Vitamin C, but also for being a humanist thinker. Already a recognised scientist in 1957, he wrote an article in *Science* entitled "Science, Ethics and Politics" [1] discussing how far should science be allowed to go in order to support politics, while commenting on the line beyond which scientists should stand up for their own ethical beliefs. As he was examining the correlation between science, morality, and politics, he should be considered one of the first Hungarian unofficial bioethicists. His recognition of the importance of ethics in education supports this argument: "we need an educational system based on the real understanding of moral, aesthetics, and spiritual values" [2] (p. 28).

In the following decades, he wrote and debated on the responsibility of science within the cold war, and whether science could or should provide any solutions for the war. One of his most famous books is entitled *The Crazy Ape*, which also inspired the title of this volume. In this short book, he especially wanted to shape the moral perspective of the world's youth. During the Vietnam War, under the shadow of the Cold War, and the threat of a possible nuclear crisis, he discussed the fundamental question of the technological progress of mankind. Are we more than technologized apes, who now have the potential to destroy the environment? "Will mankind be able to survive the machinations of present-day men who often appear to act more like crazy apes than sane human beings?" [2] (p. 4). Although this subject may seem old fashioned (and his answers on this sometimes seem a bit idealistic today), the questions raised by him still remain fundamental at the gate of the post-human world.

Contributions to the book

This book is mostly composed of selected papers from the Second International Conference in the New Age of Science (BNAS), with some contributions from additional scholars. It is divided four main parts: (1) Ethical Challenges in Enhancement Technologies; (2) Bioethics, Anthropology, Psychology, and Philosophy; (3) Globalization and Environmental Ethics; and (4) Ethical and Legal Challenges in Medicine and Research.

The book opens with a colourful article by Stefan Lorenz Sorgner ("Dignity of Apes, Humans and AI"), which also echoes the title of the book: the dignity of (the crazy) apes. Half a century ago, Albert Szent-Györgyi was focusing on the potential problem of technology in the hand of the evolutionary- (but not morally-) selected creature; by now, society is facing human-created technology (transhumans, AI) which will possibly overcome the age of apes.

In the second contribution in the volume, Vivien Szútor continues the investigation into human dignity in relation to the rapidly-evolving genetic engineering, mainly focusing on legal aspects. The special emergent point of this essay is the discussion on therapeutic cloning.

The second part of the volume gathers papers which investigate a series of bioethical questions from a broader anthropological and psychological perspective.

Attila Dobos, in his paper "Please geben vôtre consentimiento! - Informed Decision-Making in Intercultural Context" stresses the intercultural and linguistic context of moral decision-making processes. Furthermore, the author also discusses this thought provoking-issue in the context of medical informed consent.

Daniela Reisz and Alexandra Anghel introduce a challenging discussion, arguing that information (especially information provided by the most sophisticated technologies such as MRI) can create an unexpected side effect: fear. The authors thus mainly debate on autonomy and the beneficence principles in the modern medical settings. The paper includes many case reports which portray the shortcomings of excessive neuroimaging. Based on Kantian philosophy, the authors support the position that knowledge is beneficial only if one has the possibility to cope with it.

Florin Lobonţ, in his paper "Children's Bioethics, Theory of Attachment and P4C," continues the discussion on a similarly hot topic: the autonomy of children and how it could be incorporated in legal and ethical frameworks. The most engaging point of his argumentation is that it tries to extend the attachment theory to the socio-cultural level.

Lobonţ's contribution is followed by a stimulating political perspective in which Mihai Murariu highlights the possibility of a society needing simplifying solutions when it becomes complex. The author reasons that the amplification of insecurity within globalised societies opens the path towards new totalitarian movements.

Further on, the ever perplexing debate of the psy-complex (psychology, psychotherapy, psychiatry) is presented by Attila Bánfalvi. In his paper "Psy-complex: Out of the Techno-Scientific Paradigm?", the author investigates the basis of this debate, namely the ontologically different types of body and psyche perspectives.

The last contribution to this part belongs to Sorin Grigore Vulcănescu who reviews intercultural differences and solutions regarding end-of-life decisions in a board summary of the contemporary legal and ethical challenges of euthanasia. The study incorporates the most recent legislative changes worldwide, highlighting that this is rapidly-changing field and that there is a global tendency in the process of the legalisation of euthanasia.

The third part of the volume is comprised of studies which investigate ethical issues regarding globalisation and the environment.

Ferenc Hérány investigates environmental ethics in relation to Christian anthropocentrism ("A Critique of the Environmental Ethical Critique of the Christian Anthropocentrism"), while Dejan Donev steps away from anthropocentric ethics and discusses ecocentrism, a philosophical approach which considers all elements of the biotic community as integral parts of the system.

Matthew J. Crippen and John Salevurakis then investigate the ethical, political, and economical perspectives of animal hunting in Southern Africa. Following a utilitarian perspective, they put aside the usual prejudices, and expose an uncommon perspective on hunting, arguing that it can be advantageous for both preserving wildlife and for the local population.

This part is closed by Ivica Kelam who discusses the historical development and complex ethical debate on genetically-modified food, emphasizing the controversial interests and forces behind the development of complex regulations on genetically-modified food.

The final part of the volume is dedicated to healthcare. Anamika Krishnan investigates the applicability of the UDBHR's global bioethical guideline in India, a country distorted by bioethical issues. The paper presents the actual discrepancy between the legal system and UDBHR in India.

Further on, Kartina Aisha Choong and Mahmood Chandia discuss the bioethical issues surrounding patients in a permanent vegetative state especially those with different religious heritages. The authors highlight vital theological questions on consciousness and the meaning of life, while also debating on practical matters related to the issue, such as secular law and financial outcomes.

Narine Harutyunyan, in her study on reproductive autonomy and genetic technologies, covers a highly debatable topic which originates in advanced medical technologies. Should preimplantation sex-selection for obviously non-medical reasons be allowed? The current demographical crisis in China should warn us that finding the balance between the most basic bioethical principles (autonomy and non-maleficence) has never been as difficult as in today's technologized era of reproductive medicine.

Gergely Tari, Csaba Hamvai and András Perbíró make a short overview of the medicalization of childbirth, while presenting their own field investigation in Hungary. The paper includes the most relevant findings of this research focusing on the non-medical causes of negative childbirth experiences, claiming respect for reproductive autonomy.

The research is followed by the study authored by Csaba Hamvai, Gergely Tari and Melinda Csenki which deals with the issue of prophylactic mastectomy as demanded by the patient. The authors outline the most important legal and bioethical assumptions associated with prophylactic mastectomy on demand, illustrating them through a case study of a clinical case. They argue for the necessity of shared decision-making between doctor and patient and, based on this, they recommend several guidelines

Engaging in the field of medical education, Coralia Cotoraci, Alciona Sasu, Mircea Onel, and Cristina Ghib-Para investigate truth disclosure in medical settings. They try to summarise medical students' understandings of truth disclosure while presenting the differences between students' perceptions based on their various cultural backgrounds.

This study is followed by a lucid overview of the debate on advertisements in pharmaceutical marketing. Miroslav Radenkovic et al. argue that doctors should always consider evidence-based medical facts and medical scientific knowledge in order to avoid prescribing specific drugs uncritically.

The volume ends with a coherent study by Adél Tóth, who discusses whether researchers in the biomedical field should disclose research results to participants of studies on human subjects. The author provides a balanced overview of the pros and cons of the different approaches, and also summarises the common points of the different approaches.

Final remarks

The Bioethics of the "Crazy Ape" collects a wide range of bioethical topics. Bioethical questions are eternal by nature, although our technologized times transform old issues in forms never before experienced. Just like the famous scientist Albert Szent-Györgyi believed in his time, we also believe that all the contributing authors recognised their moral responsibility in adding new approaches to the continuum of each debate. Although this responsibility has became increasingly complex, we must avoid to become barriers of the scientific development. Bioethics as an applied field of philosophy should always try to establish a framework for a sustainable world: in daily clinical practice, in cases of human experiments, and (not the least) in the natural environment.

[1] Albert Szent-Györgyi, "Science, Ethics and Politics" *Science*, New Series, vol. 125, no. 3241, pp. 225-226, 1957.
[2] Albert Szent-Györgyi, *The Crazy Ape*. New York: Philosophical Library, 1970.

Oguz Kelemen and Gergely Tari
University of Szeged, Hungary

PART I

Ethical Challenges in Enhancement Technologies

The Dignity of Apes, Humans, and AI

Stefan Lorenz Sorgner
John Cabot University in Rome, Italy

Abstract:
Only humans used to be granted personhood. Given this understanding of personhood, human embryos, who possess neither a brain nor a nervous system and consequently cannot even suffer, deserve more respect than adult chimpanzees who possess self-consciousness and sentience. This is not plausible. A new foundation for personhood is needed. My counter-suggestion is that we need an empirical way of analysing the intensity of suffering of various entities. These reflections also show reasons why a certain type of suffering, namely non-conscious cognitive pain, can also be attributed to a sufficiently developed AI, which is not yet presently available.

Keywords:
Personhood; dignity; Peter Singer; AI; animals; anthropocentrism; speciesism.

The Earth was created about 4-5 billion years ago. But whether life was created on earth or on another celestial body and it afterwards reached Earth is an open question. The problem is that there was only inanimate matter at first (water, stone, gas) and then suddenly a moment came when life arose together with entities capable of self-movement. At first, everything was determined solely by causal processes. Suddenly, however, certain classes contained a principle of order by which they could develop. Scientists often speak of a primordial environment populated by water, gases, and electric lightning, which gave rise to the simplest forms of life.

In Ancient philosophy, all living entities were described as animated. In Plato's work, the soul was not yet identified exclusively with the soul of reason; animals and plants were also given certain types of souls. Everything that could move by itself had a soul and was thus alive.

However, how life could arise from (apparently) inanimate matter is astonishing – just as the fact that something is there and not nothing is confusing. Furthermore, the situation became more complicated, as these simple life forms have evolved and become more complex. Some of them even developed a special ability for emotions, perceptions, abstraction, spoken language, written language, artistic and numerical forms of communication, reason, and creativity. How could such abilities develop from water, electricity, and gases? In addition, awareness has developed in the course of these processes, i.e. the ability to feel pain and to perceive and react to the environment. Finally, living beings have developed as beings who had an understanding of time, of the past and the future, and who recognized themselves in the mirror, i.e. possess self-consciousness. At present, in addition to humans, nine other animal species are known (great apes, dolphins, elephants…), of which selected members have passed the mirror test. How could this develop on the basis of water, gases, and electricity alone?

Something must have been added from the outside, such as a divine spark: the philosophers of western cultural history (decisively influenced by Plato and then developed further through the Stoics, Descartes, and Kant) predominantly argued that both the immaterial spirit and the gift of free would be able to explain these abilities. They usually adhered to the primacy of a non-empirical understanding of reason. This idea is still anchored and valid in German law today. Animals are not things but should legally be treated as things. Only human beings possess dignity [1].

These considerations have been shaping our thoughts, actions, and the culture of the western world since Plato. However, some of these considerations exist only as relics and encrusted structures of traditional considerations. This tradition began to crumble several hundred years ago. When Darwin formulated the theory of evolution and Nietzsche proclaimed the death of God, the cultural movement away from dualistic ontological thinking gained enormous importance. Instead of assuming that something would be added from the outside in the processes described above, it is now assumed that all these developments have happened on their own as emergent processes without any magical connections between material and non-empirically accessible entities. Transhumanism also rose from the tradition of evolutionary and naturalistic thinking. This is an important reason why the conservative American political scientist and cultural critic Francis Fukuyama calls transhumanism the "most dangerous idea in the world" [2] (p. 42-43). The term transhumanism was also coined in the context of the emergence of evolutionary thought. The term was described by Julian Huxley in an article published in 1951:

> Such a broad philosophy might perhaps best be called, not Humanism, because that has certain unsatisfactorily connotations, but Transhumanism. It is the idea of humanity attempting to overcome its limitations and to arrive at fuller fruition; it is the realization that both individual and social developments are processes of self-transformation. [3] (p. 139)

I think this definition is still correct. Furthermore, Huxley also coined the term "Evolutionary Humanism." The relationship between evolutionary humanism and transhumanism represented today by the Giordano Bruno Foundation needs further clarification [4]. There are tensions between Julian Huxley's technical considerations represented in his numerous publications and those of his brother Aldous Huxley, the author of the technology-critical novel "Brave New World." He also shares the fundamental evolutionary thinking with his grandfather Thomas Henry Huxley, who stood out as Darwin's supporter, Darwin's "bulldog." Julian Huxley's half-brother, Andrew Fielding Huxley, was also active as a biologist. He was a university professor in London and even won the Nobel Prize, but is currently much less well known than the other family members already mentioned. Julian Huxley was a university professor in London too, while he was also the first UNESCO Secretary-General who made a significant contribution to the first Declaration of Human Rights, and the director of the British Eugenics Society.

The close connection between transhumanism and genetic engineering also exists in contemporary transhumanism. Furthermore, the focus on cyborg techniques and artificial intelligence has been added, because with the help of all these techniques the previous limits of our humanity can be transcended. This objective is relevant for many reasons. The central assumption is that the probability of leading a good life is increased, if we develop further through training, practice, but also with the help of other techniques. Furthermore, this approach is important, since all life is always threatened by the possibility of extinction. Life has evolved evolutionarily. Just as dinosaurs are extinct, so could human beings be. Extinction depends on how well we are adapted to our environment. However, as our environment is constantly changing, there is a perpetual need for change for us too. If we develop techniques which can help us, we can increase the probability of survival.

The two most promising techniques are genetic engineering and artificial intelligence. With the help of different genetic techniques, especially CRISPR/Cas9, people could develop in a carbon-based way, i.e. people could become organic trans- or posthumans, whereby the posthuman either still belongs to the human species, but has at least one characteristic which goes beyond the limits of current human capacities, or can become a representative of a new species. The possible alteration processes using cyborg techniques and AI are even more radical. They involve an intensified fusion of brain-computer interfaces until the development goes so far that mind uploading becomes an option and we can store our personality on a hard disk, which according to the Google futurist Ray Kurzweil should already possible in a few decades [5] [6]. In this case, the posthuman would no longer be a carbon-based being but a silicon-based one. The fact that a transmission of our personality is conceivable, at least in principle, becomes clear simply by the fact that all the cells of our body renew themselves every seven years. Nevertheless, we remain a continuous unit. However, whether what exists on a carbon basis can be transferred to a silicon basis is quite questionable. At present, we do not know of a silicon-based life form; or should self-replicating computer viruses already count as a kind of life? [7] (p. 1-54).

With this development into a silicon-based posthuman, we are facing new ethical challenges. The dominant way of assessing the moral status of entities (which is currently widespread in all parts of the world) is based on the cultural tradition which assumes that in the evolutionary development into human beings is connected to the rather decisive event of the incarnation, something that goes beyond the naturalistically-explainable evolutionary process. At this point, God's divine immaterial spark, our reason, entered into us and connected with us. This process is responsible for the fact that only we humans possess something that goes beyond the purely natural world, which is why only humans have the subject status. All human beings are subjects, persons, and bearers of dignity. All other entities are objects, non-persons, and can therefore be traded, as they possess a finite value. This categorization is both morally and legally dominant. There are only a few legal exceptions, e.g. when a court in Argentina granted the status of a person to a great ape [8].

A particular legal situation appears in Germany. People have dignity; animals are not things, but should be legally treated as things. This assessment clarifies that the idea that only human beings were given the immaterial divine spark is still legally valid. Or at least this understanding is suggested by the legal assessment.

This basic attitude is problematic in many respects. Every third German sees herself as a sceptic, naturalist, or atheist, which implies that the legally suggested understanding of animals is not shared by them. Legislation thus patronizes the majority of Germans in a morally problematic and paternalistic way. Such a situation is unacceptable in a liberal-democratic state. In my opinion, therefore, ontological implications should also be banished from the constitution and replaced by contingent-normative attitudes in order to ensure that a plurality of ontologies is lived out.

However, the question of the moral status of apes and other animals is yet to be adequately addressed. The strongest intellectual counterproposal comes from Peter Singer, who makes the following considerations regarding the concept of human dignity [9] (p. 76-77). He labels this ethical theory as speciesist. Human life is preferred in a morally illegitimate way, i.e. it is preferred even if it has no morally relevant qualities. In this context, a human embryo has human dignity, even if it does not even have a nervous system or brain and otherwise does not fulfil the prerequisites for perceiving pain. However, the property of perceiving pain should, in his view, be the basis for giving a being an elevated moral status, and the more intensively someone can feel pain, the higher the corresponding moral status should be. Out of these considerations Singer develops a counterproposal. Someone who only has consciousness lives exclusively in the present. But if a being has self-consciousness, then it perceives itself as a continuous entity, an entity that existed in the past, exists now, and will probably still exist in the future. If a being has sentience and self-consciousness,[1] then the ability to feel pain should be much higher than with beings who have consciousness only and live in the moment. At present, only a few members of nine animal species, apart from humans, have this ability. It should also be noted here that only about 65% of 2-year-old human children have the ability to recognize themselves in the mirror, which is the most important test of self-consciousness [10].

However, it may well be asked whether the mirror test is an appropriate test of self-consciousness. Dogs do not pass it. Could this be because the test is biased concerning the ability to see? Could it not be the case that for some living beings, the ability to hear or smell is more pronounced and linked to self-consciousness [11]? In any case, the consequences of Singer's ethics are enormous. If the choice was to save an elephant or a newborn with severe mental disability, then his theory would give a clear answer as to which is the morally-appropriate choice. His answer is not shared by many enlightened people, which is an important reason why I find his considerations problematic. However, his theoretical reflections are quite conclusive. So how can it be explained that the intuitive reactions to the consequences of his ethics are negative? One explanation could be that our emotions are strongly influenced by our cultural imprint, which is anthropocentric. It is also possible that our emotional response to his reflections will change, if our culture becomes less anthropocentric.

The crucial point why I am addressing this issue is that Singer's ethics do not seem to provide us with a well thought-out assessment of the moral status of AI or, for instance, of the humanoid Data from Star Trek. Let us assume that it would be possible to download the personality onto a hard disk, and self-consciousness could remain. Nevertheless, this being would not meet Singer's requirements for personal status, since (most likely) sentience would be lacking. At least at present, it is difficult to imagine that AI and robots possess emotions and can feel physical pain, since there is no organic body that seems to be necessary for such sensations. Would a person uploaded to a hard disk lose their personal status through the process of mind uploading?

[1] In the following passages, I criticize that self-consciousness ought to be seen as necessary for personhood. In my book "Schöne neue Welt" (2018) [13], I explain why sentience is not necessary for personhood either.

It is highly speculative to reflect upon an uploaded posthuman, whom we may at best know from the film *Transcendence* starring Johnny Depp. But how should we generally deal morally with computers and AI? On the basis of anthropocentric human dignity ethics, these entities would clearly be things that can be owned, destroyed, and sold. If we look at the relationship between soldiers and combat robots and between old people and care robots, which already exist to an increasing extent in East Asia, then this assessment seems too short-sighted. But on what philosophical basis could an assessment of the moral status take place? It could be argued that AIs can already pass the mirror test to some extent? Does this mean that robots with AI already have self-consciousness?

This example seems to indicate the limits of the mirror test. On the other hand, AI's with sensors only pass the test so far, if they are initially informed on how they look like. Living animals do not need this initial information. However, what we understand by self-consciousness could be nothing more than a special algorithm. In any case, it seems impossible to have physical pain without a carbon-based organism, and the ability to feel pain was decisive in giving a being a moral status based on previous considerations. But perhaps these considerations are not enough.

Avishai Margalit emphasizes the relevance of dignity in not humiliating others [12] (p. 150). What situation arises in case of humiliation? It is the situation of a relation in which one being puts itself above another and expresses contempt for the other. However, the humiliated person does not necessarily feel the physical pain he feels when he breaks his ribs. Rather, humiliation is primarily associated with the cognitive realization of not being appreciated. The process is also painful, but without having to be associated with physical pain. Perhaps it can be described as cognitive pain, which only has to be linked to cognition, but not to consciousness, if we can uncouple consciousness and pain. The possibility of decoupling seems to exist because there are signs that fetuses can feel pain without having consciousness (e.g. research indicates that from the 6th month onwards, foetuses have the physiological prerequisites for experiencing pain, but it is rather unlikely that we attribute consciousness to foetuses at this stage.Wakefulness is a post-birth phenomenon. Is pain without consciousness possible? The relationship of the central nervous system, memory, brain, consciousness, and the ability to feel pain is extraordinarily exciting).

There is also evidence that cognition seems to be possible without consciousness. One indication of this argument are the results of a selective attention test, also known as the "Invisible Gorilla." We are asked to count the number of passes of a basketball team and are then asked if we noticed anything special about the video. The kickboxing gorilla that walks through the playing field remains unnoticed by many. Nevertheless, it is in our cognitive field. Cognition and conscious perception are therefore two different phenomena. These considerations suggest that there is not only a conscious and an unconscious, but also that there may be a non-conscious cognition, as in the case of the invisible gorilla.

If both non-conscious cognition and the phenomenon of cognitive pain exist, as illustrated by the example of humiliation, then it may make sense to talk about the possibility of non-conscious cognitive pain of AI. An AI may also cognitively realize that it is not appreciated, and in this case the cognitive pain associated with humiliation may be associated with this realization process. Data from Star Trek would have to be given a moral status against the background of these considerations. Whether Data ought to receive the person or the postperson status would have to depend, among other things, on the relationship between cognitive and organic pain. I think that an empirical means of analysing the intensity of various types of pain is needed for being able to compare types of personhood. Yet, a separate article is needed to analyse these issues in more detail. In any case, I think that these traces are worth pursuing further.

References

[1] S. L. Sorgner, *Menschenwürde nach Nietzsche: Die Geschichte eines Begriffs.* Darmstadt: WBG, 2010.

[2] F. Fukuyama, "The World's Most Dangerous Ideas: Transhumanism," *Foreign Policy*, vol. 144, Sept./Oct. 2004.

[3] J. Huxley, "Knowledge, morality, and destiny—The William Alanson White Memorial Lectures," third series, *Psychiatry*, vol. 14, no. 2, pp. 127–151, 1951.

[4] S. L. Sorgner, *Transhumanismus: "Die gefährlichste Idee der Welt"!?*, Freiburg i. Br.: Herder, 2016.

[5] R. Kurzweil, *The Singularity Is Near: When Humans Transcend Biology*, London: Penguin, 2006.

[6] R. Kurzweil, T. Grossman, *Transcend. Nine Steps to Living Well Forever*, New York: Rodale Books, 2011.

[7] A. Schrauwers, B. Poolmmann, *Synthetische Biologie. Der Mensch als Schöpfer*, Berlin/Heidelberg: Springer, 2013.

[8] https://www.scientificamerican.com/article/argentina-grants-an-orangutan-human-like-rights/. Accessed: March 15th, 2019.

[9] P. Singer, *Practical Ethics*, Cambridge: Cambridge University Press, 2011.

[10] B. Amsterdam, "Mirror self-image reactions before age two," *Developmental Psychobiology*, vol. 5, no. 4, pp. 297–305, 1972.

[11] R. Cazzola Gatti, "Self-consciousness. Beyond the looking-glass and what dogs found there," *Ethology Ecology & Evolution*, vol. 28, no. 2, pp. 232–240, 2015.

[12] A. Margalit, *Politik der Würde*, Berlin: Suhrkamp, 2012.

[13] S.L. Sorgner, *Schöner Neuer Mensch*, Berlin: Nicolai Verlag, 2018.

Human Dignity in Genetic Engineering (With Some Hungarian Examples)

Vivien Szútor

Károli Gáspár University of the Reformed Church, Hungary

Abstract

The achievements of genetic engineering are rapidly expanding in everyday medical practice, as they offer solutions for diseases which have been incurable so far. However, genetic engineering also has disadvantages as it questions everything thought about humans until now, thus transforming the concept of the right of human dignity. This study searches for legal answers to the following questions: (1) how human dignity manifested? (2) in what cases may genetic engineering be harmful?

Keywords:

Genetic engineering; human dignity; self-determination; physical integrity; Oviedo Convention; human genetic examinations; life-saving siblings; cloning; pre-implantation examination; human genetic research.

I. Introduction

Around 350 BC, Aristotel already supposed that genetical information is transmitted to descendants; about two thounsand years later, in 1865, Gregor Mendel discovered these inheriting units [1] (p. 550). Since then, doctors and researchers set high hopes in genetic knowledge and technology, which may lead to curing today's uncurable hereditary diseases.

Gene technology does not only modify a human's body and define future generations, but it may also provoke changes in human rights (if people alter themselves, their rights may also be altered). Therefore, the definition of human dignity and its manifestationsmay change, as with the help of gene technology, the human body can be altered on the most basic level and with this, the entire mankind.

Whether we have the right to gene alteration is a question that cannot be answered individually. Answering such a question must comply with international standards to which countries comply and which protect human dignity. Furthermore, it is important that human dignity in case of gene editing is protected not only in individual cases, but generally worldwide.

II. The right to human dignity

A. The concept of human dignity

There are two traditional ways of interpreting human dignity. The first tradition explains human dignity as "within the doctrinal framework of the sancitity of human life" [2] (p. 40), in which humanlife is presented as God's gift. For the second tradition, human dignity is "justified by the unique abilities of normal people" [2] (p. 41). The latter implies that individuals are able to live their lives on the basis of their own moral rules; in this case, the essence of human dignity is respect.

According to Ludwig Siep, human dignity can mostly be defined in its negative meaning; that is, when human dignity is violated. For example, human dignity can be violated when an individual is seen as an object rather than a subject, but also when people are humiliated and ashamed [3] (p. 5).

B. The legal concept of the right to human dignity

In Judit Sándor's interpretation, human dignity is considered two-fold: on the one hand, it is the basis for ethical norms, a guiding principle used to define behaviour rules; while on the other hand, it is a fundamental right, included in national constitutional laws as well as in international human rights [4] (p. 47).

According to the first abortion decision of the Hungarian Constitutional Court in 1990, the right to human dignity means that "the autonomy and the self-determination of the individual has a core which prevents the disposal of others; humans are subjects, and they should not become instruments or objects" [5]. Human dignity is untouchable for every human and does not depend on what and how much people have achieved [5].

As "the practice of Hungarian constitutional court assigns the right to human dignity a general function which protects the personality" [6] (p. 104), human dignity is considered a matter of law which can be invoked before court. The right to self-identity, self-determination, and physical intergity can be derived from the general personality protection. Alongside these functions, the right to human dignity protects the equality of people's lives. The inviolability of the private sector is also protected by the right to human dignity, as the Hungarian Constitutional Court ruled in several decisions [7].

C. The appearance of gene editing in international documents

The Universal Declaration of Human Genome and Rights issued by UNESCO in 1997 rules about human dignity and the human genome in Article A. The Declaration does not just emphasize human dignity in the light of law, but it also presents several obligations for the states with regard to it. It defines the human genome as not only the symbolic inheritance of mankind, but also as the basis for the inherent dignity and diversity of humans. Furthermore, the Declaration (which does not have a binding force) establishes that every human should have their dignity and rights respected regardless of genetic attributes. It also declares that people's uniqueness and diversity must be respected on the basis of human dignity [8] (Article 1-2). The Declaration protects human diversity and takes a stand against any disadvantegeous discrimination (including genetic discrimination) which results in harm of the human dignity [8] (Article 6).

The 2003 International Declaration on Human Genetic Data (also issued by UNESCO) deals especially with the human genetic data, containing rules on its collection, application, and storage. The Human Genetic Data Declaration also refers to the issue of human dignity deeming it as a key standard in the course of any procedure involving human genetic data. Thus human dignity is one of the aims of the Declaration: it combines human dignity with the prohibition of discrimination and stigmatization by stating that human genetic data cannot be used for purposes which would differentiate one person from the other and would thus lead to harming human dignity [9].

The Universal Declaration of Bioethics and Human Rights, published by UNESCO in 2005, states that results in science and research should promote respect for human dignity. It declares that the absolute goal of science and society should primarily be human interest and well-being [10] (Article 3). Article 10 of the Bioethics Declaration deems that humans are equal in their rights and dignity, thus everbody has the right to fair and equitable treatment. Moreover, Articles 11 and 12 of the Declaration rules that although cultural diversity and pluralism must be considered, it cannot be the basis for harming human dignity and fundamental rights. Also, just like the Declaration on Human Genetic Data, the Declaration highlights that it cannot be interpreted in ways that would come against human dignity [10] (Article 28). When announcing the Declaration, the director-general of UNESCO Koichire Matsuura stated that: "By adopting this Declaration, Member States have solemnly affirmed, for the first time in the history of bioethics, the moral commitment of the international community to respect a certain number of universal principles for humanity in the development and application of science and technology, set forth within a single text" [11].

The Council of Europe issued its first binding document, the Oviedo Convention, in 1997. Judit Sándor thinks that the Convention "takes human dignity as general principle for biomedicine" [4] (p. 32). Thus, the Convention admits in the Preambulum that the document was created in order to raise awareness that the improper use of medical science and biology might harm human dignity. The unconcealed purpose of the Convention is to respect human dignity and identity in the field of biology and medicine without disctinction [12] (Article 1). It is apparent from the Explanatory Memorandum of the Convention that it only contains general principles, thus the states may decide on a wider protection for the fields of biology and medicine [12] (Article 27). The Memorandum also emphasizes that "the Convention as a whole will provide a common framework for the protection of human rights and human dignity in both longstanding and developing areas concerning the application of biology and medicine" [13] (p. 2). Furthermore, it highlights that human dignity "is at the basis of most of the values emphasised in the Convention" [13] (p. 3). The Convenction "also uses the expression 'human being' to state the necessity to protect the dignity and identity of all human beings. It was acknowledged that it was a generally accepted principle that human dignity and the identity of the human being had to be respected as soon as life began" [13] (p. 4).

Therefore, it is apparent that all these documents consider human dignity as basic and general principle. The right to human dignity embodies everything which can make humans human (subjects) and which do not degrade humans to an objects. Because of dignity, not only are all people are equal but so is the human genome, which can be harmed if biology or medicine use it wrongfully. Because of this, the freedom of research must be harmonized with human dignity. This is the reason why discrimination and stigmatization on the basis of genetical attributes must be prohibited. Thus, although the Declarations acknowledge the uniqueness and diversity in the human genome, it prohibits that human dignity be harmed because of it.

III. The right to human dignity and genetic engineering

Although genetic engineering can enhance the human body, it can also alter or compromise fundamental values, such as the right to human dignity.

If gene technology spreads and there will be more genetically modified humans than unmodified ones, will the biological concept of humans be transformed? If genetically-modified people will also be considered fully human, will the concept of human dignity need to be changed? Or would the very fact that this change is considered be discriminating between genetically modified humans and genetically unmodified people?

Would we have to legally distinguish people from people? Should genetically modified people enjoy a kind of positive discrimination, and for instance, when going through a job application process, would this fact be positively evaluated? Should we differentiate between the people who are genetically modified because of certain genetical diseases and people modified solely for enhancement? If we acknowledge diversity through genealogy, should we also recognize equality? Can an unmodified person be equal to a modified one, or will a genetically modified person be more worthy?

The declaration of human dignity as guiding principle offers people enough guarantee to freely decide on genetic modification.

IV. Human genetic examinations

Human genetic examinations are meant to reveal expected genetic diseases or dispositions for certain diseases before they develop. However, this knowledge may violate the right to human dignity by giving information on a patient's health, without – in most cases – offering a cure for it. So although a person can learn about their (expected) disease, they can neither prevent nor treat it. This way, their right to self-identity and the self-determination can be violated, as they obtain information they cannot use. In such cases, the law provides the right to non-knowledge, whereby the person in question can decide whether or not the result of the human genetic test be revealed.

A. Pre-implantation examinations and prenatal tests

Pre-implantation examination shows which of the embryos to be implanted suffers from a genetic disorder, while the prenatal test reveals whether the fetus is suffering from a genetic disorder. Although the benefit of preimplantation is that only the most viable embryos are implanted in the mother's body (thus protecting the mother from possible miscarriage), the disadvantage is that in many cases, selection can result a kind of infant design when parents choose between healthy embryos (treating them as objects), based on eye colour and intelligence. Can the right to human dignity of an already born child be violated in the event that the parents have tailored the child to their liking? Does the right to physical integrity also include the right to genetic integrity, and can the right to self-identity also include the right to genetic self-identity?

Discussing the pre-implantation examination, Michael Quante asks himself: "is it philosophically acceptable that we put the incipient human's life under the boundless protection of the guiding principle of the human dignity? Or should we conclude that, in the changed situation of life sciences and medicine, the principle of human dignity should not be applied to all stages of human life?" [2] (p. 47). However, it is important to note that in the case of embryos, it is also arguable whether the principle of progressivity is applicable in regard to the right to human dignity. Two issues should be considered: stating that the embryo is not a human being (as done in many states), and ensuring protection, so that neither research nor surgical interventions can be carried out without proper ethical considerations. In addition, design interventions could be prevented if the embryo or the foetus is given a conditional human dignity from conception which could help to avoid unnecessary genetic interventions without their knowledge and consent – because in this case, it would not be primarily about parenting, but about having a child with a certain purpose and specific characteristics [14] (p. 209). Furthermore, the baby desigining process not only involves treating the child as a thing (and thus violating human dignity), but more healthy embryos could be killed in the process.

B. Life-saving siblings

Both in the United States of America and the United Kingdom (here regulated and only permitted when the procedure fits particular conditions) it is allowed to do pre-implantation examinations on the embryos awaiting implantation in order to find out which embryo genetically suits better as donor.

Since an embryo does not have human dignity, its right to dignity cannot be discussed. However, the self-identity and the right to physical integrity of the child born with the specific goal of serving as donor can be questioned. While the child may help a sibling, they cannot freely agree to the procedure as the first treatments are done when the child is very young. Nevertheless, while it cannot be argued that through the birth of a donor child, two children may live healthy lives, the life-saving sibling procedure does not guarantee to heal the child who is suffering from a genetic disorder. If the procedure is not regulated, the parents may still play the legal system by conceiving a child naturally. If in such a case the prenatal test shows that the foetus will not be a suitable donor for their sibling, the pregnancy can terminated – thus the life-saving sibling concept is realizable in a quasi legal form.

V. Genetic Engineering

Although genetic surgery is still in its research phase, its application would most likely spread rapidly, as it offers cures for diseases which have so far been considered irreversible. However, such a procedure would jeopardize the right to physical integrity because it interferes with the human body at the most basic level: the intervention modifies or replaces a defective gene. Genetic engineering could thus on the one hand cure incurable diseases, and on the other hand enhance people's abilities which would not only redefine human dignity, but also the human biological concept.

A. The right to self-determination

The major problem of genetic engineering is that there is no knowledge yet on either its long-term consequences or its side effects. Thus professionals require the permission of their patients in order to apply procedures with no exact information on side effects or consequences. This way, the right to human dignity may be harmed, while the right to self-determination is not actually practiced as the patient agrees to a procedure whose risks are unknown.

The other crucial problem with genetic engineering is that it intervenes on the most basic level, it replaces, modifies, and designs genes. There is still insufficient knowledge as to whether such interventions may lead to gene mutations or to situations when the original disease is cured, but a new disease is developed instead.

B. Enhancement

Genetic surgery does not only aim to cure diseases, but also to enhance people's abilities. However, Siddhartha Mukherjee rightfuly asks the following questions: Are we capable to responsibly develop our genes? What could be the consequences of trying to improve the natural information encoded in our genome? Can we make our genomes more "suitable," without risking to male them more unsuitable? [1] (p. 525). Therefore, although scientists would like to intervene in the process of evolution (and speed it up) by procedures which intervene in the human genome, this should be done by knowing the consequences beforehand.

If the goal of genetic surgery is enhancement, would it mean that a person undergoing such a procedure is not "good enough"? Thus is the human dignity not harmed in this case? Furthermore, could the right to human dignity also be harmed if body enhancement is banned? Because, why should people not be able to decide whether they desire sharper vision, better hearing, or particular characteristics? Could legislation prohibit the development of abilities which endanger the life and physical integrity of others, while allowing the development of other non-harmful abilities? Or should people undergo enhancement procedures if it is not hereditary to descendants (thus leaving the right to self-determination to the descendants)? These are all questions posed by enhancement procedures which are still waiting to be answered.

C. Cloning

There are two types of cloning: reproduction cloning, which is banned in most countries, and therapeutic cloning, whose permissibility is disputed. The Preambulum of the first Additional Report of the Oviedo Conventionnoted that cloning is a practice against human dignity which degrades humans to objects, by creating new humans identical to humans who already exist or have existed in the past. The Report thus banned the intervention [15].

The Fundamental Law of Hungary disposes that human cloning is prohibited [16]. Therefore specialized literature agrees that while therapeutical cloning is permitted in Hungary, reproductive cloning is not [14] (p. 217). "According to certain opinions, as long as the entire human being is not cloned, but only tissue or organs are reproduced, human cloning has not taken place" [17] (p. 120). Some countries allow therapeutic cloning (such as the USA, Sweden, Belgium, South-Korea, or Japan), while other countries have banned all research on human embryos (Ireland, Austria, Poland, Italy or Norway).

Cloning also correlates with human dignity and human equality: "the question may arise on whom would be entitled to be cloned. If cloning will depend on money, it may result in aggravating the inequality of society" [18]. Distinguinging between cloned and un-cloned people could then lead to the violation of human dignity, thus allowing eugenics to fully manifest.

VI. Human genetic research

Human genetic research is controlled both on national levels and international level. As a general rule, the Oviedo Convention permits in vitro research in connection to embryos, while it prohibits creating embryos through genetic research [12] (Article 18). The Oviedo Convention states that "scientific research in the field of biology and medicine shall be carried out freely [12] (Article 15). The Convention does not dispose of the human genetic research so it does not define genetic research.

In spring 2015, the Chinese professor Junjiu Huang announced that he received 86 embryos affected by genetic diseases from clinics which handle artificical inseminations. The research group experimented with the embryos, creating and adding healthy DNAs, but their experiments failed in the sense that the embryos were not healed and died. This is why some believe that the 2015 moratorium by Jennifer Doudna and David Baltimore on the clinical use of gene editing and gene modification techniques in the western world should be abolished, as no state should be left behind in the genetic engineering competition [1] (p. 526-527). This competitiveness is precisely what can endanger the right to human dignity, as it can easily advance the freedom and primacy of science and research, undermining the primacy of people's well-being and dignity. The intentions to modify the human embryo's genome are becoming similar to an intercontinental race [1] (p. 527). The spirit of competition does not always respect human dignity, but it sometimes stands for individual interests, particularly in cases when one would like to match the human genetic research of a very different culture, such as China (which strives to be at the head of human genetic research although it does not yet benefit from the appropriate legal acts and laws).

Human genetic research raises a particular problem: researchers can conduct whatever clinical trials they wish in human embryos. However, the embryos used in research are prohibited to be cropped into human or animal bodies.

Although human genetic research aims to examine the genome, rather than intervening through medical and surgical procedures, it may still hold cetain dangers. It may violate the right to human dignity, because it touches the physical integrity of the research subjects – although they voluntarily accepted to be part of the research. The following question arises: did the subjects accept to be part of the research being well informed that they are joining a research whose results are doubtful and which can jeopardize physical integrity and life on the most basic level? Furthermore, another question should be posed: while human genetic research can eliminate the consequences of taking medication (with several exceptions), can it revitalize until then the human genome to its authentic status, during the period when the serious side-effects occur? As long as researchers do not answer these questions and will have no precise data on side effects and whether they are reversible, a moratoriam on human genetic research should be initiated. Until these questions can be answered properly, human genetic research can be regulated by law.

The astronaut twin

Human genetic research can not only be done on Earth, but also in space, as it happened in 2016 with Scott Kelly, who was sent on a 340-day mission on the International Space Station (ISS) in order to research how the human body responds to such long periods in space. Meanwhile, Kelly's twin brother remained on Earth, therefore they both provided a good basis for comparison.

Human genetic research is permitted if the research amounts more benefits than disadvantages for the research subject; furthermore, it is permitted if the research subject's well-being and health has priority over societal and scientific interests [12] (Article 2). Nevertheless, the research could have violated Kelly's right to human dignity as he was locked up in the space station for nearly a year; but as an astronaut, he voluntarily agreed to it. Furthermore, the research could have easily harmed his human dignity as the research did not anticipate what kind of changes the stay on the ISS would produce in Kelly's body.

Although an article claimed that over 7% of his genetic stock was changed as compared to his twin brother, "these stories are biologically impossible. If 7% of Kelly's genome was altered, he would be about as different from a human as a rhesus monkey" [19]. Therefore his gene did not change; only the way it was expressed had changed, but this can be the result of the stressfull circumstances he had experienced on the ISS.

VII. Conclusion

Can we interfere in the human genome, although this would imply affecting the lives of future generations without their consent? Is this process not harming the right to human dignity of future generations, given that it would alter their physical integrity on the most basic level? Do we have the right to "determine the next generation's fate, even if it is done with the purpose of improvement?" [14] (p. 204).

The right to human dignity is a reference point in genetic engineering, playing an important role in identifying which treatments can and cannot be performed. However, just stating the right to human dignity may not be enough, as different nations have different undestandings of human dignity. Therefore as long as there are no international standards and regulations as to which procedures are allowed and which are not, human dignity can be violated, as different rules can not only increase competition in genetic engineering, but can also create the phenomenon of genetic tourism [20]. "The danger is very real that countries are running a race to be the first, without considering the human rights perspective, bioethical analysis or possible consequences" [21] (p. 9).

References

[1] S. Mukherjee, *The Gene. An Intimate History*, London: Bodley Head, Vintage, 2016.

[2] M. Quante, *Menschenwürde und personale Autonomie – Demokratische Werte im Kontext der Lebenswissenschaften*, Frankfurt: Felix Meiner Verlag, 2010.

[3] L. Siep, "Az emberi méltóság argumentuma az őssejtkutatásról folyó etikai vitában" [The argument of human dignity in the ethical debate on stem cell research], *A személy bioetikai kontextusa* [The bioethical context of the person], Sándor Kőműves and Erzsébet Rózsa, eds., Debrecen: Debreceni Egyetemi Kiadó, 2013.

[4] J. Sándor, *Az én molekulám – Bioetika és emberi jogok a XXI. század elején* [My molecule – Bioethics and human rights at the beginning of the 21st century], Budapest: L'Harmattan Kiadó, 2016.

[5] 64/1991. (XII. 17.) AB határozat [Constitutional Court Decision].

[6] J. Sári, B. Somody, *Alapjogok – Alkotmánytan II* [Fundamental rights – Constitutional doctrine II], Budapest: Osiris Kiadó, 2008.

[7] "11/2006. (III. 23.) AB határozat" [Constitutional Court Decision] *Magyar Közlöny* [Hungarian Journal], Budapest, no. 32. 2717, 2006 http://www.kozlonyok.hu/nkonline/MKPDF/hiteles/MK06032.pdf. Accessed: April 28th, 2018.

[8] UNESCO, *The Universal Declaration of Human Genome and Rights.* 1997.

[9] UNESCO, *The International Declaration on Human Genetic Data.,* Article 7, 2003.

[10] UNESCO, *The Universal Declaration of Bioethics and Human Rights,* 2005.

[11] Koicihire Matsuura director-general of UNESCO on the occasion of the twelfth session of the International Bioethics Committee, Tokyo, 15

December 2005. http://unesdoc.unesco.org/images/0014/001428/142832e.pdf. Accessed: April 28th, 2018.

[12] The Council of Europe, *The Oviedo Convention,* 1997.

[13] Explanatory Report of the Human Rights and Biomedicine Convention, European Treaty Series, No. 164. https://rm.coe.int/16800ccde5. Accessed: April 28th, 2018.

[14] Z. Navratyil, *A varázsló eltöri pálcáját? A jogi szabályozás vonulata az asszisztált humán reprodukciótól a reproduktív klónozásig* [Does the wizard break his wand? The regulatory framework from assisted human reproduction to reproductive cloning], Budapest: Gondolat kiadó, 2012.

[15] The Council of Europe, *The Oviedo Convention's first Additional Report*, Article 1. 1. 1998.

[16] The Fundamental Law of Hungary, Article III. 3.

[17] M. Julesz, *Az orvosi jog aktualitásai. Az eutanáziától a klónozásig* [The actualities of medical law. From euthanasia to cloning], Budapest: Medicina Könyvkiadó Zrt., 2016.

[18] A. O. Homicskó, "Az új technológiák megjelenése az egészségügyben, különösen a géntechnológia és az elektronikus egészségügy" [The emergence of new technologies in health care, in particular genetic engineering and e-health], *Egyes modern technológiák etikai, jogi és szabályozási kihívásai* [Ethical, legal and regulatory challenges of modern technologies], Árpád Olivér Homicskó ed., Budapest: Acta Caroliensia Conventorum Scientiarum Iuridico-Politicarum XXII., 2018. http://www.kre.hu/ajk/images/doc4/Egyes_modern_technologiak_etikai_jogi_es_szabalyozasi_kihivasai.pdf. Accessed: April 30th, 2018.

[19] S. Kaplan, "The truth about astronaut Scott Kelly's viral space genes", *Washington Post*, March 16th 2018. https://www.washingtonpost.com/news/speaking-of-science/wp/2018/03/16/the-truth-about-astronaut-scott-kellys-viral-space-genes/?noredirect=on&utm_term=.cccdb6141946. Accessed: April 28th, 2018.

[20] http://theconversation.com/the-next-frontier-in-reproductive-tourism-genetic-modification-67132. Accessed: April 28th, 2018.

[21] P. de Sutter, "The use of new genetic technologies in human beings," *Committee on Social Affairs, Health and Sustainable Development*, April 25th, 2017, http://website-pace.net/documents/19855/3313570/20170426-recours-nouvelles-technologies-génétiques-EN.pdf/75b25d58-a122-4896-91ae-295d49d42549. Accessed: April 1st, 2018.

PART II

Bioethics, Anthropology, Psychology, and Philosophy

Please geben vôtre consentimiento! Informed Decision-Making in Intercultural Context

Attila Dobos

Semmelweis University, Faculty of Health Sciences,
Department of Social Sciences, Hungary

Abstract

The present study places the well-discussed problem of reaching informed consent in medical treatments into an intercultural perspective. In doing so, it sheds light on some features of the term 'cultural context' in the twenty-first century, as it may be seen through the lenses of philosophical hermeneutics, interpretive anthropology, and intercultural discourses. These findings will be used to support the achievements of the anthropological turn in bioethics. The theoretical considerations will be followed by instances from empirical knowledge to highlight the cultural aspects of the already explored difficulties of attaining informed consent. Some of the value differences from Hofstede's cultural dimension theory, and the importance of the nativeness of the langue used in moral decision-making will be discussed.

Keywords

Informed consent; cultural context, moral decision-making, language; anthropological turn in bioethics; super-diversity.

I. Introduction

This study focuses on the obstacles of gaining informed consent in an intercultural situation. The difficulties of achieving an informed decision are thoroughly disputed in medical literature, but most assume a monocultural context for the problem [1]. If not, the main emphasis is falls on the barriers of communication and understanding [2]. (This is obviously a considerable topic in monolingual situations as well, e.g.

problems of comprehension, psychological dispositions, mental disorders, socio-economic differences, etc.). The value differences amongst cultures and their impact on medical practices and bioethical principles represent an exciting topic, on which discussions have already begun, but are still far from reaching the end [3]. First, I will discuss the significance of 'cultural context' as such, as this is needed in order to establish an intercultural perspective, after which this topic will be addressed by using empirical knowledge. Cultural differences have been debated in relation to doctor – patient encounters, but up until recently, little attention has been paid in bioethical discussions as to how having different cultural identities affects the way we make moral decisions. [4] Additionally, the freshly-revealed relation between language and moral decision-making is offering a noteworthy consideration in better understanding the situation. With the help of these findings, the problem of giving autonomous informed consent in an intercultural context will be framed.

In doing so, this essay relies on both the achievements of the cultural critique of bioethics, or the so-called anthropological turn in bioethics [5-7], and on the results of recent psychological experiments on moral decision-making processes. Furthermore, I will use some insights of interpretive anthropology – mainly through the work of Clifford Geertz – and philosophical hermeneutics – as it was developed by Hans-Georg Gadamer, to support my claims.

II. The cultural context

Thinking and talking about culture has always been a rather complex and, at the same time, sensitive question. This difficulty also draws from the level of self-reflection it demands from the thinker in question, and from the inherent conceptual abstraction of culture. The leading 20th century German philosopher Martin Heidegger identified the notion of culture as being one of the epoch-making phenomenon of modernity [8] (p. 87-88). Culture in modern societies, in contrast to culture in traditional ones, implies that we no longer see our way of living and our interpretation of the world surrounding us as naturally preordained and unproblematic, but we rather perceive it as belonging to a certain culture, i.e. embedded in a continuation of a tradition established by our predecessors, therefore conceiving itself as *a* culture, as *one* amongst many, for which being a culture has become a theme of reflection.

This density is clearly tangible in the book of Alfred Kroeber and Clyde Kluckhohn *Culture: critical review of concepts and definitions* [9], well-known in anthropological circles, which over half a century ago has collected nearly 300 definitions of culture – including all notes and citations see [9] (p. 149, 4a footnote) – of which they thematise 164 items which were in use in scientific discussions. Additionally, according to historical-semantic research, until the mid-19th century the notion of "culture" was only used in the singular [10], thus assuming the existence of only one possible culture that is *the* human culture. This might be another argument for the importance of moral circles as emphasised by Peter Singer [11]. This also allows us to better understand our self-centred attitude we may habitually (and almost automatically) hold, and what hinders us from perceiving the transcendental function of the notion. We, quite often, take terms and their conceptual framework as axiomatic, universally applicable even within the most reflective epistemic fields, such as philosophy or cultural anthropology. For instance, the concepts of morality, personality, or culture show a vast variety of meanings in different cultures across the globe, while "western" understandings of some of these concepts is nonexistent in certain cultures. "The Western conception of the person as a bounded, unique, more or less integrated motivational and cognitive universe, a dynamic center of awareness, emotion, judgment, and action organized into a distinctive whole and set contrastively both against other such wholes and against its social and natural background, is, however incorrigible it may seem to us, a rather peculiar idea within the context of the world's cultures" [12] (p. 31).

It is important to distinguish between at least two main considerations of culture in philosophical tradition: (1) culture as a social phenomenon; and (2) culture as a concept with transcendental function. The classical philosophies of culture (Die Kulturphilosophie) deal mainly with the first understanding, while epistemological research in philosophical hermeneutics and in phenomenology has placed the latter meaning into its focus. Although significant efforts were made to draw its inference [13], the "cultural turn" in philosophy has not yet been elaborated. As Gadamer argued: "…the world-constitutive subjectivity, though it may be a manifold of such constitutive subjectivities, is a part of the world constituted by these subjectivities and therefore brings into play all the specific subjective relative characters of the personal horizon which distinguishes for example the negroes of the Congo or the Chinese farmers from Professor Husserl" [14] (p. 181).

Although this quote from Gadamer poses several problems by using sloppy examples from a social scientific point of view, it nevertheless convincingly shows the main concern here. Without further immersing into the relativism debate and accounting for it, it seems inevitable that the cultural context in general plays a much more important role in understanding human actions and thoughts than we used to presume.

A. Diversification of diversity

The problem of taking culture into account mainly emerges in two major fields of bioethics: (1) in relation to the so-called particular moral values and beliefs of a given group of people, and (2) concerning the traditional (local) understanding of (and relation to) health, illness, healing in general, or end-of-life issues, moral status of the foetus, genetic design, organ donation, etc. Fortunately, considerable important work has already been done in both fields, after several anthropologists and social scientists questioned the cross-cultural appropriateness of the "western" models of moral reasoning, which is grounded in rights-based liberal individualism [15, 16]. As for the second field mentioned above, complementary and alternative medicine (CAM) as part of traditional local systems of healing throughout the world is largely practised, although bioethical literature rarely discusses its ethical particularities [3]. Concerning moral differences – although anthropologists tend to remain in the field of medical anthropology or medical sociology, while bioethics was pushed farther mainly by philosophers, lawyers, theologians, and physicians – few qualitative researches have offered anthropological critiques of bioethics, and starting with the mid 1990s, the role of different cultural norms and ethnic differences in bioethics has been slowly acknowledged [5].

However, the pluralistic nature of our societies has been in a continuous changing state for the past decades. The main elements and conceptual schemes of multicultural discourse were shaped by experiences drawn mainly from the United States of America of the

1960s, where a relatively small number of minority groups – although each comprised of large numbers – represented cultural pluralism. Moving into the 21st century, this claim can no longer be maintained due to the massive changes in the structure of societies. As one of the most celebrated figures in the anthropology of migration, Steven Vertovec argued:

"Britain can now be characterised by 'super-diversity,' a notion intended to underline a level and kind of complexity surpassing anything the country has previously experienced. Such a condition is distinguished by a dynamic interplay of variables among an increased number of new, small and scattered, multiple-origin, transnationally connected, socio-economically differentiated and legally stratified immigrants who have arrived over the last decade" [17].

These conditions do not apply to the UK alone, but are applicable worldwide. (After coining the term, Vertovec witnessed its popularity skyrocket with newly-established research centres addressing the questions of super-diversity).

This leads to rethinking and re-evaluating the classical multicultural doctor-patient encounter, where the doctor – a member of culture A – engages in interaction with the patient from culture B. Along the discourse on interculturality, different cultural identities should also be highlighted here, with their instant correlation with each other and with their environment. Interculturality presumes interaction, process and change – in this regard, it can be used in a normative sense. Furthermore, culture should not be perceived as a closed and stable box of meanings, rather a frame of reference for the given personal identity which includes group affiliations and the sum of its beliefs, social roles, statuses, which makes a unique composition for the given situation – as we may position ourselves differently under different circumstances.

Due to this diversification and the massive changes we are facing, it is common that during the healing process, three different important actors can be identified: the institution of the service provider, the health care professional, and the patient – each of which possibly representing different cultural identities with different understandings of the meaning and importance of informed consent.

III. Achieving informed decisions and the related value differences

As one of the most significant topics of bioethics, the issue of informed consent is thoroughly disputed in related literature. After the so-called anthropological turn in bioethics, the underlying principle, i.e. 'respect of autonomy', was put into cross-cultural context: does it have the same importance and meaning in different communities? Although there still is an ongoing debate as to what extent culture represents a moral weight in ethics [18], most scholars would agree that there could be important values specific to a given community which affect moral decision-making – my arguments here tend to support these claims. The appearance and deliberation of African, Latin, or Islamic bioethics [7, 19, 20] in the millennial discourse also clearly shows the understanding of the emerging need to clarify these phenomena.

The renowned Dutch social psychologist Geert Hofstede has developed the alleged cultural dimensions theory [21], through which I will exemplify how value differences may affect informed decision making. Together with his research team, he defined six dimensions of national culture. The notion of 'national culture' might seem strange at first, but it is only used to when differentiating between countries rather than individuals. He asserts that the scores on those dimensions are relative and can only be used comparatively. He also believes that values are stable elements in culture (much more than practices), so a "comparative research on culture starts from the measurement of values" [22] (p. 28). Hofstede thus identified six dimensions: power distance index (PDI); individualism vs. collectivism (IDV); masculinity vs. femininity (MAS); uncertainty avoidance index (UAI); long-term vs. short-term normative orientation (LTO); indulgence vs. restraint (IND). I will use two of them to illustrate foreseeable difficulties in achieving informed consent. In doing so, I would like to support the already-existing knowledge and claims on these arguments. [23]

Power distance index is, in short, used to indicate how a given country is dealing with inequalities among people. Societies scoring high are likely to accept hierarchical order and large differences, while low PDI countries strive for equality and demand convincing justification when inequality occurs. Translating this into the health care scenario: how does this affect the doctor - patient power asymmetry in relation to providing the best possibility for informed decision making?

PDI seems to have a strong impact on various elements of the healing procedure: in large PDI cultures, doctor-patient consultations take less time and are more formalised with "less room for unexpected information exchange" [22] (p. 71). Blood transfusion practices have also shown differences with regards to PDI: countries with low scores have more blood donors and more blood is supplied to hospitals. Naturally, there are many more factors behind these differences, but the referring studies controlled them in a methodologically-appropriate way, making it possible to observe the significance of the correlations [24]. If evidence for these differences exist, one might claim that they pose a risk to securing informed decision making. Accepting authority and the feeling of dependence and large emotional distances are likely to appear when low and high score PDI country citizens are at the other side of the desk, e.g.: institutional policy may require strict obedience to its internal rules, while the doctor is assuming a high level of independence from the patient who draws different inferences by misunderstanding the behaviour of the doctor. *Ad hominem* kind of arguments may play another role in case of patients of low PDI countries, who take everything coming from the (superior) doctor as granted, because it was he/she who said that (therefore it must be right).

By presenting these ideas, I do not claim that these risks do not exist among members of the same community, especially as the differences in PDI also appear between classes, socio-economic statuses, occupations to some extent within the same country as well. Nevertheless, I do argue that these factors pose a significant threat to informed decision-making and, in most cases, are overlooked.

The insights identified in the individualism vs. collectivism (IVS) dimension represent well-known problems in the literature: the highly esteemed principle, the 'respect for autonomy' is clearly grounded in the ethos of individualistic liberal societies. But while this standard seems self-evident in the Euro-American culture, it has a different interpretation in many cultures across the globe: it is not the individual who ought to make decisions and bear responsibility alone. In Hofstede's model, this dimension refers to the preference towards loosely- or tightly-knit social frameworks: in the former, one is expected to take care of themselves and their close family, while in the latter, the community secures its members in exchange for obedient loyalty. It might be difficult to understand shared responsibilities in cases of life-threatening decisions made by a community or a leader instead of the patient. The same applies in the case of medical interpreters or cultural mediators: the moral and legal responsibility of the mediating third party is more than dubious. Furthermore, speaking one's mind straightforward could be perceived differently, ranging from being honest to a rude violation of the interest of the community. Therefore, the appropriateness of truth-telling in a given moment requires proper cultural assumptions [25], and is not necessarily the right thing to do as moral intuitions would suggest.

The IVS dimension plainly shows value differences that may have a direct impact on the way moral decisions and informed consent can be made, insofar as it strongly relates to the different perception of autonomy.

IV. Moral decision-making and the question of language

The last, but probably the most vivid and less debated insight so far, concerns the role one's native or second language may play in moral decision-making. One might claim that answering questions in a moral dilemma should not depend on the language they were asked in as long as it is understood properly. Recent research however suggests the opposite [26, 27]: Albert Costa and his fellows refreshed the famous trolley problem by Philippa Foot [28] and have put it into a slightly different, but very new context. (This problem appeared in many different versions in the last half-century, independent of Foot's original purpose). They wanted to see what happens if two basic versions of the same problem would be presented in a language different from the subjects' native language. They first used the footbridge version (Fig. 1), where a man stands on a footbridge overlooking a train track. A trolley is coming and is about to kill five people. The only way to stop it would be to push a heavy man off the bridge in front of the trolley – thus killing the heavy man, but saving the five people. On a utilitarian basis, we would see this as sacrificing one to save five, but at the same time, also as violating the moral prohibition of killing. For methodological reasons, they repeated the experiment in another version of the dilemma (Fig. 2), where there is no footbridge, but an alternating track with a switch, so by pulling the switch one can change the direction of the trolley to the track, thus only killing one man instead of five.

Fig. 1. Footbridge version of the trolley problem
Source: https://kevinbinz.com/tag/footbridge-dilemma/

Fig. 2. Switch version of the trolley problem
Source: https://kevinbinz.com/tag/footbridge-dilemma/

The main difference between the cases is the level of emotional involvement: while the former requires physical contact and action to kill/save people, the latter is less emotionally aversive. The hypothesis was grounded on a moral psychological model, where "moral judgment is driven by a complex interaction of at least two forces: intuitive 'automatic' processes prompted by the emotional content of a given dilemma, and rational, effortful, controlled processes driven by the conscious evaluation of the potential outcomes" [25] (p. 1]. In this model, the first intuitive processes supports decisions based on the shared basic values and norms of the community, so deontological judgements, while the rationally-controlled process favours judgements for the sake of the greater good: the utilitarian decisions. They thus hypothesise that listening to these dilemmas thinking about them in a foreign language will increase utilitarian decisions, because the usage of foreign language stimulates less intense emotional reactions compared to a native language. As shown in the charts below (Fig. 3), the switch version with less emotional involvement resulted in fairly similar (~80%) utilitarian choices in native and foreign languages, while the emotionally demanding footbridge version resulted in more than twice as many utilitarian decisions in a foreign language than in the native language. The experiment was carefully designed with control over many different factors which can play any role, but, along with the results of the few other studies [27, 29, 30] on this topic, they conclude that people's moral decisions depend on the nativeness of the language in which the question or problem is presented.

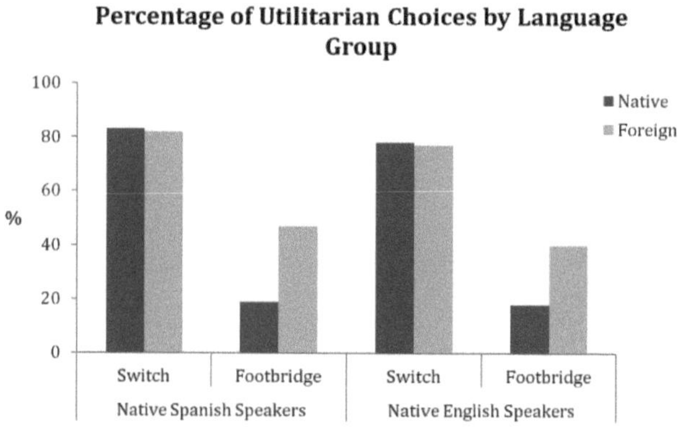

Fig. 3. Percentage of utilitarian decisions by language group. Source: [Costa] (p. 5)

Considering that in our globalised world there are hundreds of millions of people who speak and manage their businesses in a foreign language every day, including making moral decisions, this result also represents a significant finding in relation to achieving informed consent.

V. Summary

Receiving informed consent from the patient has been a central concern in the bioethical dialogue since its very beginning. Fortunately, a vast amount of research explores the difficulties of making informed decisions. Following the anthropological turn in bioethics, an increasing number of studies on cultural differences, local moralities, and culturally-affected cognitive processes were carried out, resulting in a shift in evaluating the role of culture as a context for bioethical issues. This study would like to support these efforts by giving an alternative understanding of the cultural context using insights from philosophical hermeneutics, interpretive anthropology, and intercultural discourses. Further to the theoretical findings, a few instances from empirical knowledge have been used to highlight culturally-driven difficulties in achieving informed consent, such as the different value dimensions of different societies and the hidden impact of using a foreign language during moral decision-making.

References

[1] R.R. Faden, T.L. Beauchamp, *A History and Theory of Informed Consent,* Oxford: Oxford University Press, 1986.

[2] E.E. Anderson, S.B. Newman, A.K. Matthews, "Improving informed consent: Stakeholder views," *AJOB Empirical Bioethics*, vol.8, no.3, pp. 178-188, 2017. https://doi:10.1080/23294515.2017.1362488

[3] S. Chattopadhyay, R. De Vries, "Bioethical concerns are global, bioethics is Western," *Eubios Journal of Asian and International Bioethics,* vol. 18, no.4, pp. 106–109, 2008.

[4] S. Sachdeva, P. Singh, D. Medin, "Culture and the quest for universal principles in moral reasoning," *Int J Psychology,* vol. 46, no. 3, pp. 161-176, 2011.

[5] L. Turner, "Bioethics in a Multicultural World: Medicine and Morality in Pluralistic Settings," *Health Care Analysis*, vol. 11, no. 2, pp. 99-117, 2003. https://doi.org/10.1023/A:1025620211852

[6] A. Kleinman, *Writing at the margin: Discourse Between Anthropology and Medicine.* University of California Press, 1997.

[7] *African American bioethics: culture, race, and identity.* L.J. Prograis, Jr.-E.D. Pellegrino, eds., Washington D.C.: Georgetown University Press, 2007.

[8] M. Heidegger, *Die Zeit des Weltbildes.* Gesamtausgabe. Bd. 5: Holzwege. Frankfurt (Main): Klostermann, 1977.

[9] A.L. Kroeber, C. Kluckhohn, *Culture: A Critical Review of Concepts and Definitions.* The Museum, 1952.
[10] G. Markus, *Culture, Science, Society.* Leiden: Brill, 2011.
[11] P. Singer, *The Expanding Circle*, Princeton University Press, 2011.
[12] C. Geertz, "From the Native's Point of View: On the Nature of Anthropological Understanding," *Bulletin of the American Academy of Arts and Sciences*, vol. 28, no. 1, pp. 26-45, Oct. 1974.
[13] R.A. Mall, *Intercultural Philosophy*, Rowman & Littlefield, 2000.
[14] H.G. Gadamer, "The Science of the Life-world," *Analecta Husserliana,* A-T. Tymieniecka, ed., vol. 1, pp. 173-185. Dordrecht: D. Reidel Publishing Company, 1972.
[15] A. Kleinman, *Pateints and healers in the context of culture,* University of California Press, 1980.
[16] R.C. Fox and J.P. Swazey, "Medical Morality Is Not Bioethics - Medical Ethics in China and the United States," *Perspectives in Biology and Medicine,* vol. 27, no. 3, pp. 336-360, 1984.
[17] S. Vertovec, "Super-diversity and its implications," *Ethnic and Racial Studies,* vol. 30, no. 6, pp. 1024-1054, November 2007.
[18] S. Gbadegesin, "The Moral Weight of Culture In Ethics," *African American Bioethics: Culture, Race, And Identity,* LJ. Prograis Jr. and ED. Pellegrino, eds., Georgetown University Press, pp. 25-44, 2007
[19] F. Luna, *Bioethics and vulnerability. A Latin American View.* Rodopi, 2006.
[20] A. Sachedina, *Islamic Biomedical Ethics*, Oxford: Oxford UP, 2009.
[21] G. Hofstede, *Culture's Consequences: Comparing Values, Behaviors, Institutions and Organizations Across Nations.* Thousand Oaks CA: Sage Publications, 2001.
[22] G. Hofstede, G. J. Hofstede, M. Minkov, *Cultures and organizations – software of the mind.* McGraw-Hill Education, 2010.
[23] PE. Ekmekci, B. Arda, "Interculturalism and Informed Consent: Respecting Cultural Differences without Breaching Human Rights," *Cultura (Iasi, Romania).* vol.14, no. 2, pp. 159-172, 2017.
[24] W. de Kort, E. Wagenmans, A. van Dongen, Y. Slotboom, G. Hofstede, I. Veldhuizen, "Blood product collection and supply: a matter of money?," *Vox Sang,* vol. 98(3 Pt 1):e201-8, Apr. 2010. doi: 10.1111/j.1423-0410.2009.01297
[25] J. Oliffe, S. Thorne, TG. Hislop, EA. Armstrong, "Truth Telling and Cultural Assumptions in an Era of Informed Consent," *MSN Fam Community Health,* vol. 30, no. 1, pp. 5–15, 2007.
[26] A. Costa, A. Foucart, S. Hayakawa, M. Aparici, J. Apesteguia, J. Heafner, "Your Morals Depend on Language," *PLoS ONE* vol. 9, no. 4, e94842, 2014. https://doi.org/10.1371/journal.pone.0094842

[27] J. Geipel, L. Surian, C. Hadjichristidis, "How Foreign Language Shapes Moral Judgment," *Journal of Experimental Social Psychology*, vol. 59, pp. 8–17, February 2015.
[28] P. Foot, "The Problem of Abortion and the Doctrine of the Double Effect," *Oxford Review*, no. 5, pp. 5-15, 1967.
[29] B. Keysar, S.L. Hayakawa, S.G. An, "The foreign-language effect: thinking in a foreign tongue reduces decision biases," *Psychol Sci,* vol. 23, pp. 661–668, 2012. https://doi:10.1177/0956797611432178.
[30] C. Harris, "Bilingual Speakers in the Lab: Psychophysiological Measures of Emotional Reactivity," *J Multiling Multicult Dev,* vol. 25, pp. 223–247. https://doi:10.1080/01434630408666530.

Can Kant's Position Be Guessed in a Debate on Access to New Technical Advances in Medicine?

Daniela Reisz,[1] Alexandra Anghel[2]

[1] Victor Babes University of Medicine and Pharmacy, Timisoara, Romania
[2] Neurology, SCJUT, Romania

Abstract

Neuro-ethics arises as a new field of debate in bioethics, respectively in medical ethics. The term was coined by Safire in 2002 and covers aspects linked to fundamental value questions such as what is life, when does life start, when does an embryo become human. Other subjects include questions on what the mind is; the self and the psyche; brain research; consent; brain as a source of ethics; wellbeing; eugenics; the relationship between new technologies and humans. Several pillars sustain the common theoretical background of all of these ethical enquiries. One of them is the philosophy of Immanuel Kant. MRI (Magnetic Resonance Imaging) is a fascinating tool for diagnosis but it cannot substitute the specific activity of the doctors. Even so, from the beginning, the request for MRI investigation surpassed its availability and led to the enforcement of access criteria. Who will benefit from the new technology and who will not? Sometimes the information supplied by MRI can make the difference between life and death or disability, and the selection of beneficiaries requires a strong ethical ground. The question we are trying to answer is: what would Immanuel Kant`s position be in this debate?

Keywords

MRI; prevention; neuro-ethics; human ideal; prevention.

I. Premises

According to Clausen and Levy's Handbook of Neuroethics, "neuroethics is systematic and informed reflection on and interpretation of neuroscience, and related sciences of the mind (psychology in all its many forms, psychiatry, artificial intelligence, and so on), in order to understand its implications for human self-understanding and the perils and prospects of its applications" [1] (p. vi). Among the large array of methods, techniques, and devices which contribute to the development of neuroscience and are fundamental in acquiring knowledge in neuroscience, MRI and functional MRI (fMRI) have an important place. This technology is relatively new and fast developing.

When new technology appears in medicine, the ethical issues it raised require resolution. The actual discussion on imagistic neuro-ethics refers especially to the application of functional MRI as a useful tool to study the human brain, behaviour, and cognition. Functional MRI is used for research purposes and therefore neuro-ethics debates mainly on this issue. These discussions are comprised of several issues: the accidental discovery of abnormalities which causes misinterpretations with important consequences on the subject; the ability to determine the level of consciousness in people with severe brain injury; lie detection; the attempts of mind reading in relation to privacy; the attempts to identify associated patterns with behavioural disorders and with the potential to predict future illness [2] (p. 661).

Brain and spinal cord MRI have revolutionized diagnosis and the description of diseases and thus have a relatively large impact on the quality of medical services. The high technological level, costly research, the interests of the industry and its intervention in research and development resulted in the promotion of this diagnostic tool to the public. Although until the last century, medical inventions were generally reserved for doctors and medical staff, nowadays large groups of researchers from different fields are involved in the development and production of new medical technologies. This changes the medical ethical codes (which used to belong to doctors) used for promoting technical devices and thus generate a field of advertising ethics in medicine.

Traditionally, medical advertising was considered unethical. Today, old medical codes are overcome by the industry and are subjected to new paradigms. The Pharma industry in Europe does not have permission to advertise directly to patients and to promote therapeutic solutions, but it does have permission to do so in the United States.

Depending on the regulations in use, various marketing policies exist for different Pharma actors. These differences in advertising and marketing do not operate for the new technologies and are intensely marketed directly to the public. The advertising of the new technological means has a double purpose: on the one hand, to identify potential investors for research development and, on the other hand, to identify new users and beneficiaries and create a market. The market has regulatory mechanisms through demand and supply.

Information on new technologies touch the public in various forms, while it sometimes bypasses doctors. As a result of this advertising, potential beneficiaries appear with a desire to access the new technology. Of course, new technologies bring solutions for real needs, but the marketing process generates additional needs, thus increasing the demand for a service and implicitly widening the market. Public access to new technology in medicine generates new and unexpected problems for the medical system and staff, as well as for the subject per se.

The ethics of neuroscience and the neuroscience of ethics (definition of neuro-ethics according to Adina Roskies [3] (p. 21) deals especially with good regulatory practices at institutional level, not with the individual ethical approach necessary in situations of moral relevance. Although it discusses the ethics of advertising in medicine, this essay focuses on individual approaches to new technologies and hence the relationship with Kant's ethics, utilitarian ethics, and liberalism.

II. Method

This study will address the implications of free public access to MRI with reference to traditional ethical values and from an individual perspective. Using some examples from medical practice, the benefits and shortcomings of excessive neuroimaging will be discussed. We will discuss the human reactions facing the perspective of the disease from an epistemological perspective. Humans facing illness is not be examined, but we will assert the psychosomatic pathology, as this is the great consumer of futile medical services.

Publicity makes new technology attractive. New technology thus becomes desirable (and eventually a luxury). The public who has the material resources to use new technology will do so to the detriment of those who genuinely need it but cannot afford it. In countries such as Romania, MRI devices are owned by both the public health system and by the private system. People who do not genuinely need MRI examinations put pressure on both systems and thus shorten the availability of this examination. Nevertheless, it is also possible that this desire for MRI investigations is a socio-cultural determinate and is specific to the Romanian population. The roots of this attitude are very complex and require a deep and complex analysis. They are linked to the medical system as before and after 1989 Revolution, the politics involved in the management of the medical system, the partial privatisation of medical services, mass media images of doctors and the medical system, general mistrust, and so on. Nevertheless, the reason for the increasing demand of MRIs as unrelated to the medical condition is connected to several human characteristics: the need for knowledge/curiosity, the need for explanation/causal relationship, the reference to a perfect human/ideal, the need for power/control of destiny and the continuous change in conceptions about the quality of life and good health.

A. The need for knowledge /curiosity

According to Kant, "human reason has the peculiar fate in one species of its cognitions that it is burdened with questions which it cannot dismiss, since they are given to it as problems by the nature of reason itself, but which it also cannot answer, since they transcend every capacity of human reason" [4] (p. 2343).

The patients' need for knowledge is consistent with biologically-specific features. Knowledge/curiosity is a biological attribute in large primates. Monkeys give up eating to satisfy their curiosity. So, the desire for knowledge is an intrinsic part of the human condition.

Knowledge is validated as a value in many philosophical systems. From the Kantian point of view, knowledge is supposed to be good and desirable. "The charm in expanding one's cognitions is so great that one can be stopped in one's progress only by bumping into a clear contradiction" [4] (p. 3339).

We will discuss, using a few examples, the consequences of blindly pursuing the need for knowledge in medical practice. Example 1. A young female patient with known migraine attacks is Angio-MRI examined and the main finding is a small-sized intracranial aneurysm.

The intracranial aneurysm is the main cause (80%) for sub-arachnoidian haemorrhage (SAH), a very serious condition with a high mortality rate (about 40%) [5]. If the aneurysm ruptures, its devastating consequences are difficult to control. But not all intracranial aneurysms break. In a normal population, the prevalence of intracranial aneurysms (IAs) is estimated to around 2% to 3.5% [6]. Our patient, finding about her intracranial aneurysm, starts to correlate her migraine with this vascular abnormality and infers that she has, apart from the migraine, a serious disease. Currently, technological advancement allows the obstruction of this vascular malformation by coiling or stent application. Endovascular techniques prevent aneurismal rupture, but a few years ago, the only solution to prevent the aneurysm rupture was craniectomy and clipping. In the past, patients, otherwise healthy, were subjected to brain surgery with the purpose of removing a potential risk with a low probability. The SAH incidence is 10 per 100 000 person-years, implying that many IAs never rupture [6]. Today, such a patient has several options: clipping the aneurism (a very complicated procedure, with increased risks in itself), coiling or stenting of the aneurysm (expensive, available just in specialized centres), or keeping it as it is. In small aneurysms, doctors prefer to do nothing because of the bigger risk of the procedure compared to the risk of the rupture. If not operated, the patient will live anxiously with the concern of a possible aneurysm rupture. Therefore this discovery affects the patient for life through emotional and material impact or through the secondary effect of medical interventions. As doctors, we ask ourselves: was it useful? Does MRI, in this situation, contribute to the well-being of the subject?

B. The need for an explanation/ causal relationship

Causal relationship is an *a priori* form of intuition in human knowledge. "…I really connect two perceptions in time. Now, connection is not the work of mere sense and intuition, but is here rather the product of a synthetic faculty of the imagination, which determines inner sense with regard to temporal relations. …and that is here the concept of the relation of cause and effect [4] (p.7307).

Because of this hardwired form of intuition, people are prone to establish a causal relationship between successive events. When such a relationship cannot be easily established, people are anxiously looking for one and they need an explanation for indeterministic events, even if the explanations are useless.

Example 2. A patient with minor neurological complains asks for an MRI examination because he had a head trauma many years ago. He wants to know if the head trauma is the cause of his actual distress. This particular patient does not look for a treatment, but he wants to know: maybe it is the old head trauma which starts to manifest. The effort is futile. The trauma is not reversible. We cannot change the patient history and acknowledging a possible link between the events has no actual utility. Sometimes, the need for knowing the reason of a disturbance is just speculative metaphysics, and useless from a pragmatic point of view. It is an attempt to break the *noumenal* realm.

We observe a need for knowledge in a field where knowledge is not always possible. Subjects pretend access to the reality itself. But the essence of speculative metaphysics is to transcend experience entirely in order to achieve the knowledge of the noumenal realm. Here, the faculty of reason is the only relevant one, but its most crucial speculative conclusions, its deepest convictions about the self, the world, and god, are all drawn illegitimately. Rational knowledge has access to the world of phenomena, not to reality itself. "For we have to do only with our representations; how things in themselves may be (without regard to representations through which they affect us) is entirely beyond our cognitive sphere" [4] (p.7332).

In his *Critique of Pure Reason*, Kant identifies space and time as *a priori* intuitions within which phenomenological manifestations of reality are modulated. Phenomena are the appearances which constitute our experience; noumena are the (presumed) things themselves, which constitute reality. All our synthetic *a priori* judgments apply only to the *phenomenal* realm, not the *noumenal* [7]. Kant considers it impossible to know the thing in itself, the noumenal realm. We can only know the manifestations, the appearances of reality, the phenomenal realm. Due to this limitation of reason, the access to the integral knowledge of the thing in itself is denied. The lack of access of reason to the noumenal realm leads to the fact that the speculations on God or essence or life, or sickness, are sterile as well as any other pretence of absolute knowledge. Even if inaccessible in its essence, reality exercised an irresistible fascination for the rational being and is a permanent subject of speculations. Men are incapable of believing that the world is not knowable in its essence. According to Kant, the thing in itself exists and leaves a set of traces that we can identify. We can thus infer certain characteristics, but further speculations are untenable.

C. The human ideal

Immanuel Kant proposes the existence of God as a necessity derived from moral imperative. Humans need an ideal to which any human should relate. Rational beings must act by Principle and this principle is identifiable with universal law. "Always act according to that maxim that you can will as a universal law of nature" [8]. Kant believed in the teleological evolution of morality and in inevitable historical evolution to reach a sublime. Kant considered and believed that this evolution is the duty of people.

Perfect body, skin, hair, nails, teeth, are part of today's propaganda. We are approaching the human ideal every day. We have the feeling of power and perfection. We are making changes in values, behaviour, consciousness. In this very demanding environment, a deviation from normal/ideal becomes suffering in itself.

Example 3. A man in the fourth decade of life, healthy, with good professional results, goes through stressful events which combine various professional and family threats. In response to these psychic tensions, he develops psychosomatic manifestations such as palpitations, dizziness, headaches, insomnia, malaise, the feeling of imminent death, the feeling that something is wrong – a burnout syndrome. The idea of an abnormality thus appears, and the human omnipotence is denied. He addresses a doctor. He goes through a laborious medical check-up where, even if doctors assure him of his healthiness, he denies it. Insisting, his complains are considered as potential signs of organic disease and he is directed to a cerebral MRI. A few small demyelinating lesions with no actual pathological significance are observed.

The patient reads the result and starts to study on the internet typing *demyelinating lesions*. He finds out that these occur in multiple sclerosis and this is a serious and disabling disease. Suddenly, his life changes and the prospect of suffering and disability inflicts on his psyche. The patient will live with the belief that he has a serious, yet hidden illness. The uncertainty of the diagnosis for an important period of time is a very stressful experience. Those psychological experiences (fear, depression, anxiety, obsessive claims and preoccupations) will affect others besides him. Physicians can feed this anxiety by subjecting the patient to further exploration. A lumbar puncture for dosing oligoclonal bands, measurements of potentially evoked signals (PEV), or even a longitudinal observation are stressful events. All of these interventions perpetuate long-term fears.

Of course, there is a trend to become perfect, even God-like. People with high self-esteem long cultivated by entourage develop adjusting disorders when they discover that they are not perfect and are not so close to the ideal as they use to think they are.

Achieving technological performance, mastering nature and controlling it, makes the relationship with divinity implicitly assumed and human beings feel the omnipotent master. Once the feeling of the person's invincibility has been contradicted (with minor evidence), there is a fierce battle with evidence that adds an additional moral disorder to the objective medical.

D. The need for power /control of destiny

Example 4. A young female with cervical headaches asks for a brain MRI because her grandmother has had a stroke in her seventies. The patient wants to know the cause of her headaches, in order to prevent a stroke and she wants to start early investigations to make sure it is not too late. The *a priori* intuition of causality, tends to be present in any cognitive process. As a result of this pre-set feature of our mind, there is an internal need for people to establish causal links between consecutive events. Once a symptom appears, it must be related to a cause, and to reverse that symptom, the cause must be removed. This simple thinking based on rudimentary logic is the engine for the need to know and the need to have an MRI examination as a source of knowing causality. The patients are looking for the cause of their suffering in order to remove it and to prevent greater suffering. By doing this, they feel in control of their own destiny.

People need to control their destiny in order to avoid unwanted consequences. Subsequently, prevention has become a propaganda priority in battles with unseen evil; unseen and eventually inexistent/small probability evil. Prevention is a term widely used in the media and internalized as an absolute value in the population. Better prevent than heal – is one of the constantly used slogans. The debate on vaccination is one example where an absurd militancy against vaccination manifests, because of minor (from a statistical point of view) negative effects. At the same time, vaccination is an example of how we try to avoid any of the small perils of life.

The principles that "prevention is always good" bring to mind the distinction between maxims and moral law. "Prevention is always good" is mainly a rational practical percept because the foundation of this percept is happiness, self-love and presupposes an object of pleasure or pain. "Now the moral law, which alone is truly objective (namely, in every respect), entirely excludes the influence of self-love on the supreme practical principle, and indefinitely checks the self-conceit that prescribes the subjective conditions of the former as laws" [9] (p. 1318).

III. Conclusions

Immanuel Kant notes that the only unconditionally good thing in the world, according to common understanding, is a good will. Good fortune, health, and even happiness as broadly understood are not unconditionally good, because when coupled with a bad will they become a source of condemnation for an impartial spectator [10].

Emerging knowledge from access to new technologies could have perverse effects that cannot be predicted. Sometimes unnecessary and abusive explorations create an iatrogenic pathology that impacts people's quality of life. With new technologies, brain abnormalities can be identified and treated as real diseases despite the fact there are no signs or symptoms. Medical solutions to solve brain abnormalities could have serious harmful long-term effects. We ask ourselves: is it useful to know that we have hidden genetic defects, with or without impact on our health, even if this reality cannot be changed? How does the inner or outer universe change once we get this information? Do people remain the same if they face the stigma? Must a person with a known genetic defect announce this to protect society from spreading that genetic defect?

Philosophical-religious traditions pay attention to the power of truth on unprepared persons. The knowledge/the truth is harmful if the one who discovers it does not go through an initiatory journey. Finding the truth is good just in case we have the means to cope with it.

Access to information provided by new technologies can be a source of great anxiety for those who receive them. They have to go through a long and laborious process to harmonize with the revealed truth or with the non-existence of a truth. If it is a detectable truth, nowadays access to this truth is relatively easy – but coping with the truth is another problem.

The casual discovery of certain particularities, or small defects of the human body, that offer individuality to the person, rise several problems to the medical team. There is an explicit or implicit discussion if we consider these particularities as a disease-producing potential, and we will take important steps to remedy them, or we will accept the idea of imperfection in an abstract and profound plan.

Preventing a bad thing involves considering the short, medium, or long-term effects of a present action. Prevention is more of a utilitarian value. It implies taking current measures to neutralize a long-term bad potential. Often, this prevention is done regardless of the small probability of the occurrence of the bad event and sometimes independent of the magnitude of the prevented event. A negative event deserves prevention even if it is infrequent or requires a large amount of effort.

An important aspect to discuss related to prevention is the principle of freedom. Preventive measures taken excessively challenge liberal values, constrain citizens to gestures imposed by the system in the name of an obscure good. Prevention, by its cumulative effect, ultimately restricts individual freedom and most often proves to be an unnecessary ritual. This raises the question of whether prevention is an absolute or relative value, and where is the demarcation limit that transforms prevention from a general good into a general bad.

"...That the only purpose for which power can be rightfully exercised over any member of a civilised community, against his will, is to prevent harm to others." (John Stuart Mill. *On Liberty*) [11] (p. 166). Stuart Mill refers to an actual harm, of objective physical nature and not to someone's feelings of hurt or someone's offended value systems.

By reducing to absurdity: would we rather not be born in order to prevent suffering, sickness and death? Death is the absolute fundamental bad.

Playing with ideas is a general human compulsion and we cannot preclude our patients to do so. Our duty is to answer all questions that arise regardless of the side effects of knowledge on the patients. Even if we recognize the futility of the search for inexistent answers by a patient, we must assist him during this process. The MRI does not give all the answers to our angst, but can be a step forward. Coping with consequences is a different process.

The *rational being* has a duty to pursue the *moral imperative*. Doing this, the *rational being* has the chance of liberty, the chance to avoid universal physical laws that forces us in the deterministic path, as simple parts in a complex mechanism. To have at least the sensation of liberty we must consent that we and our patients have the freedom of choice, even if the result is occasionally unwanted. Utilitarian principles, although very appealing, easy to contextualize, could have side effects and when an ethical travail arises, the Kantian view must be the larger framework of the analytic process.

"Each is the proper guardian of his own health, whether bodily, or mental and spiritual" [11].

References

[1] J. Clausen, N. Levy, "What is Neuroethics?", *Handboock of neuroethics*, J. Clausen, N. Levy, Heidelberg: Springer Reference, pp. V-VII, 2015.

[2] A. L. Roskies, "Neuroimaging Neuroethics: Introduction," *Handbook of Neuroethics*, J. Clausen, N. Levy, eds., Dordrecht: Springer Reference, pp. 659-663, 2015.

[3] A. Roskies, "Neuroethics for the New Millenium," *Neuron*, vol. 35, pp. 21–23, July 3, 2002, Cell Press.

[4] I. Kant, *Critique of Pure Reason* (The Cambridge Edition of the Works of Immanuel Kant). Cambridge: Cambridge University Press. Kindle Edition, 2009.

[5] R. C. Zebian, A. Kazzi, "Emergent Management of Subarachnoid Hemorrhage," https://emedicine.medscape.com/article/794076-overview Accessed: June 10th, 2018.

[6] T. Skodvin, L. H. Johnsen, O. Gjertsen, J. G. Isaksen, A. Sorteberg, "Cerebral Aneurysm Morphology Before and After Rupture: Nationwide Case Series of 29 Aneurysms," *Stroke*, vol. 48, no. 4, pp. 880-886, 2017. Available at https://www.ahajournals.org/doi/full/10.1161/STROKEAHA.116.015288?url_ver=Z39.88-

2003&rfr_id=ori:rid:crossref.org&rfr_dat=cr_pub%3dpubmed. Accessed: Mai 6th, 2018.
[7] G. Kemerling, "Kant: Experience and Reality," *Philosophy Pages*, 2011. http://www.philosophypages.com/hy/5g.htm. Accessed: May 6th, 2018.
[8] C. M. Korsgaard, "Kant's Formula of Universal Law," *Pacific Philosophical Quarterly*, vol 66, no. 1-2, pp. 24-47, 1985. http://nrs.harvard.edu/urn-3:HUL.InstRepos:3201869. Accessed: April 24th, 2018.
[9] I. Kant, *The Critique of Practical Reason.* University of Adelaide, 2014. Kindle Edition.
http//ebooks.adelaide.edu.au/k/kant/Immanuel/k16pra/index.html.
[10] J. Sorens, "Immanuel Kant: Philosopher of Freedom," *Learn Liberty*, 2017. https://www.learnliberty.org/blog/immanuel-kant-philosopher-of-freedom/. Accessed April 24th, 2018.
[11] J. S. Mill, *On liberty and the subjection of women*, New York: Henry Holt and Co. Kindle Edition, 1979.

Children's Bioethics, Theory of Attachment, and P4C

Florin Lobont
Department of Philosophy and Communication Sciences
West University of Timisoara, Romania

Abstract
Starting with a brief analysis of the main theories of children's rights to culture and cultural rights and their entailments in the realm of biomedical practices as culture, the study sketches a critique of these views on children as universally-dependent and separate individuals unable to make autonomous biomedical decisions for themselves. As a sound alternative, the study puts forward the ethics of care which redefines children's rights in terms of autonomy-in-relation, socio-cultural milieu, and cultural identity. In support of this tenet, the essay brings forward reputed research results associated with the attachment theory that show how an individual's current relational experience is deeply dependent on his/her belief systems, which are significantly shaped by the primary caretakers, and contribute essentially to the structuring of their future social interactions. Further, the study calls for the extension of this view to the socio-cultural level as a grounding for understanding society as built on relational selves, rather than on individual separate(d) subjects. The last section describes briefly the Philosophy for Children, a mind competency enhancement set of practices that significantly increases the chances to attain a more autonomous-in-relations identity in children and young people.

Keywords
Children bioethics; ethics of care; autonomy-in-relation; attachment theory; philosophy for children.

I. Introduction

The study's core argument is built on the ideas of children's rights to culture and cultural rights, on biomedical practices *as* culture, as well as on a number of conceptualizations of children's status,[1] and represents a critique of the views which regard them as universally dependent and separate individuals intrinsically unable to make autonomous biomedical decisions for themselves. The conventional, mainstream perspectives on children's rights, i.e. the will/power and the interest theories, overlook the life experiences of children as individuals existing *hic et nunc* (not as not-yet-actualized projections of future adults) and undergoing the process of developing their autonomy.

The feminist relational perspective, or the ethic of care, comes up with a sound alternative, which redefines children's rights in terms of relations, socio-cultural milieu, and cultural identity. Even though it has not solved all difficulties entailed by the resulting concept of autonomy-in-relation in the biomedical sphere of bodily interventions, it casts aside the conventional dichotomies like the adult/child one that compel scholars and policy makers to deny suitable space for children's expression of opinions and active participation.

Although the implications of this theory in the spheres of medical interventions and cultural identity do not and cannot find solutions to all possible biomedical conflicts, the approach it puts forward is arguably better suited to the contemporary humanistic philosophies of equality, dignity, and respect for diversity.

As a counterargument to these traditional perspectives, the study brings forth the importance of children's relation webs in general, of those defining attachment in particular, in the make-up of individuals from the earliest stages in life. In this respect, reputed attachment theory research shows how an individual's current relational experience is deeply linked to one's belief systems, which are formed in one's family of origin.

The attainment of a more autonomous-in-relations identity in children and young people (and thus, in future adults) requires a fundamental shift in children's views on knowledge, change, body, community, diversity, initiative, environment, and other related realities. Although only via a short introductory glimpse, the study closes by briefly suggesting that one of the recent interactive mind competencies,

[1] By "children" I will refer to the persons between the ages of 3 and 18, or, in case of countries where the law gives people the right to decide on their bodily intervention from the age of 16 (like in Hungary) or another age, I will implicitly consider that age as the upper limit of their childhood.

and emotional intelligence, training practices, namely Philosophy for Children (P4C), increases significantly the chances of producing these changes.

II. Some particularities of children's bioethics

The existence of a deep connection between the society members' physical bodies and cultural identity was observed and studied by scientists, including anthropologists and sociologists, who viewed the body as one of the keystones of society and a crucial building block of individual and collective identities.

One of the founders of sociology, Emile Durkheim, examined the body's use as a first site of "imagined community," and a source of the society through bodily practices (as parts of the sacred rituals), which he aptly described as "generators of social symbolism," a process through which collective emotional energies "create a [solidary] space for social life in which groups become conscious of themselves." As a conclusion, he wrote that "without the body, there can be no life and society" [1].

As Maya Sabatello justly noted, the link between body, society, and culture is most visible in case of children. Children represent the first *locus* of enculturation, the process by which individual members of a group absorb – to a great extent at subliminal level – the cultural and identity configuration of their community as their own [2] (p. 8). "Parental and others' bodily practices provide a model for children of how to relate to the world in the most basic forms—gestures, facial expressions, way of walking, postures, etc.—which children imitate and adopt as their own. In times of illness and sickness alike, children acquire the reasoning and means of healing that are used by their care givers, as these are perceived as simple acts of survival" [2] (p. 9).

One would have no difficulty understanding that many bodily-related customs of various cultures have placed – and still place – children in a particularly vulnerable situation. Children depend fundamentally on adults for survival, and this has culturally founded the conceptualization of the child-adult rapport in terms of power relations that rule children out as decision makers. This has determined a downgrading of children to an inferior status in their families and communities, which often leads to disregarding their opinions when expressed [3] (p. 5). Despite this, children are simultaneously viewed as bearers of significant power, given the concern of cultures to pass on their traditions from generation to generation, which spotlight children as the main custodians of the cultures' future. For the sake of culture's survival, communities have devised practices of inculcating in children strong commitment to the respective culture. However, observes Sabatello,

> this mutual dependency is particularly critical in the context of health-related and bodily practices. While some healing and cultural bodily practices may be harmless and mainly symbolic, others may be injurious and detrimental to the children's health. Yet, as the youngest members of the community and its cultural proprietors, children may nevertheless be exposed to either without the ability to protect themselves and regardless of the potential injurious effects of the custom [2] (p. 10).

Despite the importance given to children as gatekeepers of their group culture's survival, the most common mindset across cultures (even if the threshold of maturity differs to some extent in various cultural settings) denies them the capability of making choices for themselves (including the health-related bodily practices) due to their alleged lack of necessary wisdom and entrusts adults, especially parents, with this task [4] (p. 477).

Western societies have also developed norms and practices that try to establish when parents are the most suitable decision makers on behalf of their child (situations considered as dominant) and when they might not act in their children's best interest, situations in which states should resort to their power as *parens patriae* to intervene.

III. Autonomy-in-relation in children's bioethics

The ongoing question of children's rights existence and grounds commenced way before the "children's rights revolution." The debates around it focus mainly on the function of "rights" in relation to children and has as main opponents the traditional "power/will" theorists and the "interest" advocates.

The traditional opponents, who follow the Western human rights tradition, consider a right the "power that a person possesses to control or claim or do something" [4] (p. 19)]. In order to be a rights holder, a person needs to have a certain level of competency (i.e, rationality combined with a minimum level of agency), that is, of rational and autonomous decision capacity on their own behalf [5] (p. 101).

Although from this perspective, children's bioethical rights are not denied, in case of very young children, but also more generally in that of children viewed as a subgroup, children are attributed neither individual will, nor the capacity of rational decision making, and, consequently are denied any ability to exercise rights. As Onora O'Neil justly comments on the implications of these accounts, children are thus universally, unavoidably, and naturally dependent on their carers, who practically have power over their lives. Thus, the power/will theory deductively nullifies children's status of rights bearers. [6,] (p. 39).

At the other end of the spectrum, the interest theorists construct their argument on the idea of children's inherent moral worth in order to justify their moral and juridical condition as right holders. In short, the argument states that as children's moral rights are equal to those of adults, justifying the denial of their positive rights becomes hardly sustainable [2] (p. 165). According to T. D. Campbell, the crux of the interest theorists' argument is that essentially the rights concern first of all the "normative defense and furtherance of interests," regardless of the person(s) in question's capacity or incapacity of rational decisions. Consequently, as children fall into this category as bearers of interests, they are also bearers of rights [8] (p. 5).

When it comes to children's bioethics, both perspectives are limited for a number of reasons, the main one being that

> the evolution of others' duties and responsibilities (whether grounded in the child's interests and rights or in adults' obligations) is inherently contingent on the child's relationships with others and can thus have no meaning if taken, explored, and codified in the abstract or in isolation… Neither [view] actually captures the more complex lived experiences of children *as children* nor the informal and mutual relationships in which children are involved with and which provide them the *legal* organization of their daily lives [2] (p. 168).

Yet, precisely the web of multiple informal socio-legal structures where children live make up the *locus* where they receive their sources of attachment and relationships with the people around them – especially with their caregivers and with those children care for – that thus provides the setting for their development. The interactions and relationships of these milieus, in which children develop their autonomy and cultural identity, represent both an integral part and a prerequisite to the child's exercise of their rights and freedoms. "It is within these semilegal orders where children internalize and negotiate their cultural background in the process of constructing their identities—and hence also their duties and rights" [2] (p. 169).

Arguably, a sound alternative to the above contestable theories is offered by the feminist discourse on rights. Its core tenet is the idea that individuals develop within certain relational configurations and stresses the contextual interdependency and responsibilities. Based on this assumption, these theorists call for a debate on rights in light of these relationships.

The "relational theory of rights," which during the last three decades or so has become known in literature as the "feminist ethic of care" [e.g. 9, 10, 11], has established itself as the main alternative to the conventional human rights approach. It is grounded on the thought of Carol Gilligan, a feminist social scientist who challenges the assumptions of differences between the sexes as foundations for considering a person a rights' bearer through their rationality and autonomous decision-making capacity [2] (p. 169). According to the feminists who deal with the topic, this view rests on a distorted perspective on society, understood as consisting of detached, insulated individuals, in which one can only reach their own autonomy "by erecting a wall (of rights) between the individual and those around him" [12] (p. 12). The relational theory promoters' alternative tenet is that human beings are essentially interdependent, which makes autonomy possible only via relationships. These scholars also powerfully question the way in which the conventional discourse on human rights places reason and rational decision-making capacities at the foundation of granting rights. Such an outlook

> subjects the essence of humanity to reason and to the capacity "to analyze and calculate, to induce and deduce" although it is not the whole, nor even the paramount, of it.

> Instead, much greater attention needs to be given to contextual responsibilities, emotions, and other intuitive characteristics of human nature [13 quoted in 2] (p. 170).

Doing away with the conventional view on adults (basically males) as bearers of rights through authority, power and status, the relational perspective concentrates on the emotional and intuitive sides of relationships. As a *locus* of the latter's expression, the advocates of this approach designate the domestic (mostly feminine) private realm, epitomized by the concept of care – by far more adequate to children's condition [14] (p. 88).

In conclusion, children's rights are viewed rather as a matter of relationships and (moral) responsibilities than as a matter formal rules. This links children's rights to their informal and contextualized environment rather than to the abstract normative realm. Thus, this focus on relations as the main way to one's exercise and the realization of rights leaves behind the Western view of a world defined by dichotomies: female/male, child/adult, I/the "other," private/public, normal/abnormal, culture/individual, a.s.o. "As such, the relational approach opens a discussion on children of all ages as active participants and as citizens of the world" [2] (p. 170).

In the particular case of children's bioethics, the relational perspective has the notable advantage of overcoming difficulties inherent to dominant approaches such as principlism with its doctrine of informed consent - completely premised on the concept of rational decision making made by independent individuals. Based on the person's (defined mainly as a separate individual rather than an interdependent, connected, relational subject) inherent dignity and moral worth, children's rights to exercising informed consent under the Children's Convention [15] is inescapably restricted. Furthermore, there are no *expressis verbis* references to the child's informed consent in the Children's Convention, the European Conventions on Biomedicine or other international legal codes, which implies that children do not have this right. Even more illustratively, in the Convention on Human Rights and Biomedicine that specifically requires informed consent for any biomedical intervention, the assumption with regard to children is that they are "persons not able to consent," that right being conferred to the child's representative or guardian authorized by law, even if the child's opinion should be given a hearing as it is worth "taking it into consideration as an increasingly determining factor in proportion to his or her age and degree of maturity" [16].

Even though when given supportive environment and communication, children way below the age of ten have proven capable of giving a rationalized informed consent regarding their own bodily interventions, issues comparable to that of adults [17] (p. 190) only adults – i.e. parents, legal guardians, community, courts, and health care professionals – are the ones making the "right" biomedical choice on their behalf.

In case of children, applying the ethic of care with its emphasis on moral responsibilities leaves aside the biomedical conflict that opposes the child's guardian(s) to the medical professional, or the former's *fulfilment* duties to the state's *enforcement* attributes, and encourages all actors to join forces in finding and providing the best care available; this approach also gives the child's voice a hearing. Owing to its essentially relational character, it also favours the contextualization of bodily practices by genuinely prioritizing the assessment of risks and benefits for the child in question.

Assuming that the society's moral fabric is built on relational selves possessing extended social brains rather than individual separate(d) subjects, the relational approach views the child's guardian(s) and carers as the best experts on the child(ren) they care for and thus the bodily practices on their child have higher chances to be genuinely interested in the mental consequences and gains for their child(ren) via their sense of communitarian belonging and identity. In conclusion, writes Sabatello,

> Considering the indisputable connection between the body, society, and culture, and between these and the individual and his or her identity, children's bioethical rights cannot be seriously discussed by excluding their sociocultural sources of influence, but rather by acknowledging and involving their autonomy-in-relations with these webs around them [2] (p. 174).

In the same line of thought, Leora Bilsky convincingly argues in favour of viewing children's development as intrinsically influenced by their communities, which calls for the need to create "an explicit space for children's autonomy *as* identity" [17] (p. 145). Particularly relevant for children's bioethics, the relational approach creates the context for defining the criteria of autonomy regarding bodily interventions taking into account the intertwining of the group's and the individual's sense of identity via the webs of relations relevant for children's lives. This framework facilitates children's own voices in defining their autonomy and rights. Thus, the understanding of autonomy qua identity as far as children's bioethics is concerned rests on the paramount importance of relations, intuitions, and emotions that prevails over a traditional individualist discourse on rights. As a consequence, "a conceptualization of the entire discourse on children's bioethics needs to recognize the relational experiences of children, their dependency, and their attachments to both individuals and collectives who constitute their communities of reference" [14] (p. 91). And, in the end, these attributes need to be defined as components of the child's cultural identity.

IV. The attachment styles and their role in structuring children's relational identity

The above insights into the relational essence of children's autonomy according to the feminist relational ethics (or ethics of care) can arguably be strengthened by a few brief references to recent developments in psychological theories of attachment [18]. Our hope is that, if extended beyond the child – primary care givers realm, on cultural-relational level, the findings of the attachment theory can stand as a powerful argument against the individualist discourse on rights and its bioethical consequences. According to the attachment theory, the individual's life is essentially structured by the attachment style s/he was exposed to in early infancy and childhood. The theory's main representative, John Bowlby [19, 20] clearly stated that one of the essential components of attachment style is *the internal working models*. They are internalized on the basis of the quality of the child's experience of their primary attachment figure and will be employed in future significant relationships. In what follows I will briefly highlight the specific of the reality internalized by children, for each of the attachment styles described in the literature.

Therefore, being constituted early in a warm and responsive relationship with the significant adult, *secure attachment* acts as a protective factor for children [21]. Specifically, s/he internalizes her/his self-confidence and the ability to adapt to new and challenging situations, using the parent as secure base from which to explore the environment freely, without interruptions caused by anger or fear (Grossman, Grossman & Zimmerman, 1999, as cited in [21]).

Other studies present the current relational experience as being related to one's belief systems, which are formed in one's family of origin, and tend to be strongly and consistently reinforced after having been internalized during an extremely vulnerable period of life [22], when a child is most impressionable. In that early life context, parents and other primary caretakers appear to have a powerful influence on the development of children's belief systems [22].

According to Bowen [23], children who reached maturity may tend to imitate their parents' interaction within their own marriages and current families. Bowen also cautioned that mere exposure to individual and family functioning does not adequately explain the intergenerational transmission process, as the actual transmission process was often inconsistent and occurred at an emotional level [24].

At the same time, more studies linked perceived parental models to the attachment style [25, 26], because this is also structured in childhood, during the early experience with the caregiver. In this context, multilevel modeling results indicated that memories of childhood emotional abuse reported by students and their partners who acted as subjects in these case-studies were significantly associated with attachment strategies, as well as with romantic relationship quality. Findings supported hypothesized mediation effects of attachment anxiety and avoidance on couple interaction and satisfaction [27].

As stated above, Dattilio's research strongly suggests that belief systems are deeply rooted in experiences acquired during early childhood family life. Due to this profound grounding these systems proved very resistant to change through psychotherapy or other personal development programmes aiming to counter their dysfunctional effects [22].

Insecure attachment styles occur when the parent is unable to calm the child, due to his or her emotional unavailability or inability of decoding the child's needs. Here we are talking about the *ambivalent* (or *anxious/resistant*) style and the *avoidant* style [28, 29]. If the parent is unpredictable and inconsistent in their responses to the child's needs, the latter cannot rely on the availability of the attachment figure and will develop an *ambivalent* attachment. These children will manifest a low degree of independence in exploring the environment, due to anxiety of being away from the parent [21]. Rejective or hostile answers, as well as parental indifference lead to the development of an *avoidant* attachment style in children. The child can learn to avoid others and deny their own needs, and thus appear independent. But this is only the result of the fact that s/he has not learned to depend on others [28, 29].

More recent research [29] has shown that *disorganized/disoriented* attachment style appears in children subjected to abuse or neglect by the parent. These children are hypersensitive to abuse and appear disoriented or confused in the presence of their parent. They express a combination of strategies specific to the ambivalent and rejective styles, having moments when they try to please the parent, immediately followed by moments when they show anger or rejection towards him/her.

Back to internal working models [30] related to attachment styles, I need to stress that they are constantly reviewed and updated. Therefore, their impact on the psychosocial functioning of a child of a certain age may depend on the security of the representations developed within that period [31].

Children's early representations of their own experiences, of others and of the nature of relationships is outlined in the context of the discourse *shared with others* [32, 33]. In this respect, the influence of the main care giver upon the child's attachment structure works in two ways: through the quality of the care given, and through the interpretation of events offered during discussions with children. The security or insecurity of children's representations of relational experiences are thus generated and maintained [31]. In this context, parents' behaviour has been identified by a number of researches as the link variable between adult attachment and child attachment, securely attached children having sensitive mothers who respond promptly to their needs [3, 20].

The anxiety dimension of attachment within the couple relationship bears on the negative internal model about the self and the world [30; 35], causing partners to worry that their partner might abandon them, to become over-clinging, having the feeling they love their partner more than the partner loves them. Individuals exhibiting a high level of anxiety as adult attachment dimension need assurances from their partners that they are important and loved. The attachment theory, and the theoretical perspectives on intergenerational transmission of attachment show a repetitive pattern of attachment across generations: the attachment style in adulthood is manifested both in the couple relationships and in the romantic relationships that adults establish [36], and in the relationship with their child [37]. Concurrently, the research conducted by Mikulincer and Shaver, shows that people with anxious attachment to relationships, worried not to be abandoned by the loved ones, will adopt "hyperactive attachment strategies, being active, kind, persistent in trying to obtain love and support" [38] (p. 33).

The implication of the above account of attachment styles is that they set the individual's attitudinal and relational blueprints in the society at large. Extended, beyond the child-primary care givers realm, at cultural-relational level, these findings reinforce the argument that the society's moral fabric is built on relational selves possessing extended social brains rather than on individual separate(d) subjects. Our suggestion is that coupling the attachment theory with the relational-cultural identity approach(es) helps both the understanding of the relational core of the person's identity and possible background roots of various dysfunctions manifested by individuals within their social-cultural environments.

V. Philosophy for children as enhancer of social- and cultural-relational competences

The attainment of a more autonomous-in-relations identity of children and young people requires a fundamental shift in children's views on knowledge, change, body, community, diversity, initiative, environment, and other related realities. In our opinion, one of the recent interactive mind competencies and emotional intelligence training practices, namely Philosophy for Children (P4C), increases significantly the chances of producing these changes e.g. [39] (p. 379); [40]. At the end of a survey of a large number of research-based studies of P4C outcomes, K. J. Topping and S. Trickey conclude:

> Ten studies [of] P4C [results] in primary and high schools met the stringent criteria for inclusion, measuring outcomes by norm-referenced tests of reading, reasoning, cognitive ability, and other curriculum-related abilities, by measures of self-esteem and child behaviour, and by child and

> teacher questionnaires. All studies showed [...] a consistent [...] positive effect for P4C on a wide range of outcome measures [41] (p. 379).

Although presented below only as a short introductory glimpse, P4C's merits are summarized simply as a preliminary opening to the study of its prospective contributions to the development of the social- and cultural-relational competences and identity awareness in children.

P4C is more than a simple sum of programs of stimulating and fostering thinking abilities. In fact, it comprises a complex of specific resources and methods of tapping into and encouraging children's curiosity and natural propensity toward interconnectivity, helping them in their search for meaning, stimulates and develops their intellectual courage and helps them rigorously develop skills that enable them to make correct judgments in their day-to-day lives, but also stimulate and enhance empathy, compassion, acceptance and tolerance. In the countries where this type of skill training is already becoming a tradition, the visible result is an increasing number of assertive young people, capable to relate to knowledge in a critically-constructive way, to articulate and conceptualize their ideas and think independently and innovatively, to be tolerant toward others' ideas and beliefs, and to work within groups of likeminded people. A description of such achievements of the P4C programs in New Zealand reads:

> Doing philosophy in a community of inquiry gives all students a "voice" as well as teaching them appropriate ways to express themselves and to have their contributions heard. The voices of all students are encouraged and included in classroom dialogue. A range of co-operative skills [including] … listening to each other; thinking about and building on each other's ideas; respecting everyone's ideas…, create a space in which students can interact with both gentleness and rigor. This in turn allows students to hear, appreciate and challenge each other's thoughts and perspectives, and often leads to a new valuing of classmates [42].

The Philosophy for Children's central educational tool is the community of enquiry, a structure that can play a crucial role in "combating what is perceived to be a drift in society to the idea that opinions can't be judged and don't need to be justified" [43]. This type of community is a group of people united in the examination of an area of common interest via a process of dialogue-based enquiry. This type of community represents "a context for discussion wherein participants are challenged to justify their opinions regularly" [43]. The community of enquiry "aims at understanding the issues in question or the beliefs of other participants. Argument is seen as a collaborative effort to come to the best answer to a question" [43]. Facilitators specialized in P4C develop their class towards becoming a community of inquiry by some straightforward practices. In order to explain them, these practices are generally organized in four categories: encourage questioning, develop concepts, encourage dialogue and argument, and work for reasonableness [43]. Moreover, all four Cs of P4C's interconnected types of thinking, namely caring, collaborative, critical and creative, also include a strong and deep social skill component which encourages, cultivates and structures the social attitudes that define the relational side of children's selves.

However, the Philosophy for Children's founder, Matthew Lipman, strongly believes that the main confronting views on social learning oppose the reflective model to the standard model on the ground that the main aim of the former is the autonomy of the learner, instead of their dependency on, or reproduction of, the teaching authority figures. Yet, the independent reflective thinker has nothing in common with a "self-sufficient cognitive macho … protected by an umbrella of invincibly powerful arguments" [44] (p. 25). Instead, they are deeply social and communal. The reflective model's objective is "to articulate the friction-causing differences in the community, develop arguments in support of the competing claims, and then, through deliberation, achieve an understanding of the larger picture that will permit more objective judgment" [44] (p. 25).

According to Lipman, P4C's mission is "to help children learn how to think for themselves," via "improvement of reasoning ability; development of creativity; personal and interpersonal growth; and development of ethical understanding" [45] (p. 53, 78). Children can be helped to identify these ideas and concerns in specially written stories, as well in traditional stories, novels, or movies for children. During the philosophy for children classes, such themes are subjected to investigation procedures within the community of enquiry that resort to instruments of philosophical reasoning and imaginative exploration under the guidance of a teacher whose role is to stimulate children to think cooperatively and assume responsibility for their learning.

A community of enquiry is based on a collaborative knowledge creation concept founded on a community of philosophical enquiry theory which actively engages collectively and interconnectedly those concerned with a particular issue. As a conclusion repeatedly drawn in the studies on P4C's impact on children's social skills, D. Spiteri correctly summarizes the fact that, owing to their profound dialogic critique and internalizing reconstruction of knowledge, the communities of enquiry essentially contribute to the enhancement of intercultural and intersubjective sensitivity [46].

VI. Conclusion

By challenging the dominant views on children's rights, i.e. the will/power and the interest theories - especially their biomedical implications, the relational theory of rights exposes these traditional/conventional views' overlooking of life experiences of children as present individuals whose identities and autonomy are essentially defined by their relational and attachment contexts. As main resources, the above argument appealed to the feminist ethic of care, the theory of attachment and philosophy for children, in order to call for a reconceptualization of the entire discourse on children's bioethics, based on the integration of the relational experiences of children (including their intuitional and emotional underpinnings) as building blocks of their identity, instead of individualism. Individualism is criticized mainly for its misrepresented perspective on society, understood as consisting of detached, insulated individuals, in which one can reach their own autonomy.

As a counterargument to these traditional perspectives, the study brings forth the importance of children's relation webs in general, of those defining attachment in particular, in the make-up of individuals from the earliest stages in life. In this respect, reputed attachment theory research shows how an individual's current relational experience is deeply linked to one's belief systems, which are formed in one's family of origin. They tend to be very strongly and consistently reinforced after being internalized during an extremely vulnerable period of life, when a child is most impressionable. Parents and other primary caretakers have a powerful influence on the development of children's belief systems, which contribute essentially to the structuring of their future social interaction via the children's relational selves. Extended at cultural and relational levels, these findings reinforce the argument that the society's moral fabric is built on relational selves possessing extended social brains, rather than on individual separate(d) subjects.

As a suggested educational intervention strategy aiming to enhance in children both the autonomy-in-relation and the awareness of their cultural and relational identities, the philosophy for children's four Cs, namely caring, collaborative, critical and creative thinking, include a strong and deep social skill component which encourages, cultivates and structures the social attitudes that define the relational side of children's identities.

References

[1] E. Durkheim, *The Elementary Forms of Religious Life*, trans. J. W. Swain. New York: Free Press, 1965.

[2] M. Sabatello, *Children's Bioethics: The International Biopolitical Discourse on Harmful Traditional Practices and the Right of the Child to Cultural Identity*. Leiden: Martins Nijhoff Publishers, 2009.

[3] C. Jenks, "Many Childhoods?" *Childhood,* vol. 11, no. 1, pp. 5-8, 2004.

[4] J. Roche, "Children: Rights, Participation and Citizenship," *Childhood,* vol. 6, no. 4, pp. 475-493, 1999.

[5] J. Griffith, "Do Children Have Rights?" *The Moral and Political Status of Children*, David Archard and Colin M. Macleod, ed., Oxford: Oxford University Press, pp. 19-30, 2002.

[6] R. Noggle, "Special Agents: Children's Autonomy and Parental Authority," *The Moral and Political Status of Children*, David Archard and Colin M. Macleod, ed., Oxford: Oxford University Press, pp. 97-117, 2002.

[7] O. O'Neill, "Children's Rights and Children's Lives," in *Children, Rights and the Law,* Philip Alston, Stephen Parker and John Seymour, ed., New York: Oxford University Press, pp. 24-42, 1995.

[8] T.D. Campbell, "The Rights of the Minor: As Person, as Child, as Juvenile, as Future Adult," *Children, Rights and the Law,* Philip Alston, Stephen Parker, and John Seymour, ed., New York: Oxford University Press, pp. 1-23, 1995.

[9] J. O'Manique, *The Origins of Justice: The Evolution of Morality, Human Rights and Law,* Philadelphia: University Pennsylvania Press, 2003.

[10] M. Minow, "Rights for the Next Generation: A Feminist Approach to Children's Rights," *Women's Law Journal,* vol. 9, 1986.
[11] T. Cockburn, "Children and the Feminist Ethic of Care," *Childhood* vol. 12, no. 1, pp. 71-89, 2005.
[12] J. Nedelsky, "Re-conceiving Autonomy: Sources, Thoughts and Possibilities," *Yale Journal of Law & Feminism* vol. 1, pp. 7-36, 1989.
[13] J. O'Manique, *The Origins of Justice: The Evolution of Morality, Human Rights and Law,* Philadelphia: University Pennsylvania Press, 2003.
[14] B. Arneil, "Becoming versus Being: A Critical Analysis of the Child in Liberal Theory," *The Moral and Political Status of Children*, David Archard and Colin M. Macleod, ed., Oxford: Oxford University Press, pp. 70-96, 2002.
[15] Convention on the Rights of the Child, G.A. Res. 44/25, annex, 44th Sess., 61st plenary meeting. U.N. GAOR Supp. (No. 49), at 167, U.N. Doc. A/44/49 (1989), entered into force September 2, 1990, reprinted in 28 I.L.M. 1448 (1989).
[16] Article 6(2) of the Convention on Human Rights and Biomedicine.
[17] L. Bilsky, "Child-Parent-State: The Absence of Community in the Courts' Approach to Education," *Children's Rights and Traditional Values,* Gillian Douglas and Leslie Sebba, ed., Burlington, VT: Ashgate Publishing, pp. 134-159, 1998.
[18] M. A. Ionescu, "Intersubjectivity and relational health in mother and child attachment dimensions." *Procedia. Social and Behavioral Sciences,* vol. 163, pp. 159-168, 19 December 2014.
[19] J. Bowlby, *Attachment and Loss, Vol. 3. Loss, Sadness and Depression*, New York: Basic Books, 1980.
[20] J. Bowlby, *A secure base: Parent-child attachment and healthy human development*, New York: Basic Books, 1988.
[21] C. Andreassen, P. Fletcher, *Early Childhood Longitudinal Study, Birth Cohort (ECLS-B): Psychometric Report for the 2-Year Data Collection. Methodology Report.* NCES 2007-084. Washington, DC: National Center for Education Statistics, Institute of Education Sciences, U.S. Department of Education, 2007.
[22] F.M. Dattilio, "Restructuring schemata from family of origin in couple therapy," *Journal of Cognitive Psychotherapy* vol. 20, pp. 359-374, 2006.
[23] M. E. Kerr, M. Bowen, *Family evaluation*, New York: Norton, 1988.
[24] J. H. Larson, S. N. Wilson, "Family of origin influences on young adult career decision problems: A test of Bowenian theory," *American Journal of Family Therapy*, 26 (1998) published online at https://doi.org/10.1080/01926189808251085, pp. 39-53, 13 June 2007
[25] C. Perris, P. Andersson, "Experiences of parental rearing and patterns of attachment in adulthood," *Clinical Psychology and Psychotherapy*, vol. 7, pp. 279-288, 2000.

[26] C. Hinnen, R. Sanderman, M. A. G. Sprangers, "Adult attachment as mediator between recollections of childhood and satisfaction with life," *Clinical Psychology and Psychotherapy*, vol. 16, pp. 10-21, 2009.
[27] S. A. Riggs, A. M. Cusimano and K. M. Benson, "Childhood emotional abuse and attachment processes in the dyadic adjustment of dating couples," *Journal of Consulting Psychology*, vol. 58, pp. 126-138, 2011.
[28] M. E. Lamb, "Attachment." *Encyclopedia of Psychology*, A. E. Kazdin, ed., Washington, DC: American Psychological Association, 2000.
[29] M. Main, "Attachment theory," *Encyclopedia of Psychology*, A. E. Kazdin, ed., Washington, DC: American Psychological Association, 2000.
[30] J. Bowlby, *Attachment and Loss: Volume II: Separation*, New York: Basic Books, 1973.
[31] R. A. Thompson, "Early attachment and later development," *Handbook of attachment: Theory, research, and clinical applications*, Cassidy, J., Shaver P. R., ed., pp. 287-316. New York: Guilford Press, 1999.
[32] R. Fivush, "Constructing narrative, emotion and self in parent-child conversations about the past," *The remembering self: Accuracy and construction in the life narrative*, U. Neisser and R. Fivush, ed., pp. 136-157. NY: Cambridge University Press, 1994.
[33] D. Oppenheim, H. S. Waters, "Narrative processes and attachment representations: issues of development and assessment," *Caregiving, cultural, and cognitive perspectives on secure-base behavior and working models: New growing points of attachment theory and research. Monographs of the Society for Research in Child Development*, E. Waters, B. Vaughn, G. Posada, and K. Kondo-Ikemura, ed., Chicago: University of Chicago Press 1995.
[34] M. S. Ainsworth, M. C. Blehar, E. Waters, S. Wall, *Patterns of attachment: A psychological study of the strange situation*, Hillsdale, NJ: Lawrence Erlbaum, 1978.
[35] C. E. Hamilton, "Continuity and discontinuity of attachment from infancy through adolescence," *Child Development*, vol. 71, pp. 690-694, 2000.
[36] A. Nosko, T. T. Tieu, M. W. Pratt, "How do I love thee? Let me count the ways: Parenting during adolescence, Attachement Styles and Romantic narattivs in Emerging Adulthood" *Developmental Psychology*, vol. 47, no. 3, pp. 645-657, 2011.
[37] L. E. Crandell, H. E. Fitzgerald, E. E. Whipple, "Dyadic synchrony in parent-child interactions: A link with maternal representations of attachment relationships," *Infant Mental Health Journal*, vol. 18, pp. 247-264, 1997.
[38] M. Mikulincer, P. R. Shaver, *Attachment in adulthood: Structure, dynamics, and change*, New York: Guilford Press, 2007.

[39] A. Kizel, "Philosophy with Children as an educational platform for self-determined learning" *Cogent Education*, vol. 3, 2016. http://dx.doi.org/10.1080/ 2331186X.2016.1244026.

[40] N. Siddiqui, S. Gorard, S. B. H. See, "Non-cognitive impacts of philosophy for children," *Project Report. School of Education*, Durham University, Durham, 2016. https://www.dur.ac.uk/education/research/groups/?mode=projectid=738

[41] K. J. Topping, S. Trickey, "Philosophy for Children: A systematic review," *Research Papers in Education,* vol. 19, no. 3, pp. 365-380, 2004.

[42] http://www.p4c.org.nz/new-zealand-curriculum/p4c-and-the-nz-curriculum/

[43] http://p4c.com/teachers-guide. Accessed May 20th, 2018).

[44] M. Lipman, *Thinking in Education.* Cambridge: Cambridge University Press, 2003.

[45] M. Lipman, A. M. Sharp, F. S. Oscanyan, *Philosophy in the Classroom*, 2nd edition, Philadelphia: Temple University Press, 1980.

[46] D. Spiteri, "The Community of Philosophical Enquiry and the Enhancement of intercultural Sensitivity," *Childhood & philosophy*, vol. 6, no. 11, pp. 87-111, 2010.

Implementing Purity and Combating Impurity: Biopower and Totalist Movements

Mihai Murariu

Political Sciences Department,
West University of Timisoara, Romania

Abstract

This essay discusses the interaction between totalistic worldviews on one side and biopower and perspectives on purity and impurity, on the other side. It does so by using the concept of totalism, a system of thought defined by its pursuit of the complete reconstruction of society in accordance with its soteriological-simplifying principles. Totalistic worldviews tend to enforce the implementation of what they see as purity and the combating of what they see as impurity by enforcing their ultimate aims on their host society or beyond.

Keywords

Totalism; biopower; heterodoxy; totalitarianism; purity; modernity.

I. Biopower and Totalitarianism. Premodern Roots

This essay deals with the importance and the possible interaction of totality with biopower in premodern and modern contexts, taking into account perspectives that use categories of purity and impurity. This is achieved by making use of the concept of totalism, a system of thought that places a supreme value on totality, and which, particularly in militant variants, pursues the complete reconstruction of society in accordance with its soteriological-simplifying principles, while claiming a monopoly on interpreting the ideological truth of its own doctrinal core [1].

Before moving on, one must briefly approach Foucault's view on biopower, as well as his description and understanding of totalitarian regimes, particularly in the case of National Socialism. This is a necessary step, since it is a perspective which is not only intrinsically linked with Foucault's own concept of biopower, but, in many respects, it also represents the apparent highpoint of its application. Yet, as it will be briefly developed later in the essay, totalitarianism arguably does not represent a fundamentally new development in terms of intent, but, rather, can be defined merely as a stage in the possible evolution of regimes which may be called totalist ideocracies. It is also worth mentioning that such ideocratic polities – and the totalist ideologies which typically precede them – would obviously not be as bound by the constraints of pluralist democracies when it comes to bioethical dilemmas, particularly when the totalist ultimate aims would demand a certain course of action. In the context of an apparent contraction of liberal-democratic normative power, and the increased visibility and success of various hybrid forms of illiberal, autocratic, and even ideocratic hybrids, a look at the possible evolution and importance of such factors in totalist contexts is all the more necessary.

In his *History of Sexuality*, Foucault first mentions the "era of biopower" by referring to "an explosion of numerous and diverse techniques for achieving the subjugations of bodies and the control of populations" [2] (p. 140). While Foucault would not revisit the concept of biopower in that particular work, he would later come to persistently focus on the importance of sovereignty when it comes to the right to kill, along with regularization:

> Beneath that great absolute power, beneath the dramatic and somber absolute power that was the power of sovereignty, and which consisted in the power to take life, we now have the emergence, with this technology of biopower, of this technology of power over "the" population as such, over men insofar as they are living beings. It is continuous, scientific, and it is the power to make live. Sovereignty took life and let live. And now we have the emergence of a power that I would call the power of regularization, and it, in contrast, consists in making live and letting die [3] (p. 247).

Having said this, Foucault saw in National Socialism nothing less than the "paroxysmal development of the new power mechanism that had been established since the eighteenth century" [3] (p. 259). In his view, not only was there no state which could ever have more "disciplinary power" than the National Socialist regime, but there was also no other state which regulated the biological so tightly and insistently:

> Disciplinary power and biopower: all this permeated, underpinned, Nazi society (control over the biological, of procreation and of heredity; control over illness and accidents too). No society could be more disciplinary or more concerned with providing insurance than that established, or at least planned, by the Nazis. Controlling the random element inherent in biological processes was one of the regime's immediate objectives. But this society in which insurance and reassurance were universal, this universally disciplinary and regulatory society, was also a society which unleashed murderous power, or in other words, the old sovereign right to take life. This power to kill, which ran through the entire social body of Nazi society, was first manifested when the power to take life, the power of life and death, was granted not only to the State but to a whole series of individuals, to a considerable number of people (such as the SA, the SS, and so on). Ultimately, everyone in the Nazi State had the power of life and death over his or her neighbors, if only because of the practice of informing, which effectively meant doing away with the people next door, or having them done away with. So murderous power and sovereign power are unleashed throughout the entire social body [3] (p. 259).

In this case, the most important thing to take away from Foucault's interpretation is that he accurately points to the totalistic nature of the National Socialist regime, both in its potential and its application. Once again, it is the combination of the sovereign right to kill and the new mechanism of biopower, of managing of life that "coincide exactly," with momentous consequences:

> We have, then, in Nazi society something that is really quite extraordinary: this is a society which has generalized biopower in an absolute sense, but which has also generalized the sovereign right to kill. The two mechanisms—the classic, archaic mechanism that gave the

> State the right of life and death over its citizens, and the new mechanism organized around discipline and regulation, or in other words, the new mechanism of biopower—coincide exactly. We can therefore say this: The Nazi State makes the field of the life it manages, protects, guarantees, and cultivates in biological terms absolutely coextensive with the sovereign right to kill anyone, meaning not only other people, but also its own people. There was, in Nazism, a coincidence between a generalized biopower and a dictatorship that was at once absolute and retransmitted throughout the entire social body by this fantastic extension of the right to kill and of exposure to death [3] (p. 260).

Nevertheless, this combination of – at least apparent – absolute control over both life and death, while relatively rare in terms of historical occurrences, is not quite as "extraordinary" as Foucault suggests, unless one explicitly focuses on its scale alone and its use of industrial modernity to achieve its purpose. Indeed, the pursuit of such power is far older than the National Socialist project, with its origins already clearly identifiable in the great totalistic socio-political reconstruction projects of the premodern era. Certainly, in a premodern context, there had been no real possibility for entire societies to be organised and regimented with quite the same efficiency. However, the idea of pursuing a series of soteriological-simplifying principles and of applying them in a total manner on a local host society or even on a universal level can be encountered time and again in the case of premodern totalist heterodoxies. As such, it is in heterodoxies which placed a supreme value on totality that one finds the roots leading up to the formation of modern totalitarianism and biopower. And if a totalitarian phase represents the greatest and most dangerous use of biopolitics and biopower, as well as in organising the dealing of death, it is can only originate from totalist visions.

Rather than being born from the tensions inherent in the project of modernity, totalist projects – either in a quietist or militant variant – can be encountered far earlier, along with the formation of complex premodern societies and the implications posed by systems of thought which hold totality as a supreme value. Thus, whereas modern nomic crises can certainly contribute to the emergence of such worldviews, their roots go far deeper. Perhaps most importantly, the pursuit of a totalistic social reconstruction by militant heterodox groups or by successfully established totalist ideocracies typically leads to a distinct outlook on human life and on how it should be led. Whether they are religious, secular, or hybrid in nature, such totalistic perspectives tend to enforce the implementation of what they see as purity and the combating of what they see as impurity. To reiterate an essential point here, totalism is not to be confused with totalitarianism, despite the similarities which they share in their supreme focus on totality. Thus, totalism represents a system of thought pursuing the reconstruction of society in accordance with the soteriological-simplifying principles forming its doctrinal core, to which it claims a monopoly of interpretation.

In a Western European context, it was the overall failure of the great eschatological and utopian heterodoxies of the Middle Ages – such as those expanded upon by Cohn [4], Eisenstadt [5], and Kaplan [6], that decisively contributed to the triumph of the State, as its eventual alliance with either the established Church or with more pragmatic reformist movements eventually allowed it to move fully into the centre of political power. The case of heterodoxies mixing eschatological expectations with distinctly renovative or utopian totalism is encountered across Eurasia, with certain common features uniting them, both in success and failure – such as their readiness to take on established authority in pursuit of their own take on an ideal human existence and how it should be organised. Manifesting with a high degree of regularity, although with varying intensity in their potential for militantism, these heterodoxies form an important part of what may be called the shadow history of modernity.

To make only a brief point, three major Eurasian civilizational areas, (Christendom, Islamdom, and the Sinic cultural sphere) provide one with a large number of heterodox movements that sought the overthrow and transformation of their host societies. Nevertheless, most eventually lost their battle with the polity they were up against, with some exceptions, particularly in the Islamic case, which repeatedly saw such heterodoxies overthrow the established order successfully and begin implementing their vision upon their host society.[1] This is part of cyclical patterns that have been part of Islamdom since its beginnings, due in no small part to the great reverence held for the first three generations and for the person of Muhammad.[2]

It was only in the 20th century that European and Asian states collapsed under the onslaught of such totalist heterodoxies, whether in the case of Russian Bolshevism, Italian Fascism, German National Socialism, or Chinese Communism, to name only the most influential. With them, they brought the recurrent tendencies of the totalist mindset – the desire and necessity to totally reconstruct society according to the soteriological-simplifying principles dominating their doctrinal core. The main difference was that modernisation now granted them the means which other premodern, totalist heterodoxies and ideocracies had lacked. At least in theory, the whole of society was now unprecedentedly open to the power of the state – and, in those ideocracies which did not abandon their ideological purity to the pragmatism necessary for governing a relatively stable realm, this unleashed the revolutionary implementation phase of their ultimate aims, or what is commonly understood as totalitarianism.[3]

Lastly, when it comes to the distinct environment shaped in totalitarian regimes, Padovan is right to point to the "unique, unrepeatable, convergence between State policy, Social Science and political ideology, a convergence that makes it possible to formulate a whole system of strategic knowledge about society" or, in other words, the creation of "a 'total' system of biopsychosocial control" [7] (p. 494).

[1] Relevant premodern examples range from the Fatimid Caliphate in Egypt and the rise of the Iranian Safavid Empire to the Fulani jihad and Mahdist state in West and East Africa, respectively.

[2] To this one must add the strength of the *mujaddid* tradition, which essentially pursues religious renewal. This led to periodical attempts at "restoring" this ideal state of affairs and correcting the divergences separating the many Muslim societies.

[3] The process through which such heterodoxies were affected and, in turn, sought to affect themselves the project of modernity, will require a more detailed approach elsewhere since it does not represent the primary focus of this essay.

II. Purity and Impurity in Service of Totalism

As it has been mentioned previously, the apparently exceptional nature of totalitarianism is somewhat reduced when the modern means of control and organisation are no longer considered their primary traits. It is, after all, no coincidence that Carl J. Friedrich, one of the foremost scholars of totalitarianism, gradually came to regard the existence of a totalist ideology as the key essential aspect for such systems [8] (p. 126). Moreover, rather than being truly extraordinary, totalist worldviews – including non-militant ones – represent instead a deep-seated aspect of human culture, even though factors such as perceptions towards contradictions may influence their evolution. Nevertheless, totalism as a scientific concept is first encountered in German social science debates during the early 1920s, after which it resurfaced in the United States, most famously in a psychological context in the works of Erik Erikson and Robert Lifton, yet always tied to totalitarianism and, ultimately, to what can be called totalist ideologies. Somewhat unsurprisingly – due to their scientific backgrounds – Erikson and Lifton tend to use the term in an individualistic manner, representing the propensity of the individual for all or nothing alignments, rather than going any further.

Lifton's eight criteria which may be associated with totalism are worth mentioning here. These are as follows:

- Milieu control
- Mystical manipulation
- The demand for purity
- The cult of confession
- The "sacred science"
- Loading the language
- Doctrine over person
- The dispensing of existence

Lifton points out that each of these features possesses a totalistic quality, depending on an "equally absolute philosophical assumption", whilst mobilizing certain individual emotional tendencies which are "mostly of a polarizing nature". Moreover, Lifton argues that such features, when combined, "create an atmosphere which may temporarily energize or exhilarate, but which at the same time poses the graves of human threats" [9] (p. 420). In approaching these eight themes, Anthony and Robbins have rightly pointed out that some of the most important factors identifiable here are the presence of total commitment, the presence of duality, and an "apocalyptic" worldview [10].

Yet, it must be considered that totalism itself as a concept need not be understood merely as an individual predisposition for adherence to totalist ideologies, but may also be expanded to include a system of thought which places a supreme value on totality and which, consequently, is typically inimical to plurality and ambiguity. Moreover, totalist projects can include both religious and secular perspectives, since totality as a concept may be – at least in some cases – more important than ideological features. And while totalism can also be encountered in quietist movements, it is its potential for fuelling militancy that is the most dangerous. Lastly, the potential for conflict increases when such worldviews must interact with ambiguity and pluralist values.

The ideal evolution path of a totalist movement typically has its starting point in a charismatic heterodoxy, which, when insufficiently challenged, morphs into a movement able to achieve political hegemony, with the final and most difficult step being the transformation into an ideocracy, with the potential for undergoing a totalitarian phase. Furthermore, when looking at the common features encountered across the wide spectrum of totalist heterodoxies, premodern or modern, quietist or militant, secular or religious, the following points tend to stand out in their doctrinal core:

- Totality as supreme value
- Pure-Impure dichotomy
- Truth-Untruth dichotomy
- Ambiguity-Plurality as minimal values
- Soteriological-simplifying principles

In this respect, the relationship between purity and truth is critical, since it represents a powerful driving force for the implementation of the "good" and the combating of the "evil." Considering the importance placed by totalist heterodoxies on achieving their ultimate aims, the first focus is on the implementation of what they associate with purity – or, in other words, staying true to the doctrinal core of the original charismatic heterodoxy and its expansion as a model across its host society. The second focus is on the combating impurity, namely counterideological factors, potentially dangerous neutrals and elements providing contradictions. Totalistic social reconstruction projects may take various forms, such as concentrating on ideological adherence, on the idea of race, or of class [1] (p. 73). Unsurprisingly, in a socio-political sense, is this very attempt at total implementation of what is associated with purity and the combating of what is associated with impurity that has historically contributed to massive human mobilisation, as well as to disastrous human tolls – as the biopolitics of both totalist heterodoxies and modern totalitarian phases have made repeatedly made clear.

The pure-impure dichotomy – which tends to function alongside an equally powerful truth-untruth dichotomy – is, in turn, influenced by the intensity with which contradictions are handled. There are a variety of ways to react to and deal with contradictions in social situations, ranging from the usually extreme gesture of embracing the contradiction, to not acknowledging its existence (or reinterpreting its nature), to attempting to modify the contradiction. In the case of totalist movements, especially in the militant variants, the focus will be on a direct countering of the impure contradiction to the pure model offered by the doctrinal core. This is made clear even briefly analysing the main ideological features of two apparently different movements, such as in Anabaptist ideocratic experiment in Münster (1534-1535) and the movement known today as Islamic State. Thus, for the Anabaptists, the following features stand out:

- Pure Model: The perfect society is found in the imagined, early Biblical community.
- Impure Contradiction: Those who have not chosen re-baptism and, as a result live in a state of impurity and untruth. Essentially all who do not belong to the chosen community are included in this category.
- Ultimate Aim: Property, behaviour, and spiritual life are to be organised according to the doctrinal core of the movement, namely, the Anabaptist interpretation of the Scriptural traditions. Implementing this model on a worldwide scale, through elimination of impure, counterideological factors.

By comparison, despite being overall a much more deadly and bloody movement, Islamic State shares both the totalist vision and the ideological ambition to expand its value system and implement its ultimate aims across the world:

- Pure Model: Islam as an unchanging, perfect guide, which enables the model to appear both relevant and implementable.
- Impure Contradiction: Any existing contradiction to the ideal model must be eliminated and the purity of the original community and its model restored. Those who are part of the chosen group may wage war or eliminate those who are not, even as the impure categories have their own hierarchy, from Sunni rival movements, to Shia "heretics", to Western "Crusaders" and Jews.
- Ultimate Aim: The community must be organised solely according to the rules of the doctrinal core, namely, Islamic State's interpretation of the Qur'an and Hadiths. Ignorance must be destroyed, even if the majority of the population itself judged to be guilty of it. The pure chosen community has the task to implement purity and reconstruct society worldwide according to its doctrinal core, once again with a licence to eliminate any counterideological factors deemed impure.

While both are movements which can be associated with the renovative totalism type –that is, a type of totalism focussed on the past as an ideal implementable model – they also share a number of other features, such as attempting to maintain control over the lives of their followers, while seeking to expand the principles based on their doctrinal core beyond their political centres.

III. Conclusion

Thus, if a distinct type of biopolitics and biopower converged in an unprecedented manner with the rise of modernity and the emergence of ideocratic polities, such processes have deep roots in the premodern era, namely, in the series of socio-political projects which formed the doctrinal core of numerous totalist heterodoxies. If the totalitarian phase may be considered an overall exception in its ruthless and total attempt at implementing control over society and organising of life and death, then totalist movements, with their inherent focus on a pure-impure dichotomy, must necessarily be taken into account for any such conceptual archaeology. And if the convergence between science, policy, and ideology can be associated with totalising programs, the contemporary era has added yet another dimension to the mix, that of personal choice in the creation and customisation of the body – both of the present and of the future. This can be seen in the growing debates on issues such as reproduction autonomy, along with the possibility of effectively customising infants by the use of various enhancements. As it can be deduced from understanding the main features of totalism, if pluralist contexts themselves make the abuse of such developments far less likely, such is not the case in totalist ones, where little if any limit can be considered when it comes to the totalist ultimate aims. Lastly and most importantly in this regard, as society becomes ever more complex and fragmented, the demand for all-encompassing, soteriological-simplifying solutions increases, which turn leads to opportunities for the emergence and expansion of new totalist movements.

References

[1] M. Murariu, *Totality, Charisma, Authority. The Origins and Transformations of Totalist Movements*, Wiesbaden: Springer VS, 2017.

[2] M. Foucault, *The History of sexuality. Volume 1. An Introduction*, trans. Robert Hurley, New York: Pantheon Books, 1978.

[3] M. Foucault, *"Society Must Be Defended," Lectures at the College de France 1975-76*. Arnold 1. Davidson ed., trans. David Macey, New York: Picador, 2003.

[4] N. Cohn, *The Pursuit of the Millennium: Revolutionary Millenarians and Mystical Anarchists of the Middle Ages*, Oxford: Oxford University Press, Revised and Expanded Edition 1970.

[5] S.N. Eisenstadt, *Fundamentalism, Sectarianism and Revolution, The Jacobin Dimension of Modernity*, Cambridge: Cambridge University Press, 1999.

[6] J. Kaplan, "Nothing is True, Everything is Permitted: Premodern Religious Terrorism," *Terrorism and Political Violence*, DOI: 10.1080/09546553.2017.1314964, 2017.

[7] D. Padovan, "Biopolitics and the Social Control of the Multitude", *Democracy & Nature*, vol. 9, no. 3, pp. 473-494, Nov. 2003.

[8] C.J. Friedrich, "The Evolving Theory and Practice of Totalitarian Regimes," *Totalitarianism in Perspective. Three Views*, C.J. Friedrich, M. Curtis, and B.R. Barber, eds., New York: Praeger, pp. 123-164, 1969.

[9] R.J. Lifton, *Thought Reform and the Psychology of Totalism: A Study of "Brainwashing" in China*. New York: W. W. Norton & Co., 1989.
[10] D. Anthony, T. Robbins, "*Religious Totalism*, Violence, and Exemplary Dualism," *Terrorism and Political Violence*, vol. 7, no. 3, pp. 10–50, 1995.

The Psy-complex: Out of the Techno-Scientific Paradigm?

Attila Bánfalvi

Division of Humanities for Health Care, Institute of Behavioural Sciences, Faculty of Public Health, University of Debrecen, Hungary

Abstract

In the age of modern sciences, a basic discrepancy has been raised between the ontologically-different kinds of body and psyche on the one hand, and the scientific epistemology on the other.

a. Psychology as a presumed scientific endeavor uses mainly quantitative, statistical research methods which are based on the hypothesis that the psychic phenomena are essentially measurable, quantifiable.

b. Psychotherapies are broadly equivalent in their effectiveness. It seems that the psychotherapeutic efficacy is based mostly on common, non-technical elements.

c. The two-hundred-year-old 'pendulum' of psychiatry which has been swinging between the somatic and psychotherapeutic-social treatment approach, is now on the way to a less technically-oriented, more humane helping attitude.

Keywords

Psy-complex; psychology; psychotherapy; psychiatry; techno-scientific; humanistic.

Mottos

> Science, as its name implies, is primarily knowledge; by convention it is knowledge of a certain kind, the kind, namely, which seeks general laws connecting a number of particular facts. Gradually, however, the aspect of science as knowledge is being thrust into the background by the aspect of science as the power of manipulating nature. It is because science gives us the power of manipulating nature that it has more social importance

> than art. Science as the pursuit of truth is the equal, but not the superior, of art. Science as a technique, though it may have little intrinsic value, has a practical importance to which art cannot aspire. […]
> If, therefore, a scientific civilization is to be a good civilization it is necessary that increase in knowledge should be accompanied by increase in wisdom. I mean by wisdom a right conception of the ends of life. This is something which science in itself does not provide. Increase of science by itself, therefore, is not enough to guarantee any genuine progress, though it provides one of the ingredients which progress requires [1] (p. 10-11).
>
> *Bertrand Russell*

> …science is dogmatic to an almost unbelievable degree everywhere, i.e., it operates with preconceptions and prejudices [which have] not been reflected upon. There is the highest need for doctors who *think* and who do not wish to leave the field entirely to scientific technicians [2] (p. 103).
>
> *Martin Heidegger*

> It is not the victory of science that distinguishes our nineteenth century, but the victory of scientific method over science [3] (p. 261).
>
> *Friedrich Nietzsche*

I. The big suspicion

This narrative is not aimed to provide well-balanced 'pro and con' arguments for the knowledge-producing processes and treatment techniques of the disciplines in the so-called psy-complex (psychology, psychotherapy, psychiatry) but to raise suspicions about the epistemological status and the theoretical basis of the practices which the psy-complex disciplines provide (with serious ethical consequences). The basic arguments of the suspicions are:

1. By following the so-called empirical, quantitative methods, the mainstream psychology cannot provide valid knowledge on psychic phenomena, because it uses inadequate research methods.
2. Some psychotherapists have mistaken ideas on basic determinants of their own effectivity, as it is supposed that, in psychotherapies,

the same kind of technical attitude is applied for controlling mental problems as for controlling material things.

3. It seems that psychiatry has been falling into its latest crisis partly because its medical model of psychic/mental problems cannot generate valid and reliable knowledge on the nature of psychic suffering by looking at them as purely medical problems. Its mainstream treatment methods – in the last few decades treating mental problems by drugs – by focusing on the brain processes have not proven successful; that is, the bio-scientifically and technically oriented treatment approach was able to gain only partial successes in alleviating psychic pain and suffering.
4. As a consequence of the above-mentioned suspicions, it seems that the modern (postmodern) psy-complex – because of several culturally created, deep seated contradictions – cannot find a stable place in the contemporary disciplinary landscape.

From the beginning of modern times, a new kind of knowledge started to rise. (We take modernity as a new period in Western history partly because of the birth of this type of knowledge.) This turn was strongly connected to a fundamental change of the whole worldview, especially to a reorganisation of ontology and structure of the world. The contemplative attitude of premodern knowledge - aiming to understand God's will and decisions – was not enough to answer the issues which evolved from the new existential insecurity of the new age anymore. Who and in what way should nature be controlled if the 'dying' God is not strong enough for this task? The Baconian thesis: 'knowledge is power' refers to a new kind of knowledge that can be suitable for gaining power over the universe by way of replacing God with man. "'Savoir pour prévoir, prévoir pour pouvoir." The meaning of knowledge is in *what you can do with it*: the more you can do, the better the knowledge [4] (p. 165). This new type of knowledge has special direction and special aim – it is not knowledge for its own sake only, just because knowing is better than not knowing. In the case of this new kind of knowledge, 'good' means useful. And it is useful if it can provide tools and means for controlling nature. Natural scientific knowledge has special value and merit in our culture, because it must and can provide useful knowledge in the strict sense.

> A demarcation between pure science and technology could be based on the distinction between truth (for explanations) and usefulness (for solving practical problems). A scientific theory may be true without being useful (a theory of cosmology, for example), and it may be useful without being true (Newtonian mechanics, for example). Techniques are of course neither true nor false. They are useful (to some extent or other) or useless. And they are functional (to some extent or other) or non-functional in obtaining desired results. The aim of pure science is to produce true and explanatory theories about nature, society and man. The aim of technology is to produce functional and useful techniques [5] (p. 24).

Nevertheless, the whole modern project is clearly biased towards a kind of hierarchy of knowledge, namely the useful knowledge is better and has a higher value than the pure knowledge. This cultural bias towards useful knowledge has become stronger in the last decades.

This huge and deep modern cultural trend is clearly expressed in a concise form by Max Weber:

> Natural science gives us an answer to the question of what we must do if we wish to master life technically. It leaves quite aside, or assumes for its purposes, whether we should and do wish to master life technically and whether it ultimately makes sense to do so [6] (p. 144).

This means that the culture of natural scientific worldview and method is biased towards a useful kind of knowledge because even its most 'contemplative' basic disciplines can provide a hope that based on this kind of knowledge techniques and technology through which humans can control nature. (If one would argue that some natural sciences are not 'useful' in the above mentioned instrumental sense (e.g. astronomy), then it can be said that at the present historical moment this claim can be correct, but even so the ruling tendency is what is expressed in Weber's quote. This controlling pursuit has special importance for psychotherapy and psychiatry because the task of controlling 'mental death' is parallel with the main task of modern medicine.)

As the truth of this modern version of knowledge 'consisted in its effects,' it seemed in the 19th century that this project was working as a perfectly successful endeavour; that is, Western culture could deepen its belief that knowledge provided by natural sciences is the only valid knowledge (after God's death) about that nature.

On this basis, it seemed plausible that this project could be extended to society (cf. Marxism) or even to persons. The imperative was that if you wanted to control psychic phenomena then you had to turn to natural sciences and to technologies built on this knowledge. These were important aspects of that cultural milieu when the so-called psy-complex – psychology, psychotherapy, psychiatry – came into being. It seemed that what was important was the method itself and the differences between the ontological status of its 'objects' was secondary, or the method could transcend and mostly diminish these differences to the level of unimportance by successfully creating adequate knowledge for the controlling project. This means that in the light of scientific method, psychic phenomena should show some special traits, namely they should be predictable, law-like, measurable, and quantifiable. These traits create the conditions for useful knowledge, which can provide a basis for using techniques and technology on them. 'Thus technology is an activity that develops repeatable and transferable methods (techniques) with specified effects that are useful for solving practical problems [5] (p. 11).'

In the era of the birth of the psy-complex, the so called 'streetlight effect' was the mainstream epistemological attitude. 'A policeman sees a drunk man searching for something under a streetlight and asks what the drunk has lost. He says he lost his keys and they both look under the streetlight together. After a few minutes the policeman asks if he is sure he lost them here, and the drunk replies, no, that he lost them in the park. The policeman asks why he is searching here, and the drunk replies, "this is where the light is [8].' Or as somebody would put on special glasses through which some aspects of the world could be seen but other ones would remain hidden if one did not have other type of glasses. That is, one is sure that any other images of the world provided by other types of glasses are false or they cannot provide correct images of nature. The fundamental (culturally-ordered) scientific position was that the possible objects of scientific research are open for the scientific method, that is, there are no existing beings or attributes of beings which do not fit into the scientific method. (the aspects of human existence should be measured or should be made measurable.) From another perspective, sciences can and should play an ontological critical or cleaning task, namely what cannot be seen through the scientific glasses is not real or is less real than the scientifically-discernible beings or aspects of these beings. It can be said that

> there are many mysteries in the world about which science cannot say much: whether there is a God, whether there is life after death. what moral values are the true ones, etc. We just must accept this limitation, I think. Science is limited by its empirical methods. If there is no way to test a particular theory (as in the case of a metaphysical one), then we have to accept that too, even if the theory plays an important role in science. Science simply has to stick with what it can do: develop our knowledge of the world by developing and testing empirically testable theories [5] (p. 178).

The basic problem with any other knowledge-producing method is that it is not able to provide that kind of knowledge which is useful, that is suitable for controlling nature (inc. human being). From this perspective, the basic requirement is to 'unfold it by way of scientific method' or 'make it suitable for scientific research'. This positivist epistemology calls upon 'play the scientific game even if some of its "objects" are not proper objects for this game. 'But - as Heidegger warns - if there are matters that resist calculability due to their nature, then any attempt to measure them according to the method of an exact science is inappropriate [2] (p. 132).'

What if the 'object' and the epistemological tools and means do not meet? What if phenomena relating to the psy-complex cannot be subsumed under the classical scientific method without hurting their ontological attributes? As Chesterton argues: 'You can free things from alien or accidental laws, but not from the laws of their own nature. You may, if you like, free a tiger from his bars; but do not free him from his stripes. Do not free a camel of the burden of his hump: you may be freeing him from being a camel. [8]' In the Western culture from Descartes on an ontological difference has existed between the body and the mind (psyche) (*res extensa*, *res cogitans* respectively). The body can be opened for scientific scrutiny. It seems that methods of quantification, measurement, supposed natural laws can provide the kind of knowledge which is useful for creating technical control on nature including human beings.

> But the laws of Mind, and, in even a greater degree, those of Society, are so far from having attained a similar state of even partial recognition, that it is still a controversy whether they are capable of becoming subjects of science in the strict sense of the term: and among those who are agreed on this point, there reigns the most irreconcilable diversity on almost every other. [...] it is by generalizing the methods successfully followed in the former inquiries, and adapting them to the latter, that we may hope to remove this blot on the face of science [9] (p. 1015).

Therefore, the so-called moral sciences, including psychology, should use (adapt) the logic and method of natural sciences, because they were successful in revealing the laws of nature. The mainstream psychology and psychiatry followed this route. (And for the time being, this is the ideal route for psychotherapy as well.)

Of course, by way of this decision, the ontological gap between body and mind could not be covered but was only pushed under the surface. The 'taken-for-granted' cultural attitude could not avoid a philosophical critique:

> *Psyche* and psychology are attempts to objectify the human being. Something noncorporeal is acknowledged, but it is determined simultaneously according to the method of corporeal [physical] objectification. The method of its determination is not derived from the "psychical," from the noncorporeal itself, but occurs without [further] determination [from] within the horizon of scientific research, which alone counts as scientific.
> The justification of psychology lies in the fact that it acknowledged something noncorporeal, and its limitation lies in the fact that it wanted to determine it [noncorporeal reality] with the method of physical research [with the method] of natural science. The justification of psychology consists only in its point of departure and in its taking the

> noncorporeal seriously. But then its justification already ends because it researches this noncorporeal with inappropriate methods. It is a justification turned into something unjustified [2] (p. 216).

From this point of view, the professions of the psy-complex have suffered from the incompatibility between their 'object' and the method which they use for researching and approaching the psyche from the start. It is similar to a hidden disease which runs in the background seemingly without symptoms, but which would eventually have undesirable consequences on the validity of the outcomes of researches in these fields and also in the treatment practices. The aim of this essay is to unfold the signs of this hidden basic 'complex' in the practices of psychic disciplines, namely that the mainstream epistemological practices of psychic sphere do not correspond with some ontological traits of the human being. On these false premises, only two types of healing approaches can be built: the first one could work and could be effective but not because of its proper theory – explanatory theories of successful practices could be false (psychotherapies); the other type of treatment methods could not work properly – truly successful psychic cures cannot be shown by manipulating the so-called somatic basis of the psyche, namely the brain.

We can be suspicious about the consequences of the uncertain ontological status of the psyche, of the epistemological (scientific) worldview for unfolding its traits, of the knowledge background of the treatment methods of mental problems, of the culturally-embedded status of psychotherapy and psychiatry. As mentioned above, the existential hierarchy of beings is defined through the lenses of scientific method, the aspects of bodily phenomena can be more clearly shown, but psychic phenomena can have only a vague image, a kind of secondary and an underprivileged position on the ontological field.

> To know psychology one has to know its object, the soul. But given its peculiar nature, psychology finds itself in a unique position: it must provide the object of its study—a scientific concept of soul. In fact, psychology does not know its own object, and flatly denies the object that tradition hands down [10] (p. 1).

Previously, in Western culture, the soul was one of the key elements in the relationship between God and human. The soul was – unlike the body – man's immortal 'part,' the incorruptible, eternal element which directly relates man to God. This means that the soul would lose its existence without God. After the cultural experience of God's death (see Nietzsche), the soul lost its basis and disappeared, or rather it changed into the psyche.

But unlike the soul, the existence of the psyche is unstable and insecure. The soul was firmly connected to God, in that framework it was – if it can be said so - even more securely established than the body (e.g. immortal). In our contemporary culture, the existence of the psyche is more controversial. In folk psychology – from the point of view of the 'man on the street' – its existence is taken for granted. Everybody knows that he/she has psyche, but it is not as well-embedded in the scientific frame of thoughts. It is wrapped in different concepts such as personality, self, mind, behaviour, but the meanings of these substitutes are not well-defined and are presumably narrower than that of the psyche. Its origins, roots, and 'place' are also disputed: brain or society or culture, or all of these together.

Overall, it seems that the psyche has different names, but as we are coming closer to the sciences, its contours are being increasingly blurred, its existence is waning. It has different names but no 'real' existence.

It seems that because of this doubtful status of the psyche in modern Western culture, double discrepancies have raised concerning the research and the healing of the psyche.

The object of psychic disciplines is basically unknown: it is not a bodily one, but there are no further details on what it is; scientific knowledge (in the strict sense) on the psyche is impossible because it is not a material thing; thus scientific methods are not proper for unfolding its attributes.

II. Psychology

Today's mainstream psychology follows the idea of natural scientific experiments and especially the methods of statistical analysis. This tendency is the heir of Galileo's claim that the book of nature is written in mathematical language.

There has been a strong tendency since at least the 19th century that the method of the successful natural sciences should be expanded into social sphere or even beyond – to the psychic one. It means that by way of quantification as much valid knowledge on the psyche can be generated as in the case of natural, material objects. The huge number of publications in the last few decades seemingly proved that mainstream methods in psychology research are at least as valid concerning the psyche as the physical methods on natural objects; that is, modern mainstream psychology is as much as scientific as, for instance, physical sciences. Consequently, any other methods and approaches are less scientific and probably could provide less valid knowledge about the psyche than the quantitative one.

For most professionals for whom the measurability and quantifiability of the psychic phenomena is the unshakable foundation of a scientific psychology, it can be a shocking surprise that some of their colleagues argue in an opposite way. Namely the above outlined basic 'fault' between the ontological status of the psyche and the improper natural scientific epistemology can be translated into the methodological problems of psychology as a scientific enterprise.

If 'There are no psychological attributes that are quantitative! Or put it in a different way – *all quantitative derivations of psychological attributes are themselves qualitative in their nature* [11] (p.327)', then the psychology which uses quantitative methods cannot be scientific in a natural scientific way and the knowledge which is generated by this method cannot be valid in the same way as the outcomes of researches in physics are valid.

This problem was clearly stated (before some contemporary psychologists' critical works) by a philosopher, Martin Heidegger. He argued that the traits of the above-mentioned 'psychological attributes' are unmeasurably qualitative.

> How do we measure sadness? Evidently, one cannot measure it at all! Why not? If one approached sadness with a method of measuring, the very approach would already be contrary to the meaning of sadness. Thus, one would preclude sadness as sadness beforehand. Here, even the claim to measure is already a violation of the phenomenon as a phenomenon. But do we not also use quantitative concepts in our speech about sadness? One does not speak of an "intense" sadness, but of a "great" or a "profound"

> sadness. One can also say, "He is 'a bit sad,'" but that does not mean a small quantity of sadness. The "a bit" refers to a quality of mood. This very depth, however, is by no means measurable. Not even the "depth" of this room as experienced in my being-in-the-world is measurable. That is, when I attend to depth in order to measure it by approaching the window over there, then the depth experience moves with me as I move toward the window, and it goes right through it. I can objectify and measure this depth as little as I can traverse my relationship to this depth. Yet I am able, more or less, to estimate the distance precisely from me to the window. Certainly. Yet, in this case, I measure the distance between two bodies, not the depth opened up in each case by my being-in-the-world. Regarding the depth of a feeling of sadness, there is no reason or occasion whatsoever to estimate it quantitatively, let alone to measure it. As far as sadness is concerned, it can only be shown how a person is affected by it and how his relationship to himself and the world is changed [2] (p. 82).

This immeasurability is rooted in the above-discussed ontologically-different status of the psyche and the soma. Now we can get closer to the basic aspects of this basic disruption, if by way of using a kind of 'epokhé', we are concentrating on our lived experiences:

'But we suspend all philosophical speculations and hold to a simple principle to distinguish *soma* and *psyche*: *Psychic phenomena cannot be weighed and measured*, but only felt intuitively, whereas everything somatic can be somehow grasped by numbers. As soon as numerical values change, they indicate a change in somatic structures, a change, which of course can be conditioned emotionally. Sadness cannot be measured, but tears formed by sadness (due to psychosomatic relationships) can be investigated quantitatively in various directions. It is possible that emotional tension, by itself not measurable in terms of natural-scientific methods, can also result in a contraction of the capillaries, leading to an increase in blood pressure. Both states of tension [*Spannungszustände*] must not be equated, of course, because a person with a high degree of emotional tension does not always have the symptoms of an arterial tension (contraction).' [...] the principle of the distinction between *soma* and *psyche* involves a different comprehension of *soma* and *psyche*, which can be stated in the following way: Psychical phenomena cannot be weighed and measured, but only felt intuitively, whereas all that is somatic can somehow be comprehended by means of numbers. Therefore, the two thematic domains of *psyche* and *soma* are determined in their material content relative to accessing them [2] (p. 78-79).'

In mainstream psychological research, the logic of finding proper research methods for the special qualities of the 'object' has turned into the opposite – epistemology has overtaken ontology. It is a kind 'epistemological violence' on the quality of the object of planned research. 'In order to use statistical data analysis techniques for building explanatory theories in psychology we first should demonstrate that these analyses are appropriate for reaching the goal of explanation. I think there is clear evidence that in many situations statistical data analysis is useless because quantitative structure is imposed to qualitative psychological phenomena so that practically meaningless numbers are generated. […] the very fundamental question whether statistical data analysis can be used as a tool for building explanatory theory of mind at all must be answered before using this procedure [12].'

'The real "gold standard" in science is respect for evidence, wherever it leads us, and employment of the best and most appropriate methodology *for the particular problem being studied.* […] in any science work, we must find the best available methodology *for the problem at hand* [13].' But if psychology developed and follows a methodology which is inadequate for its object then the outcomes of research based on false premises cannot be valid.

For understanding the difference between measurable physical and unmeasurable mental qualities Joel Michell [14] (p. 58-59) refers to Collingwood's argument:

> In heat as known to the physicist there are differences of degree; so there are in the heat we feel as a bodily sensation. In physical heat, the excess of one over another is a definite amount: we can raise a pint of water from one temperature to another by adding a certain amount of heat. In heat as we feel it, this is not the case. We cannot add a slightly tepid feeling to a feeling of moderate warmth and so produce a feeling of greater warmth. An intense feeling may

> be produced by a sum of small stimuli, each of which by itself would have produced a lesser feeling; but it is not itself a sum of these lesser feelings; where it exists, they do not exist at all.
> As I move my hand nearer to the fire, I feel it, grow hotter; but every increase in the heat I feel is also a change in the kind of feeling I experience; from a faint warmth through a decided warmth it passes to a definite heat, first pleasant, then dully painful, then sharply painful; the heat at one degree soothes me, at another excites me, at another torments me.

Collingwood adds: 'Cold, as understood by the physicist, is the lack of heat; it is nothing but a name for the fact that in any given body there is no more heat present; at the zero end of the scale, it is a name for the fact that there is no heat present at all. But cold as we feel it is not mere lack of heat as we feel it, but another feeling with a positive character of its own; yet these are not two distinct feelings merely, but two opposite feelings [15] (pp. 72-75)'.

Therefore, there is a degree difference, a kind of 'more or lessness', between the degrees of the feeling of heat, but this is not quantifiably measurable because the differences between degrees are not mutually homogenous (degrees of the same kind). One of the basic differences between psychic and bodily phenomena is that the first one is not measurable therefore it cannot be fit into the method of physical sciences. In other words, the knowledge-generating method on the psyche cannot be scientific if among the essential aspects of scientific quality should be measurability.

If this provisional conclusion seems scandalous or unacceptable at first sight, then maybe this feeling of rejection mixed with surprise could be lead back to the experience that psychological researches based on quantitative methods are so 'taken for granted' nowadays. It can be said with almost certainty that, as it is a complex cultural phenomenon, its understanding should come from different directions:

> We have the possibility of a *methodological circle* where methods based on assumptions about the nature of the subject matter only produce observations which must confirm these assumptions. Within such a circle theoretical change would be limited to the set of theories which share

> the assumptions incorporated in the methodological rules. Any theoretical change beyond this would have to involve a methodological change [16].

If our worldview were so embedded in an image of a calculable world then it seems that any kind of quantifying method could prove it. If the human psyche seems to be part of the essential feature of the calculability of nature, then there is no other way leading out of this perspective but by changing the perspective itself. What if the calculability of the psyche is a hypothesis, not a proven fact; what if this 'taken-for-grantedness' is a cultural prejudice without real confirmation? 'A breakdown in critical inquiry occurs when some hypothesis is accepted as true without a serious attempt being made to test it. […] A breakdown in critical inquiry only becomes pathological when it includes a higher-order attitude, namely that of ignoring the first-order breakdown. That is, in a pathology of science not only are some hypotheses accepted within the mainstream of a discipline without a serious attempt to test it, but that fact is not acknowledged or, in extreme cases, is disguised [17].' It means that this quantitative method as a main approach in psychology is not built and supported by inherent scientific 'interests', it does not evolve from 'searching the truth' ethos, but professional blindness towards this problematic supposition is embedded in a broad cultural 'interest-system', and it is backed by stronger cultural values than professional ones coming from inside workings of psychology. This pathology serves 'foreign interests': 'The ideological interests relate to *scientism.* The term *scientism* denotes an ideologically driven, false image of science. Ironically, it was because psychology proclaimed itself a science that this pathology arose. It is still widely thought that measurement is a necessary feature of all sciences: Knowing something *scientifically* means *measuring* it. […]

The second interest sustaining this pathology is economic. […] Modem science depends for its existence on research, and research requires financial resources. […] Psychology did not actually become more rigorous, but it aped the methodological rigor of the established quantitative sciences as a way of signalling its scientific credentials to granting agencies [18].'

A tendency in psychology that quantitative method overtakes its object is much more based on ideological, social, and cultural powers than on inherent professional values:

1. 'The inferiority complex of psychology' – as a 'latecomer' discipline that did not have a special epistemological method for creating valid knowledge on an ontologically not-so-well-grounded-being like the psyche, it borrowed a successful methodology from the natural sciences (with all of its consequences e.g. universality of knowledge, predictive capacity which is based on laws and hoping that techniques can be built on this type of knowledge for controlling the psyche) and put the ontological differences of mental phenomena into brackets.
2. Group pressure, cultural pressure – accepting the values and rules of the surrounding cultural (including scientific) environment for survival with its beneficial aspects, e.g. financial support, prestige, the member of the 'high society of sciences'.
3. Postponed 'identity crisis' – they try to be disguised by way of scientifically-looking methods that the 'object' of research is not real from a modern scientific perspective (it is not a material thing), but paradoxically because supposedly scientific methods reflect reality, hence the ontological gap between the psyche and matter was successfully filled with quantifiability.

'But will psychology be ready – theoretically and methodologically – when the "normal science" of current "mainstream" collapses – this is the (open) question [19, p. x].'

III. Psychotherapy

The key contradiction between psychotherapy and the techno-scientific approach is clearly expressed by Martin Heidegger:

> If Dr. M. asserts that psychotherapy can be done only if one objectifies the human being beforehand, then what is decisive thereby is psychotherapy and not the existence of the human being. Since one can [supposedly] only do therapy, which is a concerned handling of objects, and thus something purely technical, then the outcome of such psychotherapy cannot result in a healthier human being. In such a therapy, the human being is finally eliminated. At best, such a therapy could [only] result in a more polished object [2] (p. 215).

This Heideggerian quote argues that at least two types of psychotherapies exist: one follows a kind of 'techno-scientific logic' which takes the psyche as an object, then tries to change it in a technological way. The other type tries to follow a logic based on the ontological stance that the psyche is something different than a kind of material object.

Max Scheler had already detected this difference in the early 20th century. He argues that there are two types of psychotherapists: one is the so-called 'psychic surgeon' which tries to force the patient's psyche into normalcy by way of suggestion and manipulation. In this case, the psychotherapist is a kind of 'psychic engineer' who tries to fix the patient's abnormal, 'broken' psyche. In this approach the patient's psyche is a passive object on which knowing and skilful active agent can make the necessary changes. This psychotherapy is a substitute of medical technology, a kind of surrogate drug and its efficacy should be measured the same way as the curing power of pills. The other psychotherapy is known as Socratic (psychoanalytic). The main aim of this approach is not to change the patient's psyche into a pre-given direction, but to help him/her get insight and thus free him/her from self-deception.

'The second therapeutic ideal is considerably more modest than the ideal of the psychic surgeon. It is characterized by Socratic *reserve*, instead of the "surgeon's" cynical intrusion into the way other men manage their lives [20] (p.12).'

It is interesting that even the father of psychoanalysis, Sigmund Freud, had a rather ambivalent relationship towards these two types of therapeutic ideals:

> The future may teach us to exercise a direct influence, by means of particular chemical substances, on the amounts of energy and their distribution in the mental apparatus. It may be that there are other still undreamt-of possibilities of therapy. But for the moment we have nothing better at our disposal than the technique of psycho-analysis, and for that reason, in spite of its limitations, it should not be despised [21] (p. 182).

It seems from the above perspective that psychotherapy works like a pill which is a condensed manifestation and product of technological knowledge and skills of the medical sciences. If psychotherapy is equal with - and as it is a substitution of - pills, then the logic of efficacy of psychotherapy should be the same as that of pills. But against our expectations we can find a different structure of effective factors in psychotherapy. While in the case of a pill the main effective factor should be its chemical ingredients, therefore what is created and put into it by way of technology, the so-called placebo effect – including the quality of doctor-patient relationship - can play a part in its efficacy, but only as a supplementary agent. But in psychotherapy we can find a different logic in the structure of effective components. The least technical components play the most important part in their efficacy, and the technical ones have only marginal, but indispensable importance.

Before going into details, it should be emphasized that a professional agreement on the effectiveness of psychotherapies does exist [22]. '…the average client receiving psychotherapy is better off than 79% of untreated clients... [23] (p.94).'

Beyond this fundamental verdict, two related and not less important questions can be raised concerning the effectivity-hierarchy of different types of psychotherapies:

- whether one type of psychotherapy is better than the other one for treating a psychic problem;
- whether specific distinguishing effective components exist or not in different therapeutic regimes.

It seems that the most plausible contemporary answer is the so called 'dodo verdict': 'Everybody won, and all must have prizes.' All therapies are broadly equivalent in their effectiveness. It does not matter what kind of disorder should be treated and by what kind of therapy the answer is the same: psychotherapies are effective but without effective special technical component. 'Across different kinds of therapies, technique accounts for 12%-15% of variance' [24].

The answer for the question on why there is no difference between the effectiveness of therapies which have thoroughly different theories about the human psyche and the processes of its disorder (and they have different techniques based on their theory), is that the most important effective components of psychotherapies are the so called common factors: the quality of the therapist-patient relationship and their internal and external resources. The type of therapy and technique play the role of a 'trace element' or a 'catalyst' ('healing myth'), without which a good therapy is almost impossible, but its importance falls short of common factors.

> 1. Effective psychotherapy is predicated upon the relationship of the therapist and client in combination with the inner and external resources of the client (common factors);
> 2. Type of therapy and technique add little to the effect of the relationship and client resources if not accompanied by common factors; and
> 3. Relationship variables that are most often related to effectiveness are the conditions of empathy, genuineness, and unconditional positive regard [25].

The protocols which are so important in some specialities of medicine can only play the role of the Wittgenteinian ladder: they should be thrown away for a deep understanding and for providing real help for clients.

> Doing therapy by a manual is like having sex by a manual. Perhaps the desired outcome is achieved if instructions are technically followed. But the nuances and creativity of an actual encounter flow from moment-to-moment interaction of participants, not from Step A to Step B. Experienced therapists know that psychotherapy requires the unique tailoring of any approach to a particular client and circumstance. Simply put, psychotherapist do not do therapy by the book. When they do, it does not go very well [26].

It should be mentioned that these latest developments provided by some systematic research were anticipated by Sandor Ferenczi, the psychoanalyst who had been trying to make progress on the technical side of psychoanalysis in the first third of 20th century by experimenting with almost all logically-possible technological innovations in therapeutic techniques But shortly before his premature death he had wisely realised:

> So finally one begins to wonder whether it would not be natural and also to the purpose to be openly a human being with feelings, empathic at times and frankly exasperated at other times? This means abandoning all „technique" and showing one's true colors just as is demanded of the patient. [...] As for the doctor, thus sobered from his scientific delusion, he will have a more beneficial effect in subsequent cases [27] (p. 94);

If the emphasis on the development of different types of psychotherapeutic techniques was proven to be a mirage, a dead-end, then what would be the 'effective healing component': '... no analysis can succeed if we do not succeed in really loving the patient [27] (p. 130).'

It can be argued that what Ferenczi called 'love' in technical terms is the most important aspect of the common elements in psychotherapy.

It follows from the above arguments that from the components of a healing (effective) therapy the main aspects of modern project are missing: no calculability, no prediction, no control. Almost no aspects exist (or marginal) which can have any relation to the techno-scientific attitude. What really works in psychotherapy is outside, even against the modernist (scientific) worldview. It means that regardless if psychotherapy works as an effective enterprise, it could only work if it keeps its technical features as low as possible. With the emphasis on technical differences between therapeutic schools they not only cover their real identity, but at the same time by reifying and objectifying patients' personalities with emphasis on techniques, they put them at distance from their own human capacities.

The final question is why psychotherapy is so strongly connected to or even embedded in the medical model, or more precisely in the *biomedical* model, where the objectification of the human being is an unavoidable consequence of the natural scientifically-based knowledge production and the treatment established on a technological paradigm.

> To put it simply, the medical model has remained the dominant descriptive system for psychotherapy, not because it offers the most accurate description of what actually occurs in therapy but, rather, because the model's association with medicine and science gives psychotherapy

> a level of cultural respectability and economic advantages that other descriptive systems do not [28].

Just like in the case of psychology, the ideology and attitude of psychotherapy which is based on biomedical model is not an 'organic' development, but a kind of solution for the 'homelessness' of modern psychotherapy: for gaining respect and prestige. With attaching itself to a powerful and prestigious profession (medicine) it tries to avoid being a 'no men's land' of modernity, for existing in an endless existential doubtful situation against the cultural mainstream. Since a profession whose practice is not based on scientific knowledge and wants to treat mental problems with 'love' - therefore it cannot guarantee or at least maintain the promise to control technically the human psyche - seems a strange, a 'not-professional profession' in the (post)modern culture.

From this point of view, we can see the predecessor of modern psychotherapy, F. A. Mesmer, and his animal magnetism story in a different light. It seems a superficial approach to state that mesmerism has only a historical interest, because by way of the first blind scientific research, the validity of theory of animal magnetism was disproved. But it should not be forgotten that Benjamin Franklin's commission (Bailly's report) made a more complex statement: 'Whereas magnetism appears non-existent to us, we were struck by the power of two of our astonishing faculties: imitation and imagination. Here are the seeds of a new science, that of the influence of the spiritual over the physical [29].' This means that although a scientific committee rejected animal magnetism, 'Mesmer was discredited not based on treatment effectiveness or theoretical cogency but on the observation that the proposed mechanism of illness remediation was questionable [23] (p. 6).' That is, the concrete theory of curing is false, but the practice under the false theory has beneficial consequences for the patients' health.

It is a widespread belief that what is scientifically proven and what practically works is strongly linked. Or in other words, what is scientifically incorrect cannot be used for improving (mental) health. But 'people in previous generations and most people today have beliefs about human behaviour broadly conceived that are scientifically incorrect. But the purpose of folk beliefs is to regulate social relations and internal states in order to survive, not to be scientifically correct. [...] all healing practices provide the person an explanation for their complaints and that the scientific basis of the explanation is not what is important [30]'.

What if Mesmer's legacy has been with us since his time, and we have been creating new therapy methods with new explanatory theories from psychoanalysis to cognitive behaviour therapy, but it does not matter whether these are valid theories or not, because the therapeutic efficacy of all these different psychotherapy schools is based on the common elements of these practices, and not on special traits of seemingly adequate theories – and these common elements are definitely not technical ones.

So we do not know whether the most popular therapeutic methods are supported by better theories or they are merely more plausible in an ever changing cultural environment (e.g. CBT is theoretically better than psychoanalysis) or it only fits better in a late 20th century cultural landscape which prefers forgetting to remembering, prefers present and near future to past, where people feel more at home in a language related to programming and computers [31] (p. 137), than dealing with complicated biographical narratives, and they wish for therapeutic practices which promote 'quick fix' treatments instead of painful endless self-developing processes.

If not a special technique which plays important role in psychotherapy then generally speaking, it almost does not matter what kind of specific theory or method is used in the process of alleviating suffering, psychic pain or finding new directions for one's life. From this perspective, the medical model is not the only valuable perspective from which mental suffering can be approached. Maybe it is not even the best one, because

> …the medical model has serious limitations: (a) The model fails to describe accurately what actually occurs in therapy; (b) the model continues to dominate the field not because of its accuracy but rather because of its questionable ties with medicine, science, and the health insurance industry; (c) the model obscures the fact that psychotherapy is primarily an interpersonal process rather than a medical procedure; and (d) the model cannot account for the fact that the vast majority of clients who seek psychotherapy do so for reasons other than mental illness [32] (p.50-51).

Psychotherapy is one of the enterprises which deal with human beings. Because of the ontological characteristics of being human, it cannot play the 'scientific game' properly, but it pretends (or must pretend) that it can. Their 'clumsiness' is not a genetically-based 'disorder', but a normal consequence of the human condition. Maybe it is generally valid for all professions dealing with human beings as Alasdair MacIntyre claims about managerial effectiveness:

> The expert's claim to status and reward is fatally undermined when we recognize that he possesses no sound stock of law-like generalizations and when we realize how weak the predictive power available to him is. The concept of managerial effectiveness is after all one more contemporary moral fiction and perhaps the most important of them all. The dominance of the manipulative mode in our culture is not and cannot be accompanied by very much actual success in manipulation. I do not of course mean that the activities of purported experts do not have effects and that we do not suffer from those effects and suffer gravely. But the notion of social control embodied in the notion of expertise is indeed a masquerade. Our social order is in a very literal sense out of our, and indeed anyone's, control. No one is or could be in charge [33] (p.107-108).

This inability for technical knowledge and control, that is professionals on the fields of psy-complexes cannot – or can only seemingly – use techniques is clearly expressed in Sigmund Koch's famous argument starting with 'the scales fell gradually from my eyes' expression concerning the scientific validity of the mainstream contemporary psychology. A psychologist and psychotherapist cannot be professionals in the technical sense, but they can 'only' play the role of a knowledge-bearer 'whose minds have dwelled for long periods – in disciplined and passionate ways – on certain realms of human phenomena. Assuming relevant observational sensitivities and analytical powers, such minds inevitably form distinctions, detect patterns, develop implemental skills. The owners of minds so schooled are disciplined *connoisseurs*. Much of their knowledge is tacit, some of it is explicit: all of it may prove helpful to fellow human beings whose

minds have not been comparably immersed in the same domains. The best way to provide psychological services would be to provide, where relevant, *human beings* (as consultants, therapists, etc.) in the many contexts involving human welfare; to provide psychologists as human beings who may be *disciplined connoisseurs* of certain ranges of human action and experience. The point to stress is that we do the world a great injustice if we unleash upon the world a corps of putative "technicians" who, under a cloak of scientific impeccability, have everything to offer but wisdom and understanding [34] (p. 7 and 310).'
Maybe the psychotherapist expertise is not a masquerade, but it is clearly not an engineering-like technical expertise over the psyche either. By their training, psychotherapists can be wiser, more insightful, intuitive, and authentic, but they cannot have more technical power than any other human being can have over controlling the psyche.

IV. Psychiatry

Psychiatry is a product of the period of first-wave medicalization and was established as an institution which can solve the problem of madness in the modern era. Since then, the culturally-chosen institution for solving the riddle of madness is medicine. 19th century medicine was basically biomedicine where the problems of diseases and disorders were closed into the body. After all typed of madness have pushed into the field of medicine, it became equal with mental illness. The final step in the sequence of relocating madness in modernity was projecting the biomedical worldview onto it, which redefined it as a series of brain disorders. (It must be recognised that from 'Moral Treatment' on there was a shadow-like presence of psychotherapeutic line in psychiatry, but it was always a kind of supplementary or provisional element in psychiatry's worldview, which was used in that period when somatic treatments or biological explanations did not seem plausible or successful.) It comes from the logic of medical orientation that psychiatry just right after its birth became '*enkephalatry*': medicine of the brain. The emergent psyche has always fallen back into brain. Because the existence of psyche cannot be seriously questioned in our culture, it – in Freudian terms – started to play the role of the almost repressed content in the 'half-consciousness' of psychiatric profession. It was always present in the psychiatric praxis, but outside or only on the margins of professional attention. This unsolved ontological constellation proved to be a subversive agent in the history of psychiatry: psychiatry has always been in a never-ending crisis almost in its whole history; the medical model has been maintaining some basic unsolvable problems.

Perhaps the one mostly discussed is the validity of its etiological and nosological systems. The existence of mental illness as brain disease is a philosophical supposition or a fundamental decision of Western culture. In other words, it is not the task and product of the profession of psychiatry to claim that madness is mental illness, but rather this postulate provides a basis for the existence of the medical profession of psyche. Then nosological systems of mental disorders can be based on this cultural (philosophical) decision. But the unresolved problems of the basic decision have far-reaching consequences. Creating a kind of a nosological system which has a proper etiological background and is scientifically tenable at the same time seemed unsuccessful until now; where are the roots and causes of mental illnesses in the brain or in the psyche? If they are rooted in the psyche, then only speculative etiological systems are possible from a scientific point of view (e.g. psychoanalysis); if a scientifically valid system is planned, then it can be proven to be a mirage without any scientifically valid somatically (brain) rooted, causally determined phenomena.

Let us see some typical answers for the question: Where is the 'place' from which the mental disorders originate? Thomas Insel, a former head of NIMH strongly criticised the 5th edition of DSM:

> The weakness is its lack of validity. Unlike our definitions of ischemic heart disease, lymphoma, or AIDS, the DSM diagnoses are based on a consensus about clusters of clinical symptoms, not any objective laboratory measure [35].

Insel would have liked to see a reorientation of research away from the symptom cluster logic of DSM and suggested the Research Domain Criteria (RDoC) research program which emphasizes the biology of psychopathology and researching the biomarkers. The discrepancies between the logic of DSM and of RDoC are mostly virtual, because the constructors of DSMs – from the 3rd edition – wanted to create a nosological system which was based on biological causation. But because until now no definitive biological answer has been found for the causes of the mental disorders, they had to follow a seemingly rigorous cluster logic, by way of using a scientific-like game where the clusters should follow 'objective, impartial, observational research'. Therefore, the RDoC has the same intention as the creators of DSMs, but Insel seemed to be dissatisfied with the cluster detour and wanted to follow even a more rigorous way of biological paradigm of mental disorders. As Ian Hacking detected: 'It is [...] likely that the manual will become more attuned to neurological causes as these gradually conquer more and more of psychiatry' [36] (p. 7-8).

Neuropsychiatry and neurodevelopmental disorders are only two recently-coined categories which are clear signs of the never proven, but very powerful tendency in psychiatry that the brain as the organ of psyche is the place of the causes, therefore of the treatments of mental disorders.

The inherent logic of this supposition has lead to a practical conclusion which was expressed by some neurologists:

> From our angle of vision, there are only brain disorders that psychiatrists prefer to treat and other brain disorders that neurologists (and neurosurgeons) prefer to treat. [...] We suggest that it is unscientific, misleading, and harmful to millions of people worldwide to declare that some brain disorders are not physical ailments. Neurology and psychiatry must end the 20th century schism that has divided their fields [37] (p. 937).

This is a clear logic: if mental illnesses are products of disordered brains, then psychiatry must be redefined as a medical specialty for treating some special brain problems. There is no real difference between psychiatry and neurology. Therefore, if the program of the biopsychiatry was successful, then psychiatry would disappear, as a baseless specialty, since psyche cannot be a target of treatments aiming at the causes of mental disorders. This program – which would end by the destruction of psychiatry, seemingly with the purpose of saving it by way of connecting it to brain pathology – almost necessarily raises arguments which try to find a meaning for the psyche, therefore some have been trying to keep psychiatry as a broader enterprise than a profession which simply reduces the complex biopsychosocial task of healing the psyche into manipulating the brain processes. Several

British psychiatrists [29] signed a declaration which argues against the above tendency:

> Psychiatry is not neurology; it is not a medicine of the brain. Although mental health problems undoubtedly have a biological dimension, in their very nature they reach beyond the brain to involve social, cultural and psychological dimensions. These cannot always be grasped through the epistemology biomedicine [38].

So, psychiatry cannot be part of neurology, because the brain is a necessary, albeit not a sufficient 'player' in the game named psychiatry. Or it is even more possible that the brain is not a necessary aspect in the causation and treatment of mental disorders? Can major mental disorders like schizophrenia be understood not from the perspective of the brain, but from that of culture, that these are in the strictest sense *psychic* problems, therefore the brain can be put into brackets? 'The mind is no more a product of our brain than this chapter is a product of the highly sophisticated word processor that is supporting my writing. It is produced *in* the brain *by* culture. [...] In acute psychosis it is no longer individualized culture (which is the mind), which exists by means of the individual brain, but culture as such. The brain by the means of which the process of observation and memorization happens may be completely healthy. The disturbance is not in the brain. It is in the cultural, symbolic, structures of the mind [39] (p. 84 and 158).'

As it can be seen from the above example, the culturally, poorly defined, doubtful status of the psyche causes endless identity problems for psychiatry. This can create a pendulum-like 'lifestyle': The playground for psychiatry runs throughout two other territories, namely neurology and psychotherapy: both aim to treat the brain and the psyche respectively. When psychiatry tried to make itself at home in one of these territories, it entered professional and cultural crises. The previous somatic treatments which aimed to treat mostly the brain raised more shame than pride to the profession (water treatment, rotatory machines, confining chair, purging, vomiting, electroshock, insulin coma, lobotomy, etc.). Now we live in the era of 'psychopharmacological revolution', which seems to devaluate the former physical treatments considering them rather tortures than treatments, with reinterpreting the pre-psychopharmacological era as more or less shameful or premodern, unscientific periods in the profession's history. This dark period, on the one hand, can only be looked upon as a kind of professional aberration, but on the other hand, it can play as a background which can provide a contrast for the latest genuine 'scientific' developments. But what if this latest development would be proven to be mistakes, false hopes? What if the most important healing tools in the repertoire of contemporary psychiatry, the drugs, could cause an epidemic of mental disorders or at least they do not heal their users but trigger lifelong disorders of which symptoms are more acceptable for the social environment than the original ones, but for clients they can only change one type of mental misery into another one.

> Psychiatric drugs perturb normal neurotransmitter function, and while that perturbation may curb symptoms over a short term, over the long run it increases the likelihood that a person will become chronically ill, or ill with new and more severe symptoms. A review of the scientific literature shows quite clearly that it is our drug–based paradigm of care that is fuelling this modern-day plague [40].

The 'specificity' hypothesis can be found in the background of the widespread usage of psychiatric drugs, namely that these are basically different biochemical products than harmful street drugs. They are part of the 'specificity' theory of modern medicine which means that because we have specific diseases, so they deserve specific treatments. 'Special pills for every special ills.'

> The disease-centred model of drug addiction, the myth that psychiatric drugs are curative or restorative treatments, has led to their indiscriminate prescription to millions of people often for decades on end. It is likely that many people exposed to the harmful physical and psychological effects of these drugs derive no benefit from them [41] (p. 239).

After the forty-year period of the latest turn in psychiatry from biopsychosocial towards the 'bio-bio-bio' [42] model of causation and treatment of psychic suffering, the profession following this model was not able to answer the most important questions of 'healing madness'. 'The scientific issues facing translational psychiatry—the application of basic discoveries in neuroscience, genetics, and psychology to understanding disease and to advancing therapeutics—are daunting. The molecular and cellular underpinnings of psychiatric disorders remain unknown; there is broad disillusionment with the animal models used for decades to predict therapeutic efficacy; psychiatric diagnoses seem arbitrary and lack objective tests; and there are no validated biomarkers with which to judge the success of clinical trials. As a result, pharmaceutical companies do not see a feasible path to the discovery and development of novel and effective treatments. Given the steady stream of drugs that have gained approval during recent years for treating depression, anxiety, schizophrenia, and bipolar disorder, this scientific stall may have seemed to come out of the blue. However, payers (both insurance companies and governments) and regulatory agencies have given up their willingness to accept even more expensive new drugs that, despite marketing efforts, have turned out to be no more than variations on very old themes [43].'

This conclusion implies that the somatic project that is solving the mental suffering by manipulating the brain by biochemical or technical tools and means is in crisis (again).

The pendulum is moving again towards the opposite direction, that is towards cultural, social, existential causation and treatment orientation: the Open Dialogue approach is clearly a practical consequence of a deep-seated theoretical turn from one of the main aspects of modern psychiatry, namely 'The language of psychiatry [...] is a monologue of reason *about* madness...' [44] [p. xii). This professional monologue on passive 'objects' is changing towards a *Bakhtinian*-like dialogic relationship between healers and patients:

> The consciousnesses of other people cannot be perceived, analysed, defined as objects or as things - one can only *relate to them dialogically*. To think about them means to *talk with them; otherwise they immediately turn to us their objectivized side*: they fall silent, close up, and congeal into finished, objectivized images [45] (p. 68).

The professionals in Open Dialogue take the Bakhtinian requirements very seriously, e. g. the discussions behind the patients are forbidden; every problem, topic should be put on the table in the presence of the patient, family members and professionals (who are not only psychiatrists). They use drugs only in a very limited, even marginal way.

The other important developments are the contributions of British clinical psychologists who – after several years research process – have just developed the so-called Power-Threat-Meaning Framework [46], whose aim is to understand mental suffering outside the traditional medical model and they emphasize shift of focus towards 'power imbalance' rather than 'chemical imbalance' [47]. Parallel with these two latest developments on the professional side, some important new and critical approaches have appeared on the 'consumer' side of psychiatry. One of the most appreciated is the *International Hearing Voices Network* or *Intervoice* where people - whose hallucinations from the 'monologue of reason' perspective were only meaningless signs of their brain disorders - take the role of agents of their hallucinations by redefining them as (potentially) meaningful aspects of their life. Following this logic, they try to understand their suffering from the perspective of their idiosyncratic life events and narratives. Following a dialogue with their own inside voices, they are changing the relationship between their brain and personality into the opposite of the biopsychiatric perspective: I own a brain and not my brain owns me. 'Mad people' emancipatory movements aim to redefine madness and attitudes towards madness as mental illness like suffragettes wanted to change the image and the status of femininity at that time when being a woman meant inferiority not only from a cultural but also a scientific perspective.

Some professionals were involved in this changing world of psychiatry. As one of the leading figures of *post-psychiatry* approach, Patrick Bracken claims:

> What people can make it through enormous suffering if they believe there is meaning and purpose to it. […] What people need is not technical interventions in their lives, but holdings human space where our encounters with madness, distress, and alienation can be worked out in a meaningful environment, where people are treated with dignity and respect, and their experiences are valued and listened to [48] (p. 231 and 234).

Therefore the most important aspects of the latest tendencies in psychiatry regarding scientifically-produced knowledge and technical-oriented treatments as basic features of modern approaches:

- It seems that the biological roots and causes of psychic suffering in *bona fide* 'mental illnesses' have proven to be a possible hypothesis, but so far without any *conclusive* scientific confirmation. The madness slips out of the hands of science. (Of course, the psychic sufferings are embodied phenomena, which should have cerebral, neural, or psychobiotic correlates.) The somatic – mostly biochemical, drug-based – treatments can only have partial successes, or they can even be proven to eventually become harmful. (One form of madness is treated by creating another one.)
- The social-cultural-existential triggers (causes) of mental sufferings are well-documented. (Even if their mechanisms are nor clear.) 'The anguish is deep inside me, here, / while its explanation lies out there.' [49] (p.79) – as the famous Hungarian poet Attila József wrote. It means that successful recoveries from mental malaises can be achieved mostly with communal and existential assistance. Human relationships can have healing power, technical interventions can only be their servants.
- The medical model can only be an aspect of helping arrangements. That is, psychiatrists can and should be holistic healers and therefore can play an important role in changing mainstream medicine towards a humanistic direction.

V. Conclusion

The psy-complex is only seemingly properly embedded in Western culture. It creates – in a disguised way – a professional game which can hide its strangeness in a materialistic and techno-scientifically-oriented belief-system. In this context, if we translate the manipulation and control of nature onto technical knowledge and skills for controlling other human beings then the professions of the psy-complex can be subsumed under this evaluation: they cannot be a real part of the modern techno-scientific enterprise.

Because on the one hand, if they meet the criteria of scientifically-produced knowledge, then they would not be able to understand the causes and aspects of their patients' suffering and would not able to provide them proper support for their special human recovering.

On the other hand, if their 'methods' worked, they would not meet the criteria of scientific project and technological approach. If they tried to play a scientific and technological game, then it would have dehumanizing effects by reducing the human psyche into a machinery.

If they tried to find a special 'language' for psyche and to create humane approaches for knowing and helping people who have lost some aspects of their freedom, then these approaches would not be scientific and proper for controlling the 'objects' by way of technologies.

These paradoxes or even contradictions do not let professions find a well-embedded place, a home for themselves in the modern landscape – they will always be at least a little bit strange and unfitted in the contemporary Western *Zeitgeist.*

As a branch of inherently *deviant* disciplines and institutions in modernity – by sharing the fate of their subjects - they tend to play (without choice) a provocative and disruptive role in the techno-scientific project on the human being.

References

[1] B. Russell, *The Scientific Outlook*. London: Gerge Allen and Unwin Ltd., 1954.

[2] M. Heidegger, *Zollikon Seminars.* M. Boss, eds., Evanston, Illinois: Nothwestern University Press, 2001.

[3] F. Nietzsche, *The Will to Power.* W. Kaufmann, eds, New York: Vintage Books, 1968.

[4] Z. Bauman, *Life in Fragments*. Cambridge, Massathusetts: Blackwell, 1995.

[5] I. Nordin, *Using Knowledge*. Lanham, Boulder, New York, London: Lexinton Books, 2017.

[6] M. Weber, "Science as a Vocation," *Essays in Sociology*, H.H. Gert and C.W. Mills, eds., London and New York: Routledge, pp. 129-148. 2009.

[7] Y. J. John, "*The* 'Streetlight Effect': A Metaphor for Knowledge and Ignorance," *3 Quarks Daily*, https://www.3quarksdaily.com/3quarksdaily/2016/03/the-streetlight-effect-a-metaphor-for-knowledge-and-ignorance.html. Accessed: May 9th, 2018.

[8] G.K. Chesterton, "The Suicide of Thought," Orthodoxy, *Page by Page Books*, http://www.pagebypagebooks.com/Gilbert_K_Chesterton/Orthodoxy/The_Suicide_of_Thought_p7.html. Accessed: May 9th, 2018.

[9] J.S. Mill, "A System of Logic, Ratiocinative and Inductive." *The Project Gutenberg EBook*, https://www.gutenberg.org/files/27942/27942-pdf.pdf. Accessed: May 9th, 2018.

[10] O. Rank, *Psychology and the Soul*. Baltimore, London: The John Hopkins University Press, 1998.

[11] A. Toomela, J. Valsiner, "General Conclusion, Have Sixty Years Really Gone Astray? Back to the Future," *Methodological Thinking in Psychology: Gone Astray*?, A. Toomela and J. Valsiner, eds., Charlotte, NC: Information Age Publishing, Inc., pp. 325-337, 2010.

[12] A. Toomela, "History of Methodology in Psychology: Starting Point, Not the Goal," *Integr Psych Behav*, vol. 41, pp. 75-82, 2007.

[13] P. L. Wachtel, "Beyond 'ESTs' Problematic Assumptions in the Pursuit of Evidence-Based Practice," *Psychoanalytic Psychology*, vol. 27, no. 3, pp. 251-272, 2010.

[14] J. Michell, "The Quantity/Qualioty Interchange: A Blind Spot on the Highway of Science,"*Methodological Thinking in Psychology: Gone Astray*?, A. Toomela and J. Valsiner, eds., Charlotte, NC: Information Age Publishing, Inc., pp. 45-68. 2010.

[15] R..G. Collingwood, "An Essay on Philosophical Method," *An Essay on Philosophical Method*, J. Conelly and G. D' Oro eds., Oxford: Clarendon Press, 2005.

[16] K. Danziger, "The Methodological Imperative in Psychology," *Phil. Soc. Sci.* vol. 15, pp. 1-13, 1985.

[17] J. Michell, „Normal Science, Pathological Science and Psychometrics," *Theory and Psychology*, vol. 10, no. 5, pp. 639-667, 2000.

[18] J. Michell, "Is Psychometrics Pathological Science?," *Measurement*, vol. 6, pp. 7-24, 2008.

[19] A. Toomela and J. Valsiner, "Preface,"*Methodological Thinking in Psychology: Gone Astray*?, A. Toomela and J. Valsiner, eds., Charlotte, NC: Information Age Publishing, Inc., pp. vii-xi. 2010.

[20] M. Scheler, "The Idols of Self-Knowledg," *Selected Philosophical Essays*, Evanston, Illinois: Northwestern University Press, pp. 3-97, 1973.

[21] S. Freud, *An Outline of Psychoanalysis*. New York: Norton, 1949.

[22] cf. "Recognition of Psychotherapy effectiveness The APA Resolution," *Psychotherapy*, vol. 50, no. 1, pp. 102-109, 2013.

[23] B. E. Wampold, Z.E. Imel, *The Great Psychotherapy Debate*. New York, London: Routledge, 2015.

[24] D. K. Silverman, "What Works in Psychotherapy and How do We Know?," *Psychoanalytic Psychology*, vol. 22, 306-312, 2005.

[25] J. Bozart, "The Specificity Myth: The Fallacious Premise of Mental HealthTreatment," Paper presentation at the American Psychological Association, Washington, D. C. 2000. http://personcentered.com/specificity.htm. Accessed: May 9th, 2018.

[26] B. L. Duncan, "The Legacy of Saul Rosenzweig: The Profundity of Dodo Bird," *Journal of Psychotherapy Integration*, vol. 12, no. 1, pp. 32-57, 2002.

[27] *The Clinical Diary of Sándor Ferenczi*, J. Dupont ed., Cambridge, Massachusetts, London: Harvard University Press, 1995.

[28] D. Elkins, "The Medical Model in Psychotherapy: Its Limitations and Failures," *Journal of Humanistic Psychology*, vol. 49, no. 1, pp. 66-84, 2009.

[29] C-A. Lopez, "Franklin and Mesmer: An Encounter," *Yale Journal of Biology and Medicine*, vol. 66, pp. 325-331, 1993.

[30] B. E. Wampold, "Humanism as a Common Factor in Psychotherapy,' *Psychotherapy*, vol. 49, no. 4, 445–449, 2012.

[31] 'Cognitive psychology takes computer information processing as its model of brain's function. Thereafter, this was used as the principle of the new „therapies" whose underlying postulate is something along the lines of: Russian dog + American computer = man.' A. Aflalo, *The Failed Assassination of Psychoanalysis*. London: Karnac Books, 2015.

[32] D. N. Elkins, *Humanistic Psychology: A Clinical Manifesto*. Colorado Springs, Colorado: University of the Rockies Press, 2009.

[33] A. MacIntyre, *After Virtue A Study in Moral Theory*. Notre Dame, Indiana: University of Notre Dame Press, 2007.

[34] S. Koch, *Psychology in Human Context*, D. Finkelman and F. Kessel eds., Chicago, London: University of Chicago press, 1999.

[35] Th. Insel, "Post by Former NIMH Director Thomas Insel: Transforming Diagnosis," *NIMH*, https://www.nimh.nih.gov/about/directors/thomas-insel/blog/2013/transforming-diagnosis.shtml. Accessed: May 9th, 2018.

[36] I. Hacking, Lost in the Forest, *London Review of Books*, vol. 35, no. 15, 2013.

[37] M. Baker, M. Menken, "Time to abandon the term mental illness," *BMJ*, vol. 322, 2001.

[38] P. Bracken, Ph. Thomas, Psychiatry beyond the current paradigm, *The British Journal of Psychiatry*, vol. 201, pp. 430-434, 2012.

[39] L. Greenfeld, *Mind, Modernity, Madness The Impact of Culture on Human Experience*. Cambridge, MA, London: Harvard University Press, 2013.

[40] R. Whitaker, "Anatomy of an Epidemic: Psychiatric Drugs and the Astonishing Rise of Mental Illness in America," *Ethical Human Psychology and Psychiatry*, vol. 7, no. 1, pp. 23-35, 2005.

[41] J. Moncrieff, *The Myth of the Chemical Cure*. London: Palgrave Macmillan, 2009.

[42] S.S. Sharfstein, "Big Pharma and American Psychiatry: The Good, the Bad, and the Ugly," *Psychiatric News*,

https://psychnews.psychiatryonline.org/doi/10.1176/pn.40.16.00400003. Accessed: May 9th, 2018.

[43] S. E. Hyman, "Psychiatric Drug Development: Diagnosing a Crisis," *Cerebrum*, April 2013, http://www.dana.org/news/cerebrum/detail.aspx?id=41290. Accessed: May 9th, 2018.

[44] M. Foucault, *Madness and Civilization A History of Insanity in the Age of Reason.* London and New York: Routledge, 2005.

[45] M. Bakhtin, *Problems of Dostoevsky's Poetics*, C. Emerson eds,. Minneapolis-London, University of Minnesota Press, 1999.

[46] L. Johnstone, M. Boyle, "The Power Threat Meaning Framework: Towards the identification of patterns in emotional distress, unusual experiences and troubled or troubling behaviour, as an alternative to functional psychiatric diagnosis," Leicester: *British Psychological Society*, https://www1.bps.org.uk/system/files/user-files/Division%20of%20Clinical%20Psychology/public/INF299%20PTM%20Main%20web.pdf. 2018. Accessed: May 9th, 2018.

[47] 'There is a need of a shift in investments in mental health, from focusing on "chemical imbalances" to focusing on "power imbalances" and inequalities.' D. Pūras, "Special Rapporteur on the right of everyone to the enjoyment of the highest attainable standard of physical and mental health," *UN Human Rights, Office of the High Comissioner*, http://www.ohchr.org/EN/NewsEvents/Pages/DisplayNews.aspx?NewsID=21480&LangID=E. Accessed: May 10th, 2018.

[48] P. Bracken, "Community Development," *Outside Mental Health Voices and Visions of Madness*, Will Hall, ed., Madness Radio, pp. 229-235, 2016.

[49] A. József, "Consciousness," *Winter Night*, J. Bátki, trans., Budapest, Corvina, 1997.

Euthanasia in the Contemporary World: What Role Does Faith Play in the Choice to Legalize Assisted Dying Practices?

Sorin Grigore Vulcănescu
Faculty of Orthodox Theology, University of Bucharest, Romania

Abstract
The purpose of this essay is to examine the pattern of society in the states in which euthanasia, physician-assisted suicide, and assisted suicide have been legally recognized as well as their current legal status. Moreover, this study refers to the slippery slopes that have become obvious in some cases and to the fact that euthanasia cannot be a transparent practice to be adequately controlled by law and free of abuses. An overview of the "metamorphosed" terminology, of how the teachings and morals of the Christian Church (on the basis of which these societies have developed) have been replaced by secular anthropocentric ones, the general factors leading to euthanasia, and the characteristics of pro-euthanasia laws/society constitute an attempt to identify the propulsion of this controversial subject, not an attempt to diabolize Western culture. In this sense, this essay is an approach towards understanding a fact, not a negative generalization of Western culture.

Keywords
Euthanasia; assisted dying; secularizied society; religiosity; law; slippery slopes.

I. Introduction

The term *euthanasia* stems from the ancient Greek words *eu* ("good") and *thanatos* ("death"), literally meaning a *good death*, a *gentle death*, or, more precisely, a *good dying*. In Antiquity and in the Early Church, this term did not involve the action of shortening or taking a life, but it was rather concerned with the mental or spiritual state of a person at the moment before death – a state of peace or reconciliation with the self and God [1] (p. 210). Euthanasia was a way of dying, not an intentional act of causing death, so it was not identified with taking a life [1] (p. 163). However, the practice which we would today call assisted suicide was widespread in that period [2]. It consisted of administering means by which death was speeded.

In modern terms, the Oxford English Dictionary offers three definitions for the word *euthanasia*: (1) "a gentle and easy death"; (2) "the means of bringing" this, and (3) "the action of inducing a gentle and easy death" [3] (p. 325). As Philippa Foot said, a curious fact is that none of these explanations is an adequate definition for *euthanasia* as it is understood in modern thinking because it means something more than *gentle and easy death* or even *the means* or *the action of inducing it* [4] (p. 85). A murderer can drug his victim to death and may later claim it to have been an act of euthanasia, so it is ridiculous to consider euthanasia only as a way of dying to which people use though an act of goodness [4] (p. 85). Additionally, the *euthanasia program* developed in Hitler's Germany, through which approximately 275,000 people with physical or mental disabilities, children and adults were systematically killed – *euthanized* by gas [5] (p. 158), cannot be considered a *gentle and easy death*.

As a conclusion from a multitude of definitions or interpretations, euthanasia[1] could be, with small differences, defined as the medical procedure by means of which a physician intentionally ends the life of another person suffering from an incurable disease in its terminal phase by directly administering a lethal agent, in accord with this person's explicit, voluntary, and informed request, in order to free him/her from a state of physical or mental degradation considered to be unbearable. Euthanasia can be: *voluntary* (with the patient's consent), *non-voluntary* (when the patient's will is not known), or *involuntary* (against the patient's will). Physician-assisted suicide (PAS) consists of the help offered by a qualified person (not necessarily a physician) to a terminally ill patient suffering from an incurable disease to commit suicide, which also involves the delivering of the means to achieve this purpose. Throughout this study one may notice how much relevance these definitions contained in the laws supporting assisted dying practices still have.

[1] In this essay, I refer to *euthanasia* in its active form.

Bioethicists have identified the factors that trigger the demand for euthanasia and its *surrogates* in four different contexts [6] (p. 628–629): (1) *medical* – some people choose euthanasia because of suffering and pain caused by a disease [7] (p. 730); (2) *psycho-social* – the poor condition of health accompanied by a weakened mental state greatly influences the social status of the person [8]; (3) *psycho-emotional* – patients under the burden of severe suffering develop suicidal thoughts due to underestimated factors such as depression, feelings of worthlessness, condemnation and guilt, isolation or loneliness, unbearable pain, or other forms of psychological disturbance that profoundly compromise deliberate choice [9, 10], but once treated and with improved health, suicidal intention disappears [11], and (4) *existential* – the experience of life which includes pleasant events until the disease occurs becomes the existential *burden* of the patient, caused by futility, the lack of hope for recovery, the bad intentions of relatives for inheritance, the costs of treatments, the insurers' pressures, etc. [12].

II. Pro-Euthanasia Society

A. Pro-euthanasia attitudes – a characteristic of a secularized culture

The consequences of Enlightenment, Modernism, and Postmodernism seem to raise *homo hedonicus* on a privileged pedestal in which his own person and his autonomy are the most important goals [13], which even Renascentist humanism had failed to achieve, although it covered many more areas of development and emancipation in this sense. This spectrum of consequences has amounted to the secularization of the *homo hedonicus'* society. This phenomenon has encompassed the entire society, whereby the implications of secularization are observed in many fields: medical, economic, social, cultural, religious etc. Assisted dying has gained its place, next to other controversial practices, in a space of complete openness and permissiveness. Can secularization be construed as the main factor of the openness towards assisted dying and of its acceptance in contemporary society? Is there a pattern of the society in which euthanasia and its *surrogates* were legally recognized? Can we talk about a new way of life and thinking, about a new paradigm, about a moral of an anthropology centred on the self where the *individual* dismantled himself/herself of his/her own *personhood*, rejects his/her primordial origin and purpose, lives as long as his/her life endures and then passes away into non-being when he/she dies? Why has *homo hedonicus freed himself* of the moral precepts that had respected his/her freedom until they became too strict and outdated for him/her?

Euthanasia and its *surrogates* are found in the society of the "liberal cosmopolitan ethos" [14] (p. 129, 143, 162, 311). There is a modern and postmodern way of social thinking which (in the name of unrestrained freedom, pure autonomy, hedonism, and self-interest) sacrifices the respect for life and death with great ease [15]. The progress of civilization and the laboratory pilot programmes transform the entire human perception on life and death. The perception of the *omnipotent* human being on everything that surrounds him/her has brought him/her to believe, without a real basis, that he/she is the absolute sovereign and can decide over life and death. In this *mirage of omnipotence*, the field of biomedical sciences covers the period with the most impressive discoveries and achievements. Unfortunately, some of them generate problems, challenges, and questions that require specific answers and guidelines in order to achieve a profound understanding of the specific realities involved. New scientific and technological discoveries have led from the attempt to decipher the mystery of life to "the medicalized dehumanization of the end of life" [16] (p. 177).

The moral codes established by the Church are no longer a standard for *homo hedonicus* [17]. The Christian Church's recommended rules become too strict for *homo hedonicus* because they do not live up to his lifestyle, and this produces frustration and irritation. That is why he gives up, and then challenges moral religious values. It is, in fact, a struggle with his own weaknesses. Hence, religious values have been replaced by man's desire to have his full autonomy respected: "Autonomy and personal liberty attain an almost sacrosanct status, and 'values', for many, are perceived as 'subjective' and 'personal'" [18] (p. 195). This "dominant secular culture functions as a meta-normative perspective that seeks to relocate, re-interpret, and transform large areas of traditional moralities so as to downgrade the force of claims about a wide range of issues rendering them into life- and death-style choices" [19] (p. 8). We are confronted with an individualist moral, where moral authority has shifted from religious traditions to secular traditions of the self [20, 21]; and the self is the moral authority [18] (p. 195) in a culture of morals without God's-eye perspective, moral standard or canonical moral perspective [19] (p. 6–7).

In the first place, the increasing permissiveness of contemporary society has led to specific means used to speed intentional death, this being also related to the decreasing importance that religiosity has come to have in the moral life of each person. Through an indissoluble connection to people's lives, (1) religiosity or religious orientation and (2) culture/education influence people's behaviours and attitudes with regard to the complicated problems of life. Religious traditions can contribute to the moral improvement of their adherents in cases where their message is properly understood by them, as some sociologists have indicated [22]. For example, Jews and Christians seem to be more motivated in dealing with moral and ethical challenges due to the salvation that God has performed for them [23]. Thus,

> they are no longer turned in on themselves and confined inside the constraining frontiers of their ego; in contrast to this limiting attitude, their relational self opens up to the broad horizon of the Kingdom that God has initiated – a horizon which is much larger than anything their natural mind could imagine [24](p. 618).

Along with it, the religious orientation of people can be influenced by their experiences, educational history and their fields of study. Several studies have brought up convincing evidence in this regard. On the one hand, religious guidance was significantly different for medical students and students in the humanities, whereas the higher a population's religious orientation levels were, the more they opposed euthanasia [25]. In the same context, a study made in 33 European countries showed that adults with higher academic degrees who declare themselves as less religious were open to euthanasia [26]. On the other hand, a series of recent studies demonstrates to what extent the teachings of the Church are decisive in these life-and-death decisions. Accordingly, with respect to euthanasia, one may claim that: (1) the adoption of religious views leads to non-acceptance of euthanasia [27]; (2) religious people have proved more firm in opposing euthanasia [28, 29], especially Roman Catholics whose points of view are the most restrictive [29, 30]; (3) religiosity, religious values, and the frequency of Church attendance form a vision that constantly undermines the support for euthanasia or personal justifications for suicide [31] (p. 94); (4) the main increase in openness to euthanasia is manifested in those with minimal religious dependence [28]; (5) the choice and subsequent legislation on PAS in Oregon originated, according to researchers in the field, largely in the non-religiousness of the citizens of Oregon, the state with the lowest rate of religiosity and frequency of church attendance in the nation [32](p. 30), and (6) the French have had a more positive attitude towards euthanasia than Italians and Americans, which reflects the lower religiosity of France compared to that of Italy and the USA [33]. What concerns suicide, the same studies point out that: (1) religion and spirituality reduce the risk of suicide [34], though only among conservative Christians [35]; (2) several religious factors can contribute to strengthening moral obligations against suicide [36, 37]; (3) religious views lead to non-acceptance of suicide [38, 39] and (4) in the United States alone, suicide was the tenth leading cause of death in 2016, with 44,965 people having died by suicide, and 19,362 people by homicide because of putting people before religious principles or considering the self as unique judge [40](p. 35). In this sense, one can notice how the decrease of the religiosity level has serious consequences for the moral order within society, which is to say that Church goers are probably more receptive to its moral standards and are willing to observe the traditional moral rules, whereas the people who do not go to Church "may reject the 'menu' of church beliefs, instead recomposing a religion *à la carte* – constructing their own religious patchwork" [41] (p. 115). Furthermore, it is evident that a pious believer is less permissive in his/her morality than non-religious people who are less devoted to a particular moral system [42].

Secondly, referring to the present-day European area one can speak of a bimoral Europe. On the one side, the Christian faith and especially its practice in the West have been *eroded* by other religious or pseudo-religious and secularist culture with many effects such as self-sufficiency, indifference, or selfishness encountered in various shapes: utilitarian, hedonistic, etc. Also, secularization has forced religious decline so that moral guidance coming from the Christian Church is no longer popular nowadays and participation at religious services has declined. A recent study focused on the four major themes of homosexuality, divorce, abortion, and euthanasia illustrates that in Western Europe: (1) the presence of the Church in society has clearly decreased in the last decades and (2) the reduced attendance in Church is in perfect harmony with the increased acceptance of the four mentioned themes. On the other side, in Central and Eastern Europe, as also shown by this study, the population is much more conservative than in the West [43] (p. 624–625). However, given the ongoing political, social, and religious changes taking place in these parts of Europe [44], one may ask: how long will this conservatism persist?

A comparative analysis of the European countries indicates that any form of assistance in suicide is strictly prohibited and incriminated by law in most of these states [45]. Nevertheless, a recent study regarding the approval rate of euthanasia in Europe showed a growing tendency for approval as well as a wide permissiveness of the practice, but at the same time an East-West polarization as well [46]. The predictions of researchers in this field are summed up in a quotation by the Oxford professor Julian Săvulescu when he says "Eventually all civilised countries will legalise euthanasia and assisted suicide – it is really just a matter of time." This is because the concerns of some that people might feel pressured to commit suicide were not substantiated in the aftermath of the legalization of the practices in the Netherlands and Belgium. "People want to have control over their death, and there is no reason for the state to intervene in the liberty of one individual to access assisted suicide or euthanasia when another individual is prepared to provide it," the professor adds [47] (p. 433). 67,786 citizens from 47 European countries were interviewed in the survey. The acceptance of euthanasia varied substantially, with the highest approval percentages registered in Denmark, followed closely by Belgium, France, the Netherlands, and Sweden [46] (p. 145). The lowest percentages were registered in Kosovo, Cyprus, Turkey, Georgia, and Armenia. A significant increase in the acceptance of euthanasia was recorded in Spain, Portugal, Great Britain, Germany, and Italy, while the most dramatic decrease was observed in the Russian Federation, Ukraine, Greece, Slovakia, and Belarus [46] (p. 145).

The antagonistic position between the Western and Eastern countries is, according to researchers, based on the inhabitants' positions with regard to religious values and the influence of faith among them. Two groups can be distinguished, with the first including the extremely religious countries that strongly reject euthanasia. These are Turkey, Kosovo, Malta, Georgia, and Cyprus. The second group, with an equal number of countries but with the opposite views includes the Netherlands, Belgium, Denmark, France, and Sweden [46] (p. 145). Religious characteristics are highly important descriptive factors, both on an individual and global level. On the one hand, greater acceptance of euthanasia among non-religious people is probably rooted in the influence of secular culture that puts a relatively high value on the freedom of choice and pure autonomy in issues of life and death [48]. Grounded in this secular culture, an "after God" culture emerges: "after Metaphysics" – "a culture that articulates its concerns fully within the horizon of the finite and the immanent" – and post-traditional European culture – "a culture that is no longer structured by the traditional Christian norms" [19] (p. 5), that

> re-locates traditional moral norms of sexuality, reproduction, marriage, and end-of-life decision-making within a moral perspective that would transform traditional moral norms into quasi-aesthetic norms shaping alternative life-style or death-style choices. The dominant secular morality of the West invites the demoralization and deflation of traditional Western morality, Bioethics, and palliative care ethics [19] (p. 8).

The continuous secularization process, present also in religious areas, has influenced some religions to take a more liberal path [49], even towards a religious acceptance of euthanasia [50]. In general, Protestants seem to support euthanasia more than Roman Catholics, Orthodox and Muslims. In Protestantism there exists a variety of opinions about euthanasia [51] (p. 424). Some Protestant communities seem to argue that when a person is suffering from an end-stage disease, he/she can appeal to this way of dying [52]. Additionally, Roman Catholics from Spain and France are more permissive regarding euthanasia than those in Hungary and Poland. On the other hand, one can argue that the dominant culture in a country can play a more decisive role in determining opinions on euthanasia than the religious positions of the denomination with which a particular country is commonly associated [53], a present phenomenon, as one may see below, in the Netherlands and Belgium as well.

B. The Necessity of Directives from the Church

Firstly, the current practices of hastening death draw attention to a complex set of fundamental questions which have become irrelevant for *homo hedonicus*: What is life? Where do we have it/where does it come from? To whom does it belong? What about death? For Christians, man is the image of God in the world. He receives life as a gift from God, hence his responsibility towards the gift and the Giver as a testimony of the Christian ethos [54] (p. 445). Man's life is the supreme gift of God; the beginning and the end of it depend entirely on Him [55]: "In His hand is the life of every creature" [56]. God respects us as persons with free will, which is why He knows that some can accept the gift and some can reject it. God respects each of our decisions due to the freedom that He gave us. He does not compel us to live [57]. If someone decides to commit suicide, He does not prevent him/her, which shows that He respects every person's freedom [58] (p. 138). God does not prevent us from choosing, but this does not mean that man will not be responsible for his choices [59]. Life as a gift is not simply to be disposed of at the whim of the recipient, for "this is a gift that is not a property to possess" but rather "a task to live out, a task where freedom follows upon responsibility" [54] (p. 446). Life as a gift is the gift of immortality; accordingly, man's life is the supreme gift of God. He created man immortal by grace, and immortality is the natural state of man. Death, suffering, and pain came to the world through sin: death was allowed by God so that evil did not become immortal.[2] On the one side, the Church recognizes the weaknesses of the man, embraces those in pain and suffering, prays for the release of these burdens and for life to end in peace. At the same time, the Church claims that behind every suffering there is a blessing of God. It is the

[2] According to the biblical creation narrative, the primordial humans were not submitted to dying. The Church Fathers confess that the man in the adamic state did not experience physical or emotional pain. Death appeared after man opposed the word of God wishing to be like Him. Man's main loss was the cessation of the paradisiacal existence and happiness. This resulted in serious perturbations of man's life who had become estranged from his authentic nature. Negative events, conditions, and drawbacks began to emerge, the first of which was death – an unatural event in man's life. As the first humans acquired death through their resistance, so assisted dying is the expression of the second resistance through *homo hedonicus* and thus of a second death.

limit of salvation, the dynamic boundary of the limits of existence, and the means or occasion for the preparation of eternal life. On the other side, the Church has always thought that biological life does not determine the entire existence of man. God created man with body and soul, in a unit temporarily interrupted only by death. Though the body disintegrates by death, the soul persists and continues its existence until the union with the risen body. Death is overcome through Christ, He who defeated death by His death and resurrection. Death is edifying in the process of salvation because the way someone lives and dies sets out the outline of eternal existence.

Secondly, the end-of-life decisions simultaneously constitute a social, cultural and political problem, a relational and personal moral problem with a strong religious substratum. Christian theology has approached the themes of life and death since its inception, these being grouped today under the field of medical ethics. Since the Early Church, theologians have developed moral arguments about sexuality, procreation, and the end of life [16] (p. 122, 123, 167). Because of the present-day Church's involvement in ethical education, some bioethicists have categorized believers who oppose euthanasia as *fundamentalists* [31] (p. 91). Clear directives on the part of the Church and its theologians are still necessary today as a response to the contemporary challenges. The teachings contained in these directives are fully comprehended in the light of the revelation of the Saviour Jesus Christ entrusted to His Church. Based on this revelation, the Church acquired the understanding of human nature and the responsibility to defend its dignity. The purpose of the ethical and religious directives is on this evidence double: (1) the reaffirmation of the fact that the ethical standards in the medical care field originate in the Church teaching about the dignity of the human person, and (2) the presence of an authoritative guidance in delicate moral-medical situations [60] (p. 4). These directives are addressed to both directly and indirectly involved persons: medical personnel, patients, hospital managers, administrators, sponsors, volunteers, students, or priests. The Church's point of view expressed by the clergy is vital in understanding end-of-life situations. Many people in distress call for the support of clergy, thus it plays an important role in the personal decisions of patients [61]. This also includes the shaping of the cultural viewpoint on situations related to and decisions at the end of life [62,

63]. A recent exploratory study, in which 15 clerics were interviewed on end-of-life moral deliberations, including suicide, cessation of unnecessary medical treatment, PAS, and euthanasia, testifies to support this argument of the vitality of belonging to the Church. This study resulted in a set of eight themes that summed up the moral deliberations of the clergy in the presence of those on their deathbeds: (1) *sanctity of life*; (2) *preservation of the natural course of life*; (3) *pastoral care*; (4) *support of the faith community*; (5) *referral to professional services*; (6) *end-of-life decision in community*; (7) *consultation with medical professionals*, and (8) *a shift to a hopeful narrative* [64].

Thirdly, today more than even before, one can speak of a three-step alternative for euthanasia and its *surrogates*: (1) the prevention of the disease – as a constant preoccupation throughout one's lifetime, understood as an attitude of gratitude and respect for the life received as a gift, not as an obsessive purpose;[3] (2) the support for research fields dedicated to medical technologies and pain relief – as a constant preoccupation in the service of developing solutions for the alleviation of pain other than lethal injection, and (3) palliative care – the alternative for euthanasia and its *surrogates* covering the factors that trigger the demand for assisted dying. That is a merciful alternative that truly respects the dignity of the person.[4] Palliative care emphasizes the person as a whole and the worth of life of a patient on his/her deathbed rather than the aggressive treatment against illness or death [65] (p. 2750–2751) because this new medical practice accepts the inevitability of death. The emphasis is on care rather than healing.

[3] Beside (1) each person's own care for the prevention of diseases, it is nesessary that in the near future (2) more doctors specialize in preventive medicine, as prevention impedes the occurance of diseases, whereby the work of physicians will be reduced considerably, and that (3) as many states as possible establish clear strategies for preventing diseases, as epitomized by the Swiss model.

[4] The real alternative does not imply the direct intervention by hastening death, but it allows the patient reconciliation with himself, with his neighbours and with God; in other words, a good, light death, authentically dignified. It is the loving and competent care through the management of pain and suffering for the person on the deathbed. The light/easy death is best exemplified through palliative care, not through euthanasia and its *surrogates*, nor through palliative sedation or the voluntary cessation of nutrition and hydration.

Palliative care means alleviating, reducing, and diminishing pain as well as the consideration of three main aspects [16] (p. 193): (1) focus on managing pain and suffering [66]; (2) concern, both for the medical condition and for the patient's inner/spiritual life [67], and (3) decision-making that respects the patient's freedom of choice [68]. Regarding the previous statements concerning pro-euthanasia attitudes, characteristic of a secularized culture, we move forward to approach the current legal status of assisted dying practices in the states where euthanasia and its *surrogates* have been legally recognized.[5]

III. America

A. The United States of America

The Supreme Court of the United States of America has recognized the quality of citizens to have legitimate concerns about death, and in 1997 decided that death is not a constitutionally protected right, leaving the possibility of regulations on PAS to each individual state [69]. In the USA, the number of states that allow PAS in different forms is steadily increasing. In 1994, following a referendum, the citizens of Oregon voted for the *Death with Dignity Act* (DWDA), a law that would eliminate the difference between the right of a patient in the terminal stage of a disease to refuse treatment and the right to quicken death with a lethal substance [70] (p. 154). The result was, three years after the referendum, the 1997 legalization of PAS (but not of euthanasia) in Oregon through the *Oregon Death with Dignity Act*, "a pioneering law" from the viewpoint of the other states. Some other states like Washington (2008), Montana (2009), Vermont (2013), California (approved in 2015/in effect from 2016), Colorado (2016), District of Columbia (2016/2017), and Hawaii (2018/2019) followed the same legislative process, all of them legalizing PAS, while the debate continues in several other American states, which proves that the legalization of assisted dying practices "in the United States is a train that has left the station. It will eventually reach most of the other [...]

[5] Some aspects related to the assisted dying laws in several countries were addressed in the study S.G. Vulcănescu, "The Legal States of Euthanasia and Its Surrogates around the World," *Apulia Theologica*, vol. 4, no. 1, p. 195-209, Jun. 2018.

U.S. jurisdictions where it is not yet legal" [71] (p. 301). As indicated below, the case of Montana shows some peculiarities.

1) In *Oregon*, patients who suffer from an incurable disease have the right by law to request lethal medication. The requirements that the adult patient has to fulfil in order to receive assistance are: (1) he/she has to be a resident of the State of Oregon; (2) to be in the terminal phase of an incurable disease (6 months or less until the natural death), but still capable of self-administering the lethal substances prescribed by the physician; (3) to be informed as to his/her decision and the options of PAS as well as to be able to justify his/her decision; (4) two physicians have to agree on the diagnostic, the prognosis of the disease, and the patient's capacity for decisions, and (5) the patient has to submit an oral request and after at least 15 days a written request in the presence of two witnesses who must not be his/her relatives [72].

The provisions of the Oregon Act are well known, since the controversies surrounding it are notorious, thus making this Act a model of inspiration for neighbouring American states as well as for European ones. In Oregon, unbearable suffering is not a prerequisite for PAS, and that is why the critics of this law have expressed concern and asked for more safety measures to be adopted. Another cause for concern is related to the difficulty in detecting and diagnosing depression in patients with advanced cancer [73] (p. 178). Between 15% and 50% of patients show symptoms of depression, but only between 5% and 20% of them are diagnosed [74] (p. 101). What we know about the connection between depression and the wish to hasten death is that it causes variability, inconsequence, and instability in the wish to die [75]. From 1997 until the end of 2017, when the law was adopted, a total of 1,967 people had prescriptions written under the DWDA, and 1,275 patients died from ingesting the medications [76] (p. 6). The PAS process starts with the patient ingesting the lethal agent. After taking the lethal substances secobarbital or pentobarbital, the patients become unconscious. Death generally occurs after 25 minutes. In the cases referred to, the complications were attributed to: incomplete substance consumption, medical tolerance, or vomiting. There were patients who regained consciousness after ingesting the substances, which led to a minimum level of awareness, followed by death taking place after a few days, and there was also one person who woke up 3 days after ingestion and lived for another 3 months [77] (p. 9).

2) *Washington* is the second state, after Oregon, to legalize PAS, by adopting the *Washington Death With Dignity* (WDWD) *Act* in 2008. The Washington Act is similar to the one in Oregon: (1) the resident patient must submit two oral requests and a written one at an interval of at least 15 days respectively; (2) he/she has to be of full age, and (3) to suffer from a disease in terminal phase with a life expectancy of 6 months or less [78]. Here one can also note the ongoing growth in the number of people who request and receive medical assistance to commit suicide. Immediately following the legalization in 2009 there were 64 registered deaths, while in 2017 there were 212 requests, out of which 196 persons are known to have died [79] (p. 1).

3) In *Montana*, PAS is permitted according to a decision of the State Supreme Court. Nevertheless, there is no law that recognizes PAS. In December 2009, the Montana State Supreme Court ruled in the *Baxter v. Montana* case that there is no legal provision that prohibits a physician from honouring the request of a terminally ill, but mentally competent patient to be prescribed a lethal agent that can hasten his/her death [80].

4) *Vermont* joined the states which allow PAS in 2013, by adopting the *Patient Choice at the End of Life Act*. In order to be approved for PAS, the patient (1) has to be an adult resident of Vermont; (2) has to suffer from an incurable disease in terminal phase; (3) has to be able to submit his/her request, and (4) to administer himself/herself the lethal agent, and (5) the diagnosis and the prognosis have to be confirmed by a second physician from Vermont [81]. Between May 31, 2013 and June 30, 2017 there were 52 registered requests corresponding to the PAS law, while the disease was confirmed in 48 cases [82] (p. 4).

5) *California* legalized PAS through the *End of Life Option Act*, along the lines of Oregon, in 2015, and the law came into force in June 2016. The State of California recognizes the right (1) of adults suffering from a disease in terminal phase with a life expectancy of 6 months or less (2) to require and receive a medical prescription for the purchase and self-administration of the lethal substances [83]. This law remains in force for 10 years, unless the legislators renew it. Within the first 6 months after the law came into force, 191 persons received medical prescriptions according to the law, out of which 111 died after the ingestion of the prescribed lethal substances [84] (p. 3–4). In 2017, 374

of the 577 persons who received medical prescriptions died after the ingestion of the prescribed aid-in-dying drug(s) [85] (p. 3).

6) *Colorado* was the next American state to legalize PAS. On November 8, 2016, Colorado voters decided to pass the *Access to Medical Aid In Dying Act.* This law followed the Oregon example and came into force on December 16, 2016. This law allows (1) an adult person who suffers from an incurable disease in terminal phase with a life expectancy of 6 months or less (2) to receive a medical prescription for the lethal substances for self-administration, (3) authorizes a physician to prescribe the necessary medication, and (4) establishes criminal sanctions for the case in which one's request for medical assistance or medication is deliberately not honoured [86]. During the year 2017, 50 patients of those 69 who requested medical prescription for PAS died after taking the lethal substances [87] (p. 2–3). With an over 70-percent increase compared to 2017, 125 patients in 2018 received prescriptions for aid-in-dying medications under the provisions of the law [88] (p. 1).

7) *District of Columbia*

In 2017, the District of Columbia followed the other U.S. states which passed a law granting the patients with a terminal disease the right to end their lives through medical assistance. The law permits the request for the medical prescription for PAS to (1) an adult person (2) who is in the terminal phase of an incurable disease with a life expectancy of 6 months or less, and (3) who is capable of making this decision and of deliberately requesting the assistance for suicide [89].

8) *Hawaii*

Hawaii is the most recent US state to grant PAS legal status, doing so in 2018. The law *Our Care, Our Choice Act* became effective at the beginning of 2019 and grants (1) adult patients with a disease in terminal phase and (2) decision-making capacity the right to self-administer the lethal agent at the end of life [90]. Given the short period of time elapsed since the coming into force of these laws in the District of Columbia and Hawaii as well as the special status of PAS in Montana, there are no statistics available for these three US territories.

B. Canada

Canada's resistance to the idea of assisted dying crumbled at an unprecedented speed. In less than two years, Canada changed from the nation where PAS was considered a federal crime to the nation that adopted one of the most radical systems of legalizing PAS in the world. In 2013, the National Assembly of Quebec, known to be the most liberal among the Canadian provinces, passed *Bill 52*, an Act that gives the right to receive *end-of-life care*. The novelty of this fast paradigm shift has become second to the one redefining the medical meaning of *receiving end-of-life care*. Prior to the passing of *Bill 52*, this process included hospice and palliative care provided to a terminally ill patient by qualified staff. After the law passed, these services came to include the act of administering a lethal injection or substances that expedite death. Offering medical care becomes a contradictory practice since it helps to accelerate death, an act viewed as *support*. Canadian patients must meet the following criteria to qualify for receiving *medical help* to die: (1) they must suffer from a serious and incurable disease in its terminal stage; (2) be medically insured; (3) be at the age of free and full consent; (4) be in an advanced state of irreversible decline in mental and physical capacities, and (5) have constant, unbearable physical or mental pain, from which they cannot be released in a way that they consider tolerable [91] (p. 11–12).

The Quebec law has been in effect since December 2015. After only one year, the mortality rate was three times higher than the Government had predicted. Alex Shadenberg, executive director for *Euthanasia Prevention Coalition*, remarked that euthanasia-related deaths are underreported, even though in Quebec reporting and monitoring requirements are among the strictest [92]. The Government of Quebec recently issued a report in which 262 deaths were registered in the first 7 months since legalizing this practice [93]. There are also 3 cases mentioned where the law was not observed, without any other information regarding the consequences of these transgressions. This made an enraged Aubert Martin, the executive director of the anti-euthanasia organization *Living with Dignity*, state: "We are talking about killing a human being. This is criminal" [92].

PAS became legal across Canada on June 17, 2016. The bill known as *C-14* allows *Medical Assistance in Dying* (MAID) for (1) medically assessed adult patients (2) who experience intolerable suffering as an effect of an irreversible medical condition, and (3) who face imminent natural death [94]. According to the report from Health Canada, there were 1,982 medically-assisted deaths in the first year after it became legal (803 assisted deaths in the first 6 months and 1,179 in the following 6 months) [95]. By December 31st, 2017, the number of persons who received MAID in Canada amounted to 3,714 [96]. The most common underlying medical condition was cancer, while other common reasons were neuro-degenerative disorders and circulatory or respiratory system failures [95]. An independent report on the status of *mature minors*, with focus on their potential eligibility for MAID, was required by the 2016 Act and is expected to be approved [97].

C. Colombia

Euthanasia was decriminalized in Colombia in 1997 when the Colombian Supreme Court pronounced a favourable sentence in a case where the right to euthanasia was requested. For over 15 years euthanasia was practiced in Colombia without legal regulations. The Supreme Court subsequently decided that this practice cannot be deemed criminal under certain circumstances. In a 2014 process the Court issued two orders to the Ministry of Health and Social Protection in order: (1) to offer support to the Interdisciplinary Scientific Committee which is the enabled institution to evaluate the requests submitted by patients and (2) to prepare a medical protocol that should serve as a guide for physicians and be applied as a reference for the assisted dying procedures, after it was discussed previously by a group of experts from different fields [98].

In July 2015, the Congress of the Republic of Colombia presented a bill that regulates the assisted dying practices in Colombia. This bill defines the necessary steps to be observed and establishes the monitoring and assessment mechanisms for correctly implementing the law, according to the guidelines proposed by the Ministry of Health. The preliminary conditions of the law are: (1) the patient must possess Colombian citizenship and be at least 18 years old; (2) must suffer from an incurable disease in its terminal stage; (3) must have the full capacity of mental faculties at the time of submitting the request; (4) the request must be written, voluntary, repeatedly submitted and signed in the presence of two witnesses or, in the case of the patient's inability to sign, by a person previously designated; (5) if the patient is unconscious and cannot express his/her will by any means, the family or (if he/she does not have a family) the physician can become the decision-maker; (6) the physician must be aware of the patient's serious medical condition which causes him/her great suffering, as well as the patient's capacity for decision-making. Additionally, the physician must inform him/her of the alternatives to euthanasia, the process and its consequences, and (7) after the assessment of the patient and his/her decision-making capacity, the physician sends the request to the Interdisciplinary Scientific Committee for verification [99].

According to the resolution published in March, 2018, by the Ministry of Health and Social Protection in Colombia, minors are allowed to request euthanasia when they suffer from a severe affliction for which there is no reasonable hope for healing. Whereas children under 6 years old were excluded from the euthanasia procedure, minors between 6 and 12 years old can require it if one proves a "neurocognitive development and exceptional psychological that allows them to make a free, voluntary, informed and unequivocal decision in the medical field and their concept of death reaches the expected level for a child over 12 years old" and if they have the consent of both parents. When it comes to adolescents who are 12 to 14 years of age, if there is a discrepancy in the parents' decisions, the will of the patient prevails. Finally, beginning with the age of 14 the decision of the adolescent prevails [100].

IV. Europe

A. The Netherlands

In 2002, the Netherlands was the first country in the world to legalize both euthanasia and PAS [101] (p. 197). A long process of deliberation and tolerance preceded legalizing these two practices. In the mid 1980's, the Royal Dutch Medical Association and the Attorney General came to an agreement regarding euthanasia. If a series of strict conditions were met, the physician conducting the process of euthanasia could not be put under investigation and, consequently, convicted. The first step was to define euthanasia as the ending of the patient's life at his/her own request. The second step was to establish criteria that include: (1) the voluntary request of the informed patient; (2) unbearable suffering; (3) lack of alternatives and of an improvement in health perspective, and (4) consulting a second physician [102]. In the first phase, a prosecutor was the one who evaluated whether all the criteria were fulfilled. From 1998 onwards, the prosecutor's task was taken over by a commission composed of a lawyer, a physician, and an ethicist. On January 1st, 2002, this practice became, with slight changes, formally legal.

Despite all protests and critiques, Holland has endorsed the practice of euthanasia thus becoming an emblematic case for the states that had debates and similar bills on their public agenda. The Dutch professor Theo Boer, a former member of the Evaluation Commission, sees Holland as "a laboratory from which other countries may draw lessons when considering the option of legalizing euthanasia" [101] (p. 198). In the religious circles in Holland the reaction was inconsistent and equivocal at the beginning. While Protestant theologians like Willem Velema and Jochem Douma opposed euthanasia vehemently, the liberal theologians Harry Kuitert and Heleen Dupuis played a major role in promoting euthanasia both in the Christian and in the lay community [101] (p. 198). One emphasized the individual responsibility in the interpretation of the Holy Scripture as well as in the moral capacity to make decisions. What is not mentioned very often when discussing the role of religion in this controversy is the tradition of Dutch liberalism and pluralism. The Netherlands has developed into an important maritime centre and home to many people from different social, ideological, and religious backgrounds. The liberalism of some Dutch theologians influenced by these varied worldviews has been steadily reformulated resulting in the abandonment of the precepts of traditional morality altogether and the idea that a new ethic should not be grounded on the Holy Scripture.[6]

Since the institutionalization of euthanasia in the Netherlands, the majority of euthanasia cases have been performed for people who suffered from terminal stage cancer, a few days, or a few weeks before the occurrence of natural death; these cases were known under the term *traditional cases of euthanasia* [101] (p. 198). Lately, a new category of requests has emerged, not submitted by terminally ill patients as has

[6] "Leading Reformed theologians have stressed the corruption of reason and insisted on a close link between Scripture and morality. For religious and secularized debaters alike, the rejection of religious convictions was assumed to logically imply the rejection of a traditional morality. There have been media discussions in which Reformed Christians virtually prove their opponents right in abandoning a traditional morality and embracing a liberal one: once these opponents – tragically – have rejected the Bible as the basis for ethics, what else is there left?". [101] (p. 209–210).

been the case up to this point. These *new patients* register themselves on the waiting lists justifying their resolution with age-related problems, chronic diseases, exhaustion, the loss of social status, or even loneliness [101] (p. 205). For this new category of applicants who do not receive the approval of the authorities, while fulfilling some conditions of the law, the Clinic End-of-Life offers assistance in dying [103].

The laxity with respect to the initial obligations has also brought about abuses, considering that in Holland the number of persons euthanized or medically assisted in suicide grew from 1,923 in 2006 to 6,091 in 2016, a growth namely of 317% [104]. The evaluation commissions ascertained that there were many cases in which the law was not observed or the physician did not act in conformity with the law, while some patients seemed to have the chance of improvement [105]. Dutch prosecutors are investigating more and more cases that concern suspected abuses committed by Dutch physicians who euthanized ineligible persons [106]. For example, a Dutch physician is currently being tried for the violation of the euthanasia law after having euthanized a woman with dementia without her expressing the wish to be euthanized [107].

B. Belgium

On January 20th, 2001, a committee formed of representatives of the Belgian Senate voted in favour of a law regarding euthanasia. According to this law, neither this practice nor the medical staff who performed it could be prosecuted as long as the legal requirements were met. On May 16th, 2002, after two days of harsh debate, the lower house of the Belgian Parliament passed the law with 86 votes in favour, 51 against, and 10 abstentions [108] (p. 290–291). After euthanasia has taken place, the resulting death can be classified as a *natural* death according to Article 15 in the *Euthanasia Law*, eliminating the legal duty to inform the prosecutor or the coroner. This legal duty consists of writing the patient's death certificate and sending a report to the Federal Commission for Euthanasia Control and Evaluation (CFCEE) within 4 days, after which this authority has to determine whether the procedure was carried out correctly or not. The Belgian euthanasia law, in effect since September 23th, 2002, mentions in Article 3 that the physician who carries out euthanasia is not committing a crime as long as he/she ensures that: (1) the patient has reached the age of maturity or is an emancipated minor and is conscious at the time of the submission of the request; (2) the request is voluntary, well considered, repeatedly submitted, and not the result of external pressure; (3) the patient has a serious medical condition as the result of an incurable disorder caused by disease or accident in which physical or mental suffering is unbearable and constant and can no longer be mitigated; (4) the patient has been informed in a proper manner about the available treatment options, and (5) a different independent physician was consulted [109]. This legislation took effect on September 23rd, 2002, and has since then been modified twice: on November 10th, 2005, in regard to the legal security of the pharmacist who offers the so-called *euthanatica* used by a physician who practices euthanasia, and on February 28th, 2014, in regard to euthanasia for *capable minors* [110] (p. 41). Statistical data show a constant annual increase of euthanasia cases so that their number grew with ca. 900% from 259 persons in the first year to 2,309 in 2017 [111].

In Holland, liberalism and pluralism are considered the ideological foundations on which the law of euthanasia was based. In Belgium, among many other controversial laws, the legal recognition by the Belgian Government of secular humanism alongside 6 other belief systems can be undoubtedly mentioned as the starting graft in the new legislative approach to life and death. Since 1981, the Belgian Government has paid humanitarian counselling services, the secular equivalent of clergy, to provide moral guidance to those in the army, hospitals, and prisons, and human values (autonomy, ethics based on reason and science etc.) are taught in Belgian schools as part of the non-confessional ethics curriculum. This includes the subject of euthanasia which is "part of a philosophy of taking control of one's own existence and improving the objective conditions for happiness. There is an arrow of evolution that goes toward ever more reducing of suffering and maximizing of enjoyment", says Jan Bernheim, professor emeritus of Medicine at the Free University of Brussels [69].

While being the second country in Europe to legalize euthanasia, Belgium has become the state with the highest rate of deaths through assisted dying in the world [112], but also the indisputable leader of leniency regarding the acceptance requirements for intentional death, having turned into a counterexample for the countries that legalized or are about to legalize assisted dying practices. Several amendments to the initial euthanasia law have transformed Belgium into the most open-minded country as far as euthanasia is concerned and also the most debated and controversial of all the states which recognize assisted dying. In this sense, the following aspects should be mentioned:

(1) The Belgian controlling system of euthanasia cases which functions *a posteriori* has not proved itself capable of protecting the patients against the procedures that infringe the law [110] (p. 28). The physicians [113] (p. 617–618) as well as the members of the CFCEE seem to be overwhelmed by the very high number of cases they have to assess [114]. This led to only a summary evaluation of the thousands of annual cases and to an underreporting of approximately 50% [115], a fact which confirmed many of the previous worries and warnings [116]. Moreover, this commission raised a series of suspicions as to the objectivity and independence of its members given that almost a half of the 16 commission members are affiliated to *right to die* associations [69];

(2) *The disregard of the requirement relative to a serious and incurable disorder* or the acceptance of euthanasia in the case of persons with "multiple disorders" [110, 117] – a commonly used expression which is however not found in the euthanasia law [109], nor in the parliamentary reports [110] (p. 30). The requests of those persons are approved with increasing frequency, although they cannot attest they have a severe incurable disease, and suffer from various age-related afflictions like, for example, polyarthritis, while having reduced mobility, weakened eyesight, or hearing disorders [110, 117];

(3) *The disregard of the requirement with respect to physical or mental suffering* or "the 'cure' of suicidal depression" [118] (p. 193). In this situation there are some problematic issues which tend to be ignored: (a) the usual confusion between mental suffering and mental illness; (b) the fact that there are no measurable parameters for objectivity in this regard [110] (p. 34); (c) depression and other similar forms of mental suffering consist of exactly the loss of any wish to live and of any life perspectives, which means that the feeling of having no perspectives and the wish for euthanasia do not indicate the prognosis of the mental suffering [119], and (d) the Belgian euthanasia law stipulated that psychiatric disorders, dementia and depression do not fall under the category of afflictions admitted for euthanasia [109]. However, the list of these excluded afflictions is constantly reconsidered and modified

along with the acceptance of an increasing number of euthanasia applicants with psychiatric disorders, dementia or depression [120];

(4) *Being tired of living* has become a relatively common condition for which physicians permit euthanasia [121];

(5) *Palliative care centres have become real euthanasia houses* as soon as euthanasia was officially accepted as an option within the palliative care program [122] despite the reactions of the World Health Organization [123] and the European Association for Palliative Care [124] which had opposed the introduction of euthanasia in palliative care. The Belgian model of combining euthanasia with the palliative care represents the promotion of a culture with "a high degree of tolerance and compromise" [125] (p. 524) and creates an unclear image, so that many people at the end of their days are afraid to resort to a palliative centre due to the possibility of euthanasia [126] (p. 105–106). At the same time, this integral model testifies to the negative impact that euthanasia has on the palliative care sector [127];

(6) *Permission for the transplant of viable organs from euthanized persons* [128] causes a new dilemma concerning the possible pressure exerted on a patient who could be euthanized only for the extraction of organs [129] (p. 488). Part of the debate are also the requests of some inmates in prison who, deprived of freedom, claimed to be suffering from psychological afflictions and, accordingly, asked for euthanasia and agreed to donate their organs [130]. In contrast, the most important international transplant organizations condemn the acquisition of organs from detainees in China, which is deemed as commercial exploitation, a violation of human rights, and an unacceptable practice [131];

(7) *The extension of the law to include minors* [110] (p. 41) without previous serious consultations with approval from paediatricians with experience in treating severely ill children [132], and neglecting that: (a) "emerging adulthood" – the coming of age takes places at the age from 18 to 25 years [133]; (b) a minor's capacity to fully understand the terms and medical details of the informed consent is insufficient [134], and (c) the external influences in making very difficult decisions can be sometimes exceedingly powerful, so that the freedom of decision-making is diminished [135];

(8) *The illegal practice of PAS* – although the Belgian legislator decided from the beginning that PAS has to be excluded from the application

of the euthanasia law [110] (p. 35–36), euthanasia and PAS are considered medical procedures [125], on which ground requests for PAS are regularly approved [136];

(9) *Sedation as clandestine means of euthanasia* or "underground euthanasia" – one has ascertained an increase in the use of sedation as secret means for euthanasia [137] (p. 251) as well as for causing death without the consent or request of the patient/family in 27,9% of the cases reported by specialists in the medical field [138], a fact even recognized by the representatives of the Belgian Society for Intensive Care Medicine [139] (p. 174);

(10) *Going from an exception or ultimate solution to a legal right, normality, and particular dynamic* – in 2002, euthanasia was presented as an ethical violation, an exceptional act or an ultimate solution for extreme cases [140]. Then, it was claimed that "euthanasia is neither an exception nor an ethical transgression and its practice, properly regulated, is part of end of life care" [141] (p. 24). Today, one may find websites [142], official documents [143] and informative brochures [144] that confirm *man's right to euthanasia* [145] (p. 229), something that is being stabilized in Belgian society as a rather current practice [145] (p. 230), as a valid option at the end of life [146] (p. 1180). In addition, one can notice

> there is an indication that euthanasia, once the barrier of legalisation is passed, tends to develop a dynamic of its own and extend beyond the agreed restrictions, in spite of earlier explicit reassurances that this would not happen – in Belgium, such reassurances were given when the 2002 law was being debated [145] (p. 226).

C. Luxembourg

In 2009, the Grand Duchy of Luxembourg adopted a law regarding assisted dying, becoming the third European state to legalize it after the Netherlands and Belgium. This law stipulates that the physicians who perform euthanasia or PAS will not face legal sanctions or civil suits provided they consult with a colleague first, in order to attest that the patient suffers from a severe, incurable disease [147]. The prerequisites for a request to be accepted are: (1) the patient suffers from an incurable disease that causes him/her unbearable physical or mental suffering; (2) the patient is in serious medical condition as a result of an accident or illness; (3) the patient is conscious and capable when submitting the request, and (4) the request must be written on a voluntary basis, repeatedly if necessary and should not be the result of external pressure [148]. The law was voted by the members of the small Luxembourgian Parliament with high support of its citizens, despite the firm opposition of the Roman-Catholic Church. The Church, by means of a powerful campaign against this law, succeeded to provoke one of the biggest debates in the history of the Grand Duchy. The euthanasia law even led to the need to change the Constitution following the refusal of the Grand Duke, Henri, to sign the Act, thus forcing the Parliament to add an amendment to the Constitution that reduces the monarch's power to a mostly ceremonial role [149].

D. Switzerland

In Switzerland, assisted suicide (AS) does not have a special legal status as is the case in the other states which recognize this practice. The assistance in suicide is based on Art. 115 of the Criminal Law that decriminalizes this practice as long as one does not act in one's own interest [150]. Different from other legislation, the preliminary condition of the existence of a terminal disease is not specified in Art. 115, nor are other conditions. Thus, the logic behind the decriminalization of AS resulted from the lack of incrimination. According to the principle *nulla poena sine lege*, a person must not be punished for something that is not prohibited by law. Since the parliamentarians did not have a medical perspective when they elaborated Art. 115 in 1918 [151], the Federal Supreme Court of Switzerland recognized an individual's right to request AS when this person (1) is aware of the procedure, (2) does not act under the influence of a third party or out of impulse, (3) the wish to die persists, and (4) he/she commits the act with his/her own hand [152]. In Switzerland, there is no obligation to report AS cases in a central register, and so the registration of the cases may be incomplete. However, the Federal Statistical Office makes considerable efforts to track the cases of AS in collaboration with the *right to die* organizations, the institutes of forensic medicine and physicians [153] (p. 1).

Contrary to the situation in the Netherlands, Belgium, and Luxembourg, the Swiss law does not authorize physicians to perform AS [154]. Switzerland is the only country in the world where the alternative practice to PAS, AS, is performed. This process allows a non-resident patient to obtain a lethal substance (barbiturates) from the *right to die* organizations. Thus, suicide assistance can be given to a person who: (1) is aware of this procedure; (2) does not act by impulse or under the influence of another person; (3) has a persistent desire to die, and (4) commits the act himself/herself [152].

Non-governmental organizations play a central role in facilitating AS in Switzerland and in the US State of Oregon. In Switzerland, the so-called *right to die* organizations offer assistance to their members who want to commit suicide [155] (p. 810). The most influential are: Exit Deutsche Schweiz [156], Exit ADMD (Association pour le droit de mourir dans la dignité) [157], ExInternational [158], Dignitas [159], SPIRIT [160] (p. 612), and StHD + SterbeHilfe Deutschland [161] (p. 612). Four of these organizations provide suicide assistance to non-Swiss citizens, and this situation has led to the emergence of the term *death tourism*. The most well-known organization that offers this service is Dignitas, which is also the most controversial organization. The particularities of the organization have led to intense discussions on *suicide/death tourism*. Media investigations revealed that the suicides committed by some Dignitas members were carried out in cars, by using plastic bags or inhaling helium, all of which are anything but assistance with the aim of an *easy death* [161, 162]. Former volunteers of the organization argued that applicants did not even have enough time to reconsider their wish to hasten their deaths. In addition, there were cases in which patients received lethal drugs on the same day they arrived after only a short interview [163]. At the same time, the Dignitas organization is constantly accused of a lack of transparency in financial terms, and it is further suspected that many members leave them their inheritance, contrary to the Swiss law that compels the respect of altruistic behaviour [154] (p. 249). The online news outlet the *Mirror* has reported that

> Healthy people are travelling abroad for assisted suicide simply *weary of life*, research has revealed. [...] Around 16 percent of the people who use *right-to-die* organizations such as Dignitas have no underlying health problems listed on their death certificates. And women, highly educated, divorced and rich people are more likely to die from assisted suicide [165].

According to the law, members must repeatedly request assistance to commit suicide because of unbearable suffering; then, the person is examined by a physician who decides whether the patient has the ability to decide for him/herself and if he/she can be prescribed with barbiturates. In case of approval, a volunteer within the organization purchases the barbiturates from a pharmacy and then deposits them at the clinic until they are used. On the day the suicide is scheduled, a volunteer will re-evaluate the member's ability to make decisions, and if it is determined that the patient reaffirms his/her wish to die, the volunteer will mix the substances in drinks or food and will then hand them over to the patient. If the patient is unable to swallow, a feeding tube will be inserted into his/her stomach through the abdominal wall or intravenously, so as for him/her to be able to administer the lethal substances by him/herself [166] (p. 295). In the presence of friends and family, the patient will self-administer the lethal substance and then he/she will say goodbye to his/her loved ones and fall into a deep sleep. Death occurs in a very short time as the result of cardiac arrest. After the patient is dead, the Swiss police is notified, which, together with a coroner and a district attorney, assess the legality of the file and of the procedure [155, 166, 167].

Because there is no central registration system, as mentioned above, the existing data about the frequency and amplitude of AS are limited. There are however several international studies that used the official data from the Institute of Forensic Medicine in Zürich, the organization responsible for the investigation of all the *extraordinary deaths* in this region where the centres of Exit and Dignitas are located [168]. In the interval January 1st, 2008 – December 31st, 2012, 950 cases were registered, of which 611 cases of *suicide tourism* were analysed – the cases of *suicide tourism* doubled from 2009 to 2012 [160] (p. 613–614). The most recent study regarding the rate of AS in Switzerland shows a tripling from 2003 to 2014 [153]. The increasing incidence of AS cases in Switzerland is based on a large acceptance by the Swiss population confirmed through referenda and social inquiries [169].

E. Germany

The German legal and political elites have a long history of ethical reflection on the topic of euthanasia. In order to avoid any approach that could lead to similarities to Nazi *euthanasia*, those who supported the legalization of assisted dying wanted to clarify the non-involvement of the state in the decisions of the patients, with it being considered a decriminalization rather than a right. Prior to the current law, assisted dying existed in a *grey* legislative zone of the German legal system. Physicians were allowed to speed up death for terminally ill patients by providing high doses of painkillers or by withdrawing treatment following patient requests. This legislative vacuum also existed because of the historical connection of the practice of euthanasia with the Nazi regime. During the National Socialist government, *clandestine euthanasia* began as a eugenic measure in 1939, and approximately 275,000 people with physical or mental disabilities, children and adults were systematically killed [5] (p. 158). Hence, today's avoidance of the term *Euthanasie* ("euthanasia") and the use of the terms *Sterbehilfe* and *assistierter Suizid* ("euthanasia" and "assisted suicide").

The German Parliament adopted a bill on November 6th, 2015 which allows assisted dying provided it takes place *on an individual basis, for altruistic reasons* [170], while it prohibits this practice if it has a profit as its purpose [171]. The Parliament of Germany modified Art. 217 of the Criminal Law in order to decriminalize it, although only under certain circumstances. The law (1) prohibits and punishes any attempt to commercialize assisted dying services; (2) any kind of material benefits and *death tourism*, as in the Swiss case, shall be punished by imprisonment of up to three years, and (3) the patient's relatives and close friends as well as the physicians who act out of altruism in individual cases of assistance to the persons who wish to hasten their own death will not face prosecution [172]. Nevertheless, the law remains unclear and inaccurate because of the type or the specific mean of assisted dying is not mentioned [173] (p. 691).

In the process of deciding on the law with regard to assisted dying, a moderate bill was accepted by the German Parliament that aims to indicate a middle way between complete incrimination and a total liberalization of such practices [174, 175]. Thus, the German law appears as different from other similar laws in the sense that, on the one hand, it protects assisted dying from being performed for financial reasons, and, on the other hand, it allows assisted dying not as a right of the citizen, but rather as a practice that does not contravene the law. In other words, the emphasis is not placed on the person and his/her rights, but rather on whether this undertaking is or is not against the law [176] (p. 23).

V. Australia

On March 25th, 1996, the Northern Territory of Australia became the first jurisdiction in the world to legally allow euthanasia and PAS by adopting the *Terminal Act III 1996* [177] (p. 540). This law permitted adult persons suffering from a disease in terminal stage to request the assistance of a physician in hastening death. In 1997, the Australian Government abrogated the euthanasia law by limiting the legislating role of the regions of Northern Territory, Australian Capital Territory, and Norfolk Island [178]. Consequently, after 9 months and 4 suicide cases, the law was abrogated [137] (p. 239).

In June 2016, the Standing Committee on Legal and Social Issues of the Parliament of Victoria recommended the legalization of PAS in the Australian state of Victoria [179]. The final report of the Committee made several recommendations relating to the improvement in providing end-of-life care based on the laws in force in the state of Victoria as well as the legislation of the states where this practice had been legalized. In order to be eligible for PAS, the patients (1) have to be residents of the state of Victoria, (2) with a severe and incurable disease (3) that causes an unbearable and long-term suffering, and (4) have to be at the end of life, i.e. the last weeks or months of life [179] (p. 238). Following this legislative proposal, the Government had 6 months to respond to this project [180]. In December 2016, the Government announced that it accepted this bill [181]. The vote was given in November 2017 and in the middle of 2019 Victoria will become the only Australian state which legally recognizes PAS [182]. In this sense, Premier Daniel Andrews declared himself "proud today that we have put compassion right at the centre of our parliamentary and our political process" and he considered the day of the vote "a day of reform, a day of compassion, a day of giving control to those who are terminally ill" [182]. Victorian Health Minister Jill Hennessy said that the long debate gave the Parliament the chance to properly analyse what a *good death* means and added that "We've had some frustrating moments, but ultimately we have landed in a place where Victorians who are confronted with terminal illnesses, that are enduring unbearable pains, will have a safe and compassionate option around assisted dying" [182].

VI. Conclusions

The meaning of the term *euthanasia* has undergone two major changes since its usage in Antiquity. The first change consisted in the contemporary re-appropriation of the ancient concept of a *good death* within the framework of the debates about the legalization of assisted dying in some countries. The second shift has taken place in the most recent years and represents the contrast between the initial definitions given by the first laws on euthanasia, assisted suicide, or physician-assisted suicide that were passed in several states of the world, and the actual effects of these laws on a practical level. With reference to the latter transformation, the discussion above has shown that euthanasia creates a dynamic and a direction of its own which can be measured not only by the worrying increase in the number of the euthanized persons, but also by the constantly expanding spectrum of afflictions approved for euthanasia or by the reduction of its restrictions. The momentum of euthanasia proves that it cannot be fully controlled by the law and that the slippery slopes associated with euthanasia transform it into a slippery slope itself.

It is obvious that euthanasia and its surrogates have gained ground in a pattern of society characterized by a high degree of secularization. Within this type of dominant culture *homo hedonicus* affirms his wish (that has reached an almost sacrosanct status) for pure autonomy and boundless sovereignty over his own life and death. Moreover, one rewrites traditional morality by replacing it with the self as the centre of moral authority. On the contrary, studies have pointed out that religion and spirituality reduce the risk of suicide, that religious values, and the frequency of Church attendance form a vision that constantly undermines support for euthanasia or creates the feeling of non-acceptance of euthanasia. The main increase in openness to euthanasia is manifested in those with minimal religious dependence. A pious believer is less permissive in his/her morality than non-religious people who are less devoted to a particular moral system. The Christian Church, its doctrine and faith, can truly and comprehensively contribute to understanding the meaning and purpose of human life. The Church can maintain a scientific and clinical direction in accordance with the adjacent rules, guaranteeing the relevance of the basic principles of Christian anthropology.

References

[1] T.A. Boer, "Euthanasia, Ethics and Theology: A Dutch Perspective," *RES*, vol. 6, no. 2, Aug. 2014. J.F. Drane, *A Liberal Catholic Bioethics*. Berlin: Lit, 2010.

[2] V. Zimmermann, "Die «Heiligkeit des Lebens» – Geschichte der Euthanasie in Grundzügen," *»Euthanasie« und die Aktuelle Sterbehilfe-Debatte. Die Historichen Hintergründe Medizinischer Ethik*, ["The «Holiness of Life» – An Outline of the History of Euthanasia," «Euthanasia» and the Current Euthanasia Debate. The Historical Backgrounds of Medical Ethics], A. Frewer and C. Eickhoff, eds., Frankfurt: Campus, pp. 27, 2000. J.F. Drane, *A Liberal Catholic Bioethics*. Berlin: Lit, pp. 162-163, 2010.

[3] *The Oxford English Dictionary*, vol. III. Oxford: Oxford University Press, 1963.

[4] P. Foot, "Euthanasia," *Philos. Public Aff.*, vol. 6, no. 2, pp. 85-112, Win. 1977.

[5] F. Wertham, *Sign for Cain. An Exploration of Human Violence*. New York: The Macmillan Company, pp. 153-191, 1966.

[6] See S. Shekhar and A. Goel, "India's Position in the Global Scenario," *Am. J. Hosp. Palliat. Med.*, vol. 30, no. 7, pp. 628-631, Nov. 2012.

[7] Y. Van Wesemael, J. Cohen, J. Bilsen et al., "Process and outcomes of euthanasia requests under the Belgian act on euthanasia: A nationwide survey," *J. Pain Symptom Manag.*, vol. 42, no. 5, pp. 721-733, Nov. 2011.

[8] Public Health Division, *Oregon Death with Dignity Act 2017 Data Summary*, 2018, https://www.oregon.gov/oha/ph/providerpartnerresources/evaluationresearch/deathwithdignityact/documents/year20.pdf. Accessed: January 04, 2019. Washington State Department of Health, *Washington State Department of Health 2017 Death with Dignity Act Report*, 2018, https://www.doh.wa.gov/Portals/1/Documents/Pubs/422-109-DeathWithDignityAct2017.pdf. Accessed: January 04, 2019.

[9] E.H. Cassem, "Depressive Disorders in the Medically Ill," *Psychosomatics*, vol. 36, no. 2, pp. 2-10, Mar.-Apr. 1995.

[10] R.-G. Miller, "Hospice Care as an Alternative to Euthanasia," *JLME*, vol. 20, no. 1-2, pp. 127-132, Spr.-Sum. 1992.

[11] R. Twycross, "A View from the Hospice," *Euthanasia Examined: Ethical Clinical and Legal Perspectives*, J. Keown, ed., Cambridge: Cambridge University Press, pp. 141-168, 1997.

[12] C.D. Ruijs, A.J. Kerkhof, G. van der Wal, B.D. Onwuteaka-Philipsen, "Symptoms, unbearability and the nature of suffering in terminal cancer patients dying at home: A prospective primary care study," *BMC Family Practice*, vol. 14, no. 201, pp. 1-17, Dec. 2013. R. Lamerton, *Care of the Dying*. London: Priority Press, pp. 99, 204, 1973. K. Ohnsorge, C. Rehmann-Sutter, N. Streeck, and H. Gudat, "Wishes to die at the end of life and subjective experience of four different typical dying trajectories. A qualitative interview study," *PloS one*, vol. 14, no. 1, Jan. 2019, doi:10.1371/journal.pone.0210784.
[13] L. Turner, "Euthanasia and distinctive horizons of moral reason," *Mortality*, vol. 2, no. 3, pp. 195, 1997. J. Christman, "Autonomy," *The Oxford Handbook of the History of Ethics*, R. Crisp, ed., Oxford: Oxford University Press, p. 691, 2013.
[14] H.T. Engelhardt, *The Foundations of Christian Bioethics*. Lisse: Swets & Zeitlinger, 2000.
[15] B. Pilkington, "Distinguishing deference from deferment: Assisted suicide is the wrong response," *Christ. Bioethics*, vol. 24, no. 1, pp. 59-78, Apr. 2018.
[16] J.F. Drane, *A Liberal Catholic Bioethics*. Berlin: Lit, 2010.
[17] L. Halman, E. Van Ingen, "Secularization and Changing Moral Views: European Trends in Church Attendance and Views on Homosexuality, Divorce, Abortion, and Euthanasia," *Eur. Sociol. Rev.*, vol. 31, no. 5, pp. 624, Oct. 2015. J. Casanova, "The Religious Situation in Europe," *Secularization and the World Religions*, H. Joas, K. and Wiegandt, eds., Liverpool: Liverpool University Press, p. 206-227, 2009.
[18] L. Turner, "Euthanasia and distinctive horizons of moral reason", *Mortality*, vol. 2, no. 3, pp. 191-205, 1997.
[19] H.T. Engelhardt, "Suffering, dying, and death: Palliative care ethics «after God»," *EJST*, vol. 8, no. 2, pp. 5-13, Sep. 2012.
[20] R.N. Bellah, R. Madsen, W.M. Sullivan et al., *Habits of the heart: individualism and commitment in American life*. New York: Harper & Row, 1985. M.J. Cherry, "Physician-Assisted Suicide and Voluntary Euthanasia: How Not to Die as a Christian," *Christ. Bioethics*, vol. 24, no. 1, pp. 9, Apr. 2018.
[21] P. Rieff, *The triumph of the therapeutic: uses of faith after Freud*. Chicago: University of Chicago Press, 1987.
[22] R. Stark, *Exploring the Religious Life*. Baltimore: Johns Hopkins University Press, 2004. S. Sharot, *A Comparative Sociology of World Religions: Virtuosos, Priests, and Popular Religion*. New York: New York University Press, 2001.
[23] S. Coakley, *Powers and Submissions: Spirituality, Philosophy and Gender*. Oxford: Blackwell, 2002.
[24] L. Roy, "Does Christian Faith Rule out Human Autonomy?," *Heythrop Journal*, vol. 53, no. 4, pp. 606-623, Jul. 2012.

[25] A. Farid, R.H. Kaleybar, L. Ghobadi, S.R. Mousavi, "Prediction of Students' Attitudes toward Euthanasia Using Their Religious Orientation, Self-Esteem and Death Anxiety," *Health, Spirituality and Medical Ethics*, vol. 4, no. 3, pp. 2-7, Aug. 2017.

[26] J. Cohen, I. Marcoux, J. Bilsen et al., "European public acceptance of euthanasia: Socio-demographic and cultural factors associated with the acceptance of euthanasia in 33 European countries," *Soc Sci Med*, vol. 63, no. 3, pp. 743-756, Aug. 2006.

[27] U. Cavlak, U.B. Aslan, S. Gurso et al., "Attitudes of physiotherapists and physiotherapy students toward euthanasia: a comparative study,"*Adv Ther*, vol. 24, no. 1, pp. 135-145, Jan.-Feb. 2007. J. Cohen, I. Marcoux, J. Bilsen et al., "European public acceptance of euthanasia: Socio-demographic and cultural factors associated with the acceptance of euthanasia in 33 European countries," *Soc Sci Med*, vol. 63, no. 3, pp. 743-756, Aug. 2006.

[28] A. Danyliv, C. O'Neill, "Attitudes towards legalising physician provided euthanasia in Britain: The role of religion over time," *Soc Sci Med*, vol. 128, pp. 52-56, Mar. 2015.

[29] E. Verbakel, E. Jaspers, "A comparative study on permissiveness toward euthanasia: Religiosity, slippery slope, autonomy, and death with dignity," *Public Opin. Q.*, vol. 74, no. 1, p. 109-139, Jan. 2010. I. Dowbiggin, "From Sander to Schiavo: Morality, Partisan Politics, and America's Culture War over Euthanasia, 1950–2010," *J. Policy Hist.*, vol. 25, no. 1, pp. 30, Jan. 2013.

[30] E. Tomlinson, J. Stott, "Assisted dying in dementia: A systematic review of the international literature on the attitudes of health professionals, patients, carers and the public, and the factors associated with these," *Int J Geriatr Psychiatry*, vol. 30, no. 1, pp. 10-20, Jan. 2015.

[31] R.M. Alvarez, J. Brehm, *Hard Choices, Easy Answers*. New Jersey: Princeton University Press, 2002.

[32] I. Dowbiggin, "From Sander to Schiavo: Morality, Partisan Politics, and America's Culture War over Euthanasia, 1950–2010," *J. Policy Hist.*, vol. 25, no. 1, pp. 12-41, Jan. 2013.

[33] P. Peretti-Watel, M.K. Bendiane, J.P. Moatti, "Attitudes toward palliative care, conceptions of euthanasia and opinions about its legalization among French physicians," *Soc Sci Med*, vol. 60, no. 8, pp. 1781-1793, Apr. 2005.

[34] B.H. Amit, A. Krivoy, I. Mansbach-Kleinfeld et al., "Religiosity is a protective factor against self-injurious thoughts and behaviors in Jewish adolescents: Findings from a nationally representative survey," *Eur Psychiatry*, vol. 29, no. 8, pp. 509-513, Oct. 2014.

[35] D. O'Reilly, M. Rosato, "Religion and the risk of suicide: Longitudinal study of over 1 million people," *Br J Psychiatry*, vol. 206, no. 6, pp. 466-470, Jun. 2015.

[36] K. Dervic, J.J. Carballo, E. Baca-Garcia et al., "Moral or religious objections to suicide may protect against suicidal behavior in bipolar disorder," *J Clin Psychiatry*, vol. 72, no. 10, pp. 1390-1396, Oct. 2011. A. Osman, J. Gifford, T. Jones et al., "Psychometric evaluation of the reasons for living inventory," *Psychol. Assess.*, vol. 5, no. 2, pp. 154-158, Jan. 1993.

[37] S. Hamdan, N. Melhem, I. Orbach et al., "Protective factors and suicidality in members of Arab kindred," *Crisis*, vol. 33, no. 2, pp. 80-86, Jan. 2012.

[38] G. Domino, K. Miller, "Religiosity and attitudes toward suicide,"*OMEGA–Journal of Death and Dying*, vol. 25, no. 4, pp. 271-282, Dec. 1992.

[39] W. Leane, R. Shute, "Youth suicide: The knowledge and attitudes of Australian teachers and clergy," *Suicide Life Threat Behav*, vol. 28, no. 2, pp. 165-173, Sum. 1998.

[40] J. Xu, S.L. Murphy, K.D. Kochanek et al., "Deaths: Final Data for 2016," *NVSS*, vol. 67, no. 5, pp. 1-76, Jul. 2018.

[41] W. Jagodzinski, K. Dobbelaere, "Secularization and Church Religiosity," *The Impact of Values*, J. Van Deth and E. Scarbrough, eds., Oxford: Oxford University Press, 1995.

[42] P. Scheepers, M. Te Grotenhuis, F. Van Der Slik, "Education, Religiosity and Moral Attitudes: Explaining Cross-National Effect Differences," *Sociology of Religion*, vol. 63, no. 2, pp. 157-176, Sum. 2002. S. Bruce, *Politics and Religion.* Cambridge: Polity Press, 2003.

[43] L. Halman, E. Van Ingen, "Secularization and Changing Moral Views: European Trends in Church Attendance and Views on Homosexuality, Divorce, Abortion, and Euthanasia," *Eur. Sociol. Rev.*, vol. 31, no. 5, pp. 616-627, Oct. 2015.

[44] M. Tomka, *Church, State and Society in Eastern Europe.* Washington: The Council for Research in Values and Philosophy, 2005. L.M. Hormel, "Atheism and secularity in the former soviet union," *Atheism and Secularity*, P. Zuckerman, ed., Santa Barbara: Praeger, 2010. B. Ančić and S. Zrinščak, "Religion in Central European Societies. Its social role and people's expectations," *RASCEE*, vol. 5, no. 1, pp. 21-38, Jan. 2012.

[45] C. Grosse, A. Grosse, "Assisted suicide: Models of legal regulation in selected European countries and the case law of the European Court of Human Rights," *Med. Sci. Law*, vol. 55, no. 4, pp. 246-258, Oct. 2015.

[46] J. Cohen, P. Van Landeghem, N. Carpentier et al., "Public acceptance of euthanasia in Europe: a survey study in 47 countries," *J Public Health.*, vol. 59, no. 1, pp. 143-156, Feb. 2014.

[47] T.K. Burki, "Trends in euthanasia and assisted suicide", *Lancet Oncol.*, vol. 16, no. 9, pp. 433, Sep. 2015.

[48] A. Danyliv, C. O'Neill, "Attitudes towards legalising physician provided euthanasia in Britain: The role of religion over time," *Soc Sci Med*, vol. 128, pp. 52-56, Mar. 2015. L. Halman and E. Van Ingen, "Secularization and Changing Moral Views: European Trends in Church Attendance and Views on Homosexuality, Divorce, Abortion, and Euthanasia," *Eur. Sociol. Rev.*, vol. 31, no. 5, pp. 616-627, Oct. 2015.
[49] B.E. Moulton, T.D. Hill, A. Burdette, "Religion and trends in euthanasia attitudes among U.S. adults, 1977–2004," *Sociol. Forum*, vol. 21, no. 2, pp. 249-272, Jun. 2006.
[50] C.A. Hains, N.J. Williams, "Attitudes toward euthanasia and physician-assisted suicide: A study of the multivariate effects of healthcare training, patient characteristics, religion and locus of control," *J Med Ethics*, vol. 39, no. 11, pp. 713-716, Nov. 2013.
[51] H. Bulow, C. Sprung, K. Reinhart et al., "The world's major religions' points of view on end-of-life decisions in the intensive care unit," *Intensive Care Med*, vol. 34, no. 3, pp. 423-430, Mar. 2008.
[52] C. Campbell, "Euthanasia and religion," *UNESCO Courier*, vol. 53, no. 1, p. 37-39, 2000. H. Bulow, C. Sprung, K. Reinhart et al., "The world's major religions' points of view on end-of-life decisions in the intensive care unit," *Intensive Care Med*, vol. 34, no. 3, pp. 424, Mar. 2008.
[53] See R. De Moor, *Values in Western Societies*. Tilburg: Tilburg University Press, 1995.
[54] S. Hauerwas, R. Bondi, "Memory, community and the reasons for living: Theological and ethical reflections on suicide and euthanasia," *J. Am. Acad. Religion*, vol. 44, no. 3, pp. 439-452, Sep. 1976.
[55] S. Harakas, "Bioethics in Eastern Orthodox Christianity," *Encyclopedia of Bioethics*, vol. 1, S.G. Post, ed., New York: Macmillan, pp. 691-697, 2004.
[56] Job 12:10.
[57] R. Anselm, "Leben als Gut, nicht als Pflicht. Der Beitrag der evangelischen Ethik in der Debatte um den assistierten Suizid," [Life as Good, Not as Duty. The Contribution of Evangelical Ethics in the Assisted Suicide Debate], *ZEE*, vol. 59, no. 2, pp. 104-113, Apr.-Jun. 2015.
[58] R. Banks. "C. S. Lewis on Pain," *The Consolations of Theology*, B.S. Rosner, ed., Michigan: Eerdmans, pp. 131-157, 2008.
[59] About the responsibilities, not the obligations that life as a gift creates See J. Gustafson, *Theology and Christian Ethics*. Philadelphia: Pilgrim Press, p. 170, 1974. J. Gustafson, *Can Ethics be Christian?*. Chicago: University of Chicago Press, pp. 101-102, 1975. S. Hauerwas and R Bondi, "Memory, community and the reasons for living: Theological and ethical reflections on suicide and euthanasia," *J. Am. Acad. Religion*, vol. 44, no. 3, p. 445-446, Sep. 1976. D.S. Long and T. York, "Remembering: Offering Our Gifts,"

The Blackwell Companion to Christian Ethics, S. Hauerwas and S. Wells, eds., Oxford: Blackwell Publishing, pp. 398, 2011.

[60] USCCB, *Ethical and Religious Directives for Catholic Health Care Services*, Washington, 2018.

[61] A. Goodhead, P. Speck, L. Selman, "«I think you just learnt as you went along»-community clergy's experiences of and attitudes towards caring for dying people: A pilot study," *Palliat Med*, vol. 30, no. 7, pp. 674-683, Jul. 2016. K. Mason, M. Geist, and M. Clark, "A developmental model of clergy engagement with suicide: A qualitative study," *OMEGA–Journal of Death and Dying*, Jun. 2017, doi: 30222817713289.

[62] J. Hanford, "The pastor and bioethics," *Pastoral Psych.*, vol. 55, no. 2, pp. 175-182, 2006.

[63] L. Jylhänkangas, T. Smets, J. Cohen et al., "Descriptions of euthanasia as social representations: Comparing the views of Finnish physicians and religious professionals," *Sociol. Health Illn.*, vol. 36, no. 3, pp. 354-368, Aug. 2014.

[64] K. Mason, E. Kim, W. Blake Martin, R.J. Gober, "The Moral Deliberations of 15 Clergy on Suicide and Assisted Death: A Qualitative Study," *Pastoral Psych.*, vol. 66, no. 3, pp. 335-351, Dec. 2017.

[65] D.E. Meier and O.W. Brawley, "Palliative care and the quality of life," *J. Clin. Oncol.*, vol. 29, no. 20, pp. 2750-2753, Jul. 2011.

[66] E. Kozlovb, M.C. Reid, B.D. Carpenter, "Improving patient knowledge of palliative care" *Patient Educ. Couns.*, vol. 100, no. 5, pp. 1010, May 2017. D.J. Wilkie and M.O. Ezenwa, "Pain and symptom management in palliative care and at end of life", *Nurs Outlook*, vol. 60, no. 6, pp. 357-364, Nov. 2012.

[67] S.M. Holmes, M.W. Rabow, S.L. Dibble, "Screening the soul: communication regarding spiritual concerns among primary care physicians and seriously ill patients approaching the end of life," *Am J Hosp Palliat Care*, vol. 23, no. 1, pp. 25-33, Jan.-Feb. 2006. L. Selman, R. Harding, M. Gysels et. al., "The measurement of spirituality in palliative care and the content of tools validated cross-culturally: a systematic review," *J. Pain Symptom Manag.*, vol. 41, no. 4, pp. 728-753, Apr. 2011. M. Anderson and C. Faull, "Spirituality in Palliative Care," *Handbook of Palliative Care*, C. Faull, S. de Caestecker, A. Nicholson, and F. Black, eds., Hoboken: Wiley-Blackwell, pp. 356, 2012.

[68] R. Pool, "«You're not going to dehydrate mom, are you?»: euthanasia, versterving, and good death in the Netherlands," *Soc Sci Med*, vol. 58, no. 5, pp. 955-966, 2004. B. McNamara, "Good enough death: autonomy and choice in Australian palliative care," *Soc Sci Med*, vol. 58, no. 5, pp. 929-938, Mar. 2004.

[69] R. Aviv, "The death treatment," *The New Yorker*, Jun. 2015, www.newyorker.com/magazine/2015/06/22/the-death-treatment. Accessed: November 29, 2018.
[70] L. Ganzini, E.R. Goy, S.K. Dobscha, "Why Oregon patients request assisted death: family members' views," *J Gen Intern Med*, vol. 23, no. 2, pp. 154-157, Feb. 2008.
[71] T. Pope, "Legal History of Medical Aid in Dying: Physician Assisted Death in U.S. Courts and Legislatures," *Faculty Scholarship*, vol. 432, pp. 267-301, 2018, https://open.mitchellhamline.edu/facsch/432 . Accessed: January 30, 2019.
[72] *The Oregon Death with Dignity Act, Oregon Revised Statutes*, http://public.health.oregon.gov/ProviderPartnerResources/EvaluationResearch/DeathwithDignityAct/Documents/statute.pdf. Accessed: January 30, 2017.
[73] G.A. Van Norman, "Physician aid-in-dying: Cautionary words," *Curr. Opin. Anesthesiol.*, vol. 27, no. 2, pp. 177-182, Apr. 2014.
[74] D.L. Rosenstein, "Depression and end-of-life care for patients with cancer,"*Dialogues in Clin Neurosci.*, vol. 13, no. 1, pp. 101-108, Mar. 2011.
[75] E. Emanuel, D. Fairclough, "Attitudes and desires related to euthanasia and physician-assisted suicide among terminally ill patients and their caregivers,"*JAMA*, vol. 284, no. 19, pp. 2460-2468, Nov. 2000.
[76] Public Health Division, *Oregon Death with Dignity Act 2017 Data Summary*, 2018, https://www.oregon.gov/oha/ph/providerpartnerresources/evaluationresearch/deathwithdignityact/documents/year20.pdf. Accessed: January 04, 2019.
[77] L. Ganzini, "Legalized Physician Assisted Death in Oregon-Eighteen Years' Experience," *Assistierter Suizid: Der Stand der Wissenschaft*, G.D. Borasio, R.J. Jox, J. Taupitz, and U. Wiesing, eds., Berlin: Springer Verlag, p. 7-20, 2017.
[78] Washington State Legislature, *Death With Dignity Act Requirements*, http://app.leg.wa.gov/wac/default.aspx?cite=246-978&full=true. Accessed: February 04, 2017.
[79] Washington State Department of Health, *Washington State Department of Health 2017 Death with Dignity Act Report*, 2018, https://www.doh.wa.gov/Portals/1/Documents/Pubs/422-109-DeathWithDignityAct2017.pdf. Accessed: January 04, 2019.
[80] Supreme Court of the State of Montana, *Robert Baxter v State*, http://law.justia.com/cases/montana/supreme-court/2009/50c59956-3100-468d-b397-4ab38f6eda4d.html. Accessed: February 04, 2017.

[81] The Vermont Statutes Online, *Patient Choice at End of Life*, http://legislature.vermont.gov/statutes/fullchapter/18/113. Accessed: February 04, 2017.
[82] Vermont Department of Health, *Report Concerning Patient Choice at the End of Life*, 2018, https://legislature.vermont.gov/assets/Legislative-Reports/2018-Patient-Choice-Legislative-Report-12-14-17.pdf. Accessed: January 04, 2019.
[83] Death with Dignity National Center, *Full Text of AB-15, California End of Life Option Act*, http://californiadeathwithdignity.org/resources/california-eol-act/. Accessed: February 04, 2017.
[84] California Departament of Public Health, *California End of Life Option Act 2016 Data Report*, 2017, https://www.cdph.ca.gov/Programs/CHSI/CDPH%20Document%20Library/CDPH%20End%20of%20Life%20Option%20Act%20Report.pdf. Accessed: January 04, 2019.
[85] California Departament of Public Health, *California End of Life Option Act 2017 Data Report*, 2018, https://www.cdph.ca.gov/Programs/CHSI/CDPH%20Document%20Library/2017EOLADataReport.pdf. Accessed: January 04, 2019.
[86] Colorado Department of Public Health and Environment, *Medical Aid in Dying*, https://www.colorado.gov/pacific/cdphe/medical-aid-dying#Reporting. Accessed: July 09, 2017.
[87] Colorado Department of Public Health and Environment, *Colorado End-of-Life Options Act, Year One 2017 Data Summary*, 2018, https://drive.google.com/file/d/1-kBXgAFzHl6kcfsvtLHfOQ94Unk9mDa-/view. Accessed: January 04, 2019.
[88] Colorado Department of Public Health and Environment, *Colorado End-of-Life Options Act, Year Two 2018 Data Summary, with Updates to 2017 Data*, 2019, https://drive.google.com/file/d/1FmoyCcL2gHopDO9rCJ2lGFEMUye8FQei/view. Accessed: February 08, 2019.
[89] R. Collier, "Assisted death gaining acceptance in US," *Can. Med. Assoc. J.*, vol. 189, no. 3, Jan. 2017, doi: 10.1503/cmaj.109-5366.
[90] State of Hawaii, *A Bill For An Act*, 2018, https://www.capitol.hawaii.gov/session2018/bills/HB2739_HD1_.HTM. Accessed: December 18, 2018.
[91] The Parliament of Québec, *Bill 52. An Act Respecting End-of-Life Care*, Québec: Québec Official Publisher, 2013, http://www.assnat.qc.ca/en/travaux-parlementaires/projets-loi/projet-loi-52-40-1.html. Accessed: February 04, 2017.

[92] CNS, "Assisted Suicide Count In Canada Challenged," *America. The National Catholic Review*, https://www.americamagazine.org/politics-society/2016/11/17/assisted-suicide-count-canada-challenged. Accessed: February 04, 2017.

[93] I. Peritz, "Quebec's assisted-death requests to top 300 by 20," *The Globe and Mail*, Oct. 2016, http://www.theglobeandmail.com/news/national/quebecs-assisted-death-requests-to-top-300---three-times-governments-forecast/article32561209/. Accessed: February 04, 2017.

[94] Parliament of Canada, *Bill C-14*, 2016, http://laws-lois.justice.gc.ca/PDF/2016_3.pdf. Accessed: April 10, 2018.

[95] Health Canada, *Second Interim Report on Medical Assistance in Dying in Canada*, 2017, https://www.canada.ca/content/dam/hc-sc/documents/services/publications/health-system-services/medical-assistance-dying-interim-report-sep-2017/medical-assistance-dying-interim-report-sep-2017-eng.pdf. Accessed: April 10, 2018.

[96] Health Canada, *Third Interim Report on Medical Assistance In Dying in Canada*, 2018, https://www.canada.ca/content/dam/hc-sc/documents/services/publications/health-system-services/medical-assistance-dying-interim-report-june-2018/medical-assistance-dying-interim-report-june-2018-eng.pdf. Accessed: January 04, 2019.

[97] D. Davies, *Medical Assistance in Dying: A Paediatric Perspective*, https://www.cps.ca/en/documents/position/medical-assistance-in-dying. Accessed: April 10, 2018.

[98] Ministerio de Salud, *Protocolo para la Aplicación del Procedimiento de Eutanasia en Colombia 2015*, [Protocol for the Application of the Euthanasia Procedure in Colombia 2015], https://www.minsalud.gov.co/sites/rid/Lists/BibliotecaDigital/RIDE/DE/CA/Protocolo-aplicacion-procedimiento-eutanasia-colombia.pdf. Accessed: January 31, 2017.

[99] Ministerio de Salud, *Protocolo para la Aplicación del Procedimiento de Eutanasia en Colombia 2015*, [Protocol for the Application of the Euthanasia Procedure in Colombia 2015], https://www.minsalud.gov.co/sites/rid/Lists/BibliotecaDigital/RIDE/DE/CA/Protocolo-aplicacion-procedimiento-eutanasia-colombia.pdf. Accessed: January 31, 2017. Senado de la república de Colombia, *Proyecto de ley 030/2015*, [Bill 030/2015], http://190.26.211.102/proyectos/index.php/textos-radicados-senado/pl-2015-2016/462-proyecto-de-ley-030-de-2015. Accessed: January 31, 2017. J.M Mendoza-Villa and L.A. Herrera-Morales, "Reflections on euthanasia in Colombia", *Revista Colombiana de Anestesiología*, vol. 44, no. 4, pp. 328, Oct.-Dec. 2016.

[100] L.L. Benavides, "The right to die with dignity in Colombia," *FRCIJ*, vol. 6, no. 6, pp. 428, Nov. 2018. M.A. Triviño, "Colombia has regulated euthanasia for children and adolescents," *Latin American Post*, Mar. 2018, https://latinamericanpost.com/20090-colombia-has-regulated-euthanasia-for-children-and-adolescents. Accessed: December 19, 2018.

[101] T.A. Boer, "Euthanasia, Ethics and Theology: A Dutch Perspective," *RES*, vol. 6, no. 2, pp. 197-213, Aug. 2014.

[102] Upper House, *The Dutch Termination of Life on Request and Assisted Suicide (Review Procedures) Act*, http://www.ethical-perspectives.be/viewpic.php?LAN=E&TABLE=EP&ID=58. Accessed: February 13, 2018.

[103] See M.C. Snijdewind, D.L. Willems, L. Deliens et al., "A Study of the First Year of the End-of-Life Clinic for Physician-Assisted Dying in the Netherlands," *JAMA Intern. Med.* vol. 175, no. 10, pp. 1633-1640, Oct. 2015.

[104] S. Murphy, *Euthanasia reported in Netherlands: statistics compiled from the Regional Euthanasia Review Committees'Annual Reports. Protection of Conscience Project*, 2017, http://www.consciencelaws.org/background/procedures/assist019.aspx. Accessed: December 19, 2018. European Institut of Bioethics, *Report 2016 euthanasia in Netherlands*, 2017, https://www.ieb-eib.org/en/document/report-2016-euthanasia-in-netherlands-488.html. Accessed: December 19, 2018.

[105] Dutsch News, *Number of official cases of euthanasia rise 10% in the Netherlands*, Apr. 2017, https://www.dutchnews.nl/news/2017/04/number-of-official-cases-of-euthanasia-rise-10-in-the-netherlands/. Accessed: December 19, 2018.

[106] D. Boffey, "Dutch prosecutors to investigate euthanasia cases after sharp rise," *The Guardian*, Mar. 2018, https://www.theguardian.com/world/2018/mar/12/dutch-prosecutors-investigate-euthanasia-cases-sharp-rise-docter-assisted-deaths-netherlands. Accessed: December 19, 2018.

[107] D. Boffey, "Doctor to face Dutch prosecution for breach of euthanasia law," *The Guardian*, Nov. 2018, https://www.theguardian.com/world/2018/nov/09/doctor-to-face-dutch-prosecution-for-breach-of-euthanasia-law. Accessed: January 19, 2019.

[108] J. Griffiths, "Belgium", *Euthanasia and Law in Europe*, J. Griffiths, H. Weyers, and M. Adams, eds., Oxford: Hart Publishing, pp. 259-329, 2008.

[109] Le Ministre de la Justice, *Loi relative à l'euthanasie*, [Law on Euthanasia], http://www.ejustice.just.fgov.be/cgi_loi/change_lg.pl?language=fr&la=F&cn=2002052837&table_name=loi. Accessed: January 10, 2017.

[110] E. Montero, "The Belgian Experience of Euthanasia Since Its Legal Implementation in 2002," *Euthanasia and Assisted Suicide. Lessons from Belgium*, D.A. Jones, C. Gastmans, and C. Mackellar, eds., Cambridge: Cambridge University Press, pp. 26-48, 2017.

[111] Commission fédérale de Contrôle et d'Évaluation de l'Euthanasie (CFCEE), *Huitième rapport aux Chambres législatives, années 2016-2017*, [The Eighth Report to the Legislative Chambers, Years 2016-2017], 2018, https://organesdeconcertation.sante.belgique.be/sites/default/files/documents/8_rapport-euthanasie_2016-2017-fr.pdf. Accessed: December 19, 2018.

[112] K. Chambaere, J. Cohen, L. Robijn et al., "End-of-life decisions in individuals dying with dementia in Belgium," *JAGS*, vol. 63, no. 2, pp. 290-296, Feb. 2015. K. Chambaere, R. Vander Stichele, F. Mortier, et al., "Recent trends in euthanasia and other end-of-life practices in Belgium," *NEJM*, vol. 372, no. 12, pp. 1179-1181, Mar. 2015. S. Dierickx, L. Deliens, J. Cohen, and K. Chambaere, "Comparison of the expression and granting of requests for euthanasia in Belgium in 2007 vs 2013," *JAMA Intern. Med.*, vol. 175, no. 10, pp. 1703-1706, Oct. 2015.

[113] R. Deschepper, W. Distelmans, J. Bilsen, "Request for euthanasia/physician-assisted suicide on the basis of mental suffering: vulnerable patients or vulnerable physicians?," *JAMA Psychiatry*, vol. 71, no. 6, pp. 617-618, Jun. 2014.

[114] CFCEE, *Premier rapport aux Chambres législatives (22 septembre 2002 - 31 décembre 2003)*, [The First Report to the Legislative Chambers (22 September 2002 - 31 December 2003)], pp. 14, 2004. CFCEE, *Deuxième rapport aux Chambres législatives (années 2004 et 2005)*, [The Second Report to the Legislative Chambers (Years 2004 and 2005)], pp. 22, 2006. CFCEE, *Troisième rapport aux Chambres législatives (années 2006 et 2007)*, [The Third Report to the Legislative Chambers (Years 2006 and 2007)], pp. 22, 2008. CFCEE, *Quatrième rapport aux Chambres législatives (années 2008 et 2009)*, [The Fourth Report to the Legislative Chambers (Years 2008 and 2009)], pp. 22, 2010. CFCEE, *Cinquième rapport aux Chambres législatives (années 2010 et 2011)*, [The Fifth Report to the Legislative Chambers (Years 2010 and 2011)], p. 14, 2012. CFCEE, *Sixième rapport aux Chambres législatives (2012–2013)*, [The Sixth Report to the Legislative Chambers (2012-2013)], pp. 14, 2014, https://organesdeconcertation.sante.belgique.be/fr/organe-d%27avis-et-de-concertation/commission-federale-de-controle-et-devaluation-de-leuthanasie. Accessed: December 21, 2018.

[115] T. Smets, J. Bilsen, J. Cohen et al., "Reporting of euthanasia in medical practice in Flanders, Belgium: cross sectional analysis of reported and unreported cases," *BMJ*, vol. 341, pp. 1-8, 2010. E. Montero, "The Belgian Experience of Euthanasia Since Its Legal Implementation in 2002,"

Euthanasia and Assisted Suicide. Lessons from Belgium, D.A. Jones, C. Gastmans, and C. Mackellar, eds., Cambridge: Cambridge University Press, p. 28, 2017.

[116] T. Smets, J. Bilsen, J. Cohen et al., "Reporting of euthanasia in medical practice in Flanders, Belgium: cross sectional analysis of reported and unreported cases," *BMJ*, vol. 341, pp. 1-8, 2010. R. Cohen-Almagor, "First do no harm: pressing concerns regarding euthanasia in Belgium," *Int. J. Law Psychiatry*, vol. 36, no. 5-6 , pp. 515-521, Sep.-Dec. 2013.

[117] E. Saget, "L'euthanasie, ma mère et moi," [Euthanasia, My Mother and I], *L'Express*, Apr. 2008, www.lexpress.fr/actualite/societe/l-eutha nasie-ma-mere-et-moi_472327.html. Accessed: September 03, 2018.

[118] H. Hendin, "Assisted suicide, euthanasia, and suicide prevention: the implications of the Dutch experience," *Suicide Life-Threat. Behav.*, vol. 25, no. 1, pp. 193-204, Spring 1995.

[119] E. Emanuel, D. Fairclough, "Attitudes and desires related to euthanasia and physician-assisted suicide among terminally ill patients and their caregivers,"*JAMA*, vol. 284, no. 19, pp. 2460-2468, Nov. 2000. A. Bazan, G. Van de Vijver, W. Lemmens et al., "Euthanasie pour souffrance psychique: un cadre légal discutable et des dommages sociétaux," [Euthanasia for Mental Suffering: A Questionable Legal Framework and Societal Damages], *Le Soir*, Sep. 2015, https://www.euthanasiestop.be/telecharger-article.php?id=411&lng=fr. Accessed: September 03, 2018.

[120] L. Thienpont, M. Verhofstadt, T. Van Loon et al., "Euthanasia requests, procedures and outcomes for 100 Belgian patients suffering from psychiatric disorders: a retrospective, descriptive study," *BMJ*, 2015, doi:10.1136/bmjopen-2014-007454. Le Point, *Un Belge euthanasié après une opération de changement de sexe ratée*, [Belgian Man Dies by Euthanasia after Failed Sex Change Operation], Oct. 2013, http://www.lepoint.fr/monde/un-belge-euthanasie-apres-une-operation-de-changement-de-sexe-ratee-02-10-2013-1738312_24.php. Accessed: September 04, 2018. É. Sensique, "Une interview à coeur ouvert de Wim Distelmans," [An interview with Wim Distelmans], *ADMD*, May 2018, http://www.admd.be/une-interview-a-coeur-ouvert-de-wim-distelmans/.Accessed: 13.02.2019.

[121] R. Cohen-Almagor, "Euthanizing People Who Are «Tired of Life»," *Euthanasia and Assisted Suicide. Lessons from Belgium*, D.A. Jones, C. Gastmans, and C. Mackellar, eds., Cambridge: Cambridge University Press, p. 188-201, 2017. S. Dierickx, L. Deliens, J. Cohen, and K. Chambaere, "Comparison of the expression and granting of requests for euthanasia in Belgium in 2007 vs 2013," *JAMA Intern. Med.*, vol. 175, no. 10, pp. 1703-1706, Oct. 2015.

[122] J.L. Bernheim, K. Raus, "Euthanasia embedded in palliative care. Responses to essentialistic criticisms of the Belgian model of integral end-of-life care," *J. Med. Ethics*, vol. 43, no. 8, pp. 489-494, Aug. 2017.

[123] World Health Organization, *WHO Definition of Palliative Care*, http://www.who.int/cancer/palliative/definition/en/. Accessed: February 01, 2017.

[124] L. Radbruch, C. Leget, P. Bahr et al., "Euthanasia and physician-assisted suicide: A white paper from the European Association for Palliative Care," *Palliat Med*, vol. 30, no. 2, pp. 104-116, Feb. 2016. L. Materstvedt, D. Clark, J. Ellershaw et al., "Euthanasia and physician-assisted suicide: a view from an EAPC Ethics Task Force," *Palliat Med*, vol. 17, no. 2, p. 97-101, Mar. 2003. L. Materstvedt, "Misconstrual of EAPC's position paper on euthanasia," *J Med Ethic*, vol. 41, no. 8, pp. 655-656, Aug. 2015.

[125] J.L. Bernheim, W. Distelmans, A. Mullie, M.A. Ashby, "Questions and answers on the Belgian model of integral end-of-life care: Experiment? Prototype?," *J. Bioethical Inq.*, vol. 11, no. 4, pp. 507-529, Aug. 2014.

[126] B. Beuselinck, "2002–2016: Fourteen Years of Euthanasia in Belgium," *Euthanasia and Assisted Suicide. Lessons from Belgium*, D.A. Jones, C. Gastmans, and C. Mackellar, eds., Cambridge: Cambridge University Press, pp. 105-106, 2017.

[127] C. Dopchie, "L'euthanasie tue-t-elle les soins palliatifs?," [Is Euthanasia Killing Palliative Care?], *Les Cahiers francophones de soins palliatifs*, vol. 13, no. 2, pp. 28-41, 2013. L'Institut Européen de Bioéthique, *Le modèle belge des «soins palliatifs intégraux» dénature-t-il la pratique des soins palliatifs?*, [Does the Belgian Model of «Integrated Palliative Care» Distort Palliative Care Practices?], https://www.ieb-eib.org/en/pdf/20150904-dossiers-euthanasie-et-soins-palliatifs.pdf. Accessed: December 20, 2018.

[128] T. Stammers, "«A Last Act of Grace»? Organ Donation and Euthanasia in Belgium," *Euthanasia and Assisted Suicide. Lessons from Belgium*, D.A. Jones, C. Gastmans, and C. Mackellar, eds., Cambridge: Cambridge University Press, pp. 114-130, 2017.

[129] J. Bollen, R. Ten Hoopen, D. Ysebaert et al., "Current controversy: Legal and ethical aspects of organ donation after euthanasia in Belgium and the Netherlands," *J Med Ethics*, vol. 42, pp. 486-489, 2016.

[130] U. Bacchi, "Frank Van Den Bleeken: Belgian serial rapist and murderer wins right to die," Sep. 2014, *International Business Times*. www.ibtimes.co.uk/frank-van-den-bleeken-belgian-serial-rapist-murderer-winsright-die-1465667. Accessed: December 20, 2018.

[131] J. Huang, "Ethical and legislative perspectives on liver transplantation in the People's Republic of China," *Liver transplantation*, vol. 13, no. 2, pp. 193-196, Feb. 2007. J.D. Briggs, "The use of organs from executed prisoners in China," *Nephrol. Dial. Transplant.*, vol. 11, no. 2, pp. 238-240,

1996. M. Wang and X. Wang, "Organ donation by capital prisoners in China," *J. Med. Philos.*, vol. 35, no. 2, pp. 197-212, Apr. 2010.

[132] S. Van Gool, J. De Lepeleire, "Euthanasia in Children," *Euthanasia and Assisted Suicide. Lessons from Belgium*, D.A. Jones, C. Gastmans, and C. Mackellar, eds., Cambridge: Cambridge University Press, pp. 173-187, 2017. A. Hovine, "Il faut reporter le vote sur l'euthanasie des enfants," [Postponing the Vote on Child Euthanasia], *La Libre Belgique*, Feb. 2014, http://www.lalibre.be/actu/belgique/il-faut-reporter-le-vote-sur-l-euthanasie-des-enfants-52fa4b633570516ba0b8edd3. Accessed: Sep. 05, 2018.

[133] J.J. Arnett, "Emerging adulthood: a theory of development from the late teens through the twenties," *Am. Psychol.*, vol. 55, no. 5, pp. 469-480, May 2000.

[134] A. Iltis, "Parents, adolescence, and consent for research participation," *J. Med. Philos.*, vol. 38, no. 3, pp. 332-346, Jun. 2013.

[135] R. Barina, J. Bishop, "Maturing the minor, marginalizing the family: on the social constitution of the mature minor, sexual politics, and the family," *J. Med. Philos.,* vol. 38, no. 3, pp. 300-314, Jun. 2013.

[136] CFCEE, *Premier rapport aux Chambres législatives (22 septembre 2002 - 31 décembre 2003)*, [The First Report to the Legislative Chambers (September 22, 2002 - December 31, 2003)], pp. 17, 2004. CFCEE, *Deuxième rapport aux Chambres législatives (années 2004 et 2005)*, [The Second Report to the Legislative Chambers (Years 2004 and 2005)], pp. 24, 2006. CFCEE, *Troisième rapport aux Chambres législatives (années 2006 et 2007)*, [The Third Report to the Legislative Chambers (Years 2006 and 2007)], pp. 24, 2008. CFCEE, *Quatrième rapport aux Chambres législatives (années 2008 et 2009)*, [The Fourth Report to the Legislative Chambers (Years 2008 and 2009)], pp. 24, 2010. CFCEE, *Cinquième rapport aux Chambres législatives (années 2010 et 2011)*, [The Fifth Report to the Legislative Chambers (Years 2010 and 2011)], pp. 17, 2012, https://organesdeconcertation.sante.belgique.be/fr/organe-d%27avis-et-de-concertation/commission-federale-de-controle-et-devaluation-de-leuthanasie. Accessed: December 21, 2018.

[137] D.A. Jones, "Euthanasia and Assisted Suicide in Belgium," *Euthanasia and Assisted Suicide. Lessons from Belgium*, D.A. Jones, C. Gastmans, and C. Mackellar, eds., Cambridge: Cambridge University Press, pp. 235-257, 2017.

[138] E.E. Papavasiliou, K. Chambaere, L. Deliens et al., "Physician-reported practices on continuous deep sedation until death: A descriptive and comparative study," *Palliat. Med.*, vol. 28, no. 6, pp. 491-500, 2014.

[139] J.L. Vincent, M. Schetz, J.J. De Waele et al., "«Piece» of mind: end of life in the intensive care unit – statement of the Belgian Society of Intensive Care Medicine," *J. Crit. Care*, vol. 29, no. 1, pp. 174-175, Feb. 2014.

[140] J. Masson, "L'exception euthanasique en droit belge", [The Euthanasia Exception in Belgian Law], *Louvain Médical*, vol. 124, no. 7, pp. 238-245, 2005.

[141] D. Lossignol, "Soins palliatifs et euthanasie: la fin du conflit," [Palliative Care and Euthanasia: The End of the Conflict], *La revue des soins palliatifs en Wallonie*, vol. 14, pp. 24, 2012.

[142] Association pour le Droit de Mourir dans la Dignité, www.admd.net/international/la-belgique.html. Accessed: September 04, 2018.

[143] Belgium.be, www.belgium.be/fr/sante/. Accessed: September 04, 2018.

[144] Solidaris, http://www.solidaris.be/MonsWP/Pages/accueil.aspx?choixregion=true. Accessed: September 04, 2018.

[145] C. Mackellar, "Some Possible Consequences Arising from the Normalisation of Euthanasia in Belgium," *Euthanasia and Assisted Suicide. Lessons from Belgium*, D.A. Jones, C. Gastmans, and C. Mackellar, eds., Cambridge: Cambridge University Press, pp. 219-234, 2017.

[146] K. Chambaere, R. Vander Stichele, F. Mortier et al., "Recent trends in euthanasia and other end-of-life practices in Belgium," *N. Engl. J. Med.*, vol. 372, no. 12, pp. 1179-1180, Mar. 2015.

[147] Le Gouvernement du Grand-Duché de Luxembourg, *Loi du 16 mars 2009 sur l'euthanasie et l'assistance au suicide*, [Act of 16 March 2009 on Euthanasia and Assisted Suicide], http://legilux.public.lu/eli/etat/leg/loi/2009/03/16/n2/jo. Accessed: January 27, 2017.

[148] Le Gouvernement du Grand-Duché de Luxembourg, *S'informer sur une demande d'euthanasie ou d'assistance au suicide*, [Learning about a Request for Euthanasia or Assisted Suicide], http://www.guichet.public.lu/citoyens/fr/famille/euthanasie-soins-palliatifs/fin-de-vie/euthanasie-assistance-suicide/index.html. Accessed: January 27, 2017.

[149] R. Watson, "Luxembourg is to allow euthanasia from 1 April," *BMJ*, 2009, doi: 10.1136/bmj.b1248.

[150] Swiss Criminal Code, https://www.admin.ch/opc/en/classifiedcompilation/19370083/201701010000/311.0.pdf. Accessed: January 24, 2017. Schweizerische Eidgenossenschaft, *Schweizerisches Strafgesetzbuch*, http://www.admin.ch/ch/d/sr/311_0/a115.html. Accessed: January 24, 2017.

[151] G. Bosshard, S. Fischer, W. Bär, "Open regulation and practice in assisted dying," *Swiss Med Wkly*, vol. 132, no. 37-38, pp. 527-534, Oct. 2002. N. Steck, C. Junker, and M. Zwahlen, "Increase in assisted suicide in Switzerland: did the socioeconomic predictors change? Results from the Swiss National Cohort," *BMJ Open*, vol. 8, no. 4, pp. 1, 2018.
[152] N. Steck, C. Junker, M. Zwahlen, "Increase in assisted suicide in Switzerland: did the socioeconomic predictors change? Results from the Swiss National Cohort," *BMJ Open,* vol. 8, no. 4, pp. 1, 2018. EXIT, *https://www.exit.ch/en/en/faq/ https://exit.ch/startseite/*. Accessed: January 24, 2017.
[153] N. Steck, C. Junker, M. Zwahlen, "Increase in assisted suicide in Switzerland: did the socioeconomic predictors change? Results from the Swiss National Cohort," *BMJ Open,* vol. 8, no. 4, pp. 1-11, 2018.
[154] G. Bosshard, "Switzerland," *Euthanasia and Law in Europe*, J. Griffiths, H. Weyers, and M. Adams, ed., Oxford: Hart Publishing, pp. 463-481, 2008. N. Steck, C. Junker, M. Zwahlen, "Increase in assisted suicide in Switzerland: did the socioeconomic predictors change? Results from the Swiss National Cohort," *BMJ Open,* vol. 8, no. 4, pp. 1, 2018.
[155] S. Fischer, C.A. Huber, L. Imhof et al., "Suicide assisted by two Swiss right-to-die organisations," *J Med Ethics*, vol. 34, no. 11, pp. 810-814, Nov. 2008.
[156] Exit Deutsche Schweiz was founded 03 April 1982; membership count: around 100,000. Exit, *EXIT at a Glance*, https://www.exit.ch/en/en/exit-at-a-glance/. Accessed: January 26, 2018.
[157] Exit ADMD was founded in 1982, Geneva; membership count: around 29,000. Exit ADMD, *Comment est né le mouvement EXIT?*, http://www.exit-geneve.ch/Buts.htm. Accessed: February 15, 2019.
[158] ExInternational was founded in 1996, Bern; membership count: around 800. ExInternational, *Offener Brief an nachdenkliche Menschen*, http://www.exinternational.ch/frame.html. Accessed: July 07, 2018.
[159] Dignitas was founded in 1998, Forch; membership count: around 7,100. Dignitas, *Who is DIGNITAS* http://www.dignitas.ch/index.php?option=com_content&view=article&id=4&Itemid=44&lang=en. Accessed: January 26, 2018.
[160] SPIRIT was founded in 2011, Basel. S. Gauthier, J. Mausbach, T. Reisch et al., "Suicide tourism: a pilot study on the Swiss phenomenon," *J Med Ethics*, vol. 41, no. 8, pp. 611-617, Aug. 2015.
[161] K. Konnolly, "Dignitas attacked for new assisted-suicide method," *The Guardian*, Mar. 2008, https://www.theguardian.com/world/2008/mar/21/germany. Accessed: July 11, 2017. Daily Mail Online, *Euthanasia group Dignitas films gas and plastic bag deaths*, Mar. 2008, http://www.dailymail.co.uk/news/article-

539929/Euthanasia-group-Dignitas-films-gas-plastic-bag-deaths.html. Accessed: July 11, 2017.
[162] M. Leidig, H. Samuel, "Swiss suicide group operates in car parks," *The Telegraph*, Nov. 2007, http://www.telegraph.co.uk/news/worldnews/1568874/Swiss-suicide-group-operates-in-car-parks.html. Accessed: July 11, 2017.
[163] B. Pancevski, "Swiss suicide clinics «helping depressives die»," *The Telegraph*, Jun. 2007, http://www.telegraph.co.uk/news/worldnews/1553490/Swiss-suicide-clinics-helping-depressives-die.html. Accessed: July 11, 2017.
[164] R. Adorno, "Nonphysician-Assisted Suicide in Switzerland," *CQH*, vol. 22, no. 1, pp. 246-253, Jul. 2013.
[165] R. Flint, "Healthy people are travelling abroad for assisted suicide because they grow 'weary of life'," *Mirror*, Feb. 2014, https://www.mirror.co.uk/news/uk-news/assisted-suicide-healthy-people-travelling-3164625. Accessed: July 11, 2017.
[166] S.J. Ziegler, G. Bosshard, "Role of non-governmental organizations in physician assisted suicide," *BMJ*, vol. 334, no. 7588, pp. 295-298, Mar. 2007.
[167] EXIT, www.exit.ch. Accessed: July 11, 2017.
[168] G. Bosshard, E. Ulrich, W. Bär, "748 cases of suicide assisted by a Swiss right-to-die-organisation," *Swiss Med Wkly*, vol. 133, pp. 310-317, May 2003. S. Fischer, C.A. Huber, L. Imhof et al., "Suicide assisted by two Swiss right-to-die organisations," *J Med Ethics*, vol. 34, no. 11, pp. 810-814, Nov. 2008. S. Gauthier, J. Mausbach, T. Reisch et al., "Suicide tourism: a pilot study on the Swiss phenomenon," *J Med Ethics*, vol. 41, no. 8, pp. 611-617, Aug. 2015.
[169] Kanton Zürich, *Kantonale Volksinitiative zur Einreichung einer Standesinitiative "Stopp der Suizidhilfe!"*, [The Initiative of Cantonal Population for the Submission of the Project "Stop Suicide Aid!"], *Kantonale Volksinitiative "Nein zum Sterbetourismus im Kanton Zürich!"*, [The Initiative of the Cantonal Population "No to the Death Tourism in the Canton of Zurich!"], https://wahlenabstimmungen.zh.ch/internet/justiz_inneres/wahlenabstimmungen/de/abstimmungen/abstimmungsarchiv.html. Canton Vaud, *Assistance au suicide,* https://www.elections.vd.ch/votelec/results.html?scrutin=VDVO20120617. Accessed: December, 21, 2018.
[170] Deutscher Bundestag, *Geschäftsmäßige Hilfe zum Suizid wird bestraft*, [Commercilized Help to Suicide Is Punished], 2015, https://www.bundestag.de/dokumente/textarchiv/2015/kw45_de_sterbebegleitung/392450. Accessed: October 18, 2016. Strafgesetzbuch, *Geschäftsmäßige Förderung der Selbsttötung*, [Commercializing Assisted Suicide],

https://dejure.org/gesetze/StGB/217.html. Accessed: October 18, 2017. Deutscher Bundestag, *Drucksache 18/537. Entwurf eines Gesetzes zur Strafbarkeit der geschäftsmäßigen Förderung der Selbsttötung*, [Printed Issue 18/537. Bill on the Criminal Liability of Commercialized Assited Suicide], http://dip21.bundestag.de/dip21/btd/18/053/1805373.pdf. Accessed: April 12, 2018.

[171] Deutscher Bundestag, *Geschäftsmäßige Hilfe zum Suizid wird bestraft*, [Commercilized Help to Suicide Is Punished], 2015, https://www.bundestag.de/dokumente/textarchiv/2015/kw45_de_sterbe begleitung/392450. Accessed: October 18, 2016. T. Weigend and E. Hoven. "§ 217 StGB – Bemerkungen zur Auslegung eines zweifelhaften Tatbestandes," [§ 217 StGB – Remarks on the Interpretation of a Doubtful Fact], *Zeitschrift für Internationale Strafrechtsdogmatik*, vol. 10, pp. 681-691, 2016.

[172] Deutscher Bundestag, *Geschäftsmäßige Hilfe zum Suizid wird bestraft*, [Commercilized Help to Suicide Is Punished], 2015, https://www.bundestag.de/dokumente/textarchiv/2015/kw45_de_sterbe begleitung/392450. Accessed: October 18, 2016. T. Weigend and E. Hoven. "§ 217 StGB – Bemerkungen zur Auslegung eines zweifelhaften Tatbestandes," [§ 217 StGB – Remarks on the Interpretation of a Doubtful Fact], *Zeitschrift für Internationale Strafrechtsdogmatik*, vol. 10, pp. 681-691, 2016.

[173] T. Weigend, E. Hoven. "§ 217 StGB – Bemerkungen zur Auslegung eines zweifelhaften Tatbestandes," [§ 217 StGB – Remarks on the Interpretation of a Doubtful Fact], *Zeitschrift für Internationale Strafrechtsdogmatik*, vol. 10, pp. 681-691, 2016.

[174] J. Römer, "Neues Sterbehilfe-Gesetz. Die Grauzone bleibt," [New Euthanasia Law. The Grey Area Remains], *Der Spiegel*, http://www.spiegel.de/gesundheit/diagnose/sterbehilfe-geschaeftsmaessige-sterbebeihilfe-in-der-grauzone-a-1061369.html. Accessed: February 06, 2017.

[175] *Deutsche Welle*, "German parliament opens debate on assisted suicide," http://www.dw.com/en/german-parliament-opens-debate-on-assisted-suicide/a-18558626. Accessed: February 06, 2017.

[176] H. Feichtinger, "Death Rights," *First Things*, vol. 267, pp. 21-23, Nov. 2016.

[177] L. Bartels, M. Otlowski, "A Right to die? Euthanasia and the law in Australia," *JLM*, vol. 17, no. 4, pp. 532-555, Feb. 2010.

[178] Australian Government, *Euthanasia Laws Act 1997*, https://www.legislation.gov.au/Details/C2004A05118. Accessed: February 06, 2017.

[179] Parliament of Victoria, *Inquiry into end of life choices*, 2016, https://www.parliament.vic.gov.au/file_uploads/LSIC_pF3XBb2L.pdf. Accessed: February 11, 2019.

[180] R. Spooner, R.Willingham, "Dying with dignity: Parliamentary report recommends euthanasia laws," *The Age*, Jun. 2016, http://www.theage.com.au/victoria/dying-with-dignity-parliamentary-report-recommends-euthanasia-laws-20160608-gpeycy. Accessed: January 16, 2017.

[181] G. Kenyon, "Assisted dying legislation in Victoria, Australia," *Lancet Oncol.*, vol. 18, 2017, doi: 10.1016/S1470-2045(16)30657-X.

[182] J. Edwards, "Euthanasia: Victoria becomes the first Australian state to legalise voluntary assisted dying," *ABC*, Nov. 2017, https://www.abc.net.au/news/2017-11-29/euthanasia-passes-parliament-in-victoria/9205472. Accessed: December 22, 2018.

PART III

Globalization and Environmental Ethics

A Critique of the Environmental Ethical Critique of Christian Anthropocentrism

Ferenc Hérány

Eötvös József College, Baja, Hungary

Abstract

The trends in environmental-ethical thinking that oppose anthropocentrism formulate criticism directed toward ideologies that have been defining the Western ethical mentality up to our days and that are, at the same time, paradigms that obstruct sustainability. Such criticism facilitates the development of an ethical system that is more likely to be able to generate a sustainable social-economic structure. With my research, I wish to contribute to this endeavour. However, there is another aspect that I consider to be of crucial importance, namely that numerous individuals in the Judeo-Christian, respectively in Antiquity-based Western civilizations are unable to identify with this type of criticism grounded in environment-ethical thinking. Consequently, the objective proposed for the 'criticism of the criticism' is to reveal new perspectives of interpretation which can reconcile ecologic sustainability and Western Christian ideologies that are, from certain points of view, rightfully criticized. What I propose in my essay is not to repudiate environment-ethical criticism, but to amend it in order to serve a common goal: ecologic sustainability.

Keywords

Anthropocentrism; christianism; cartesianism; panexperientialism; malthusianism; overpopulation; care.

I. The background of the theses and the main critiques in opposition the anthropocentrism

Environmental philosophy emerged in the 1970s to reveal the spiritual and ideological foundations of the arrogant human behaviour which was and is inducing the environmental crisis. The movement was relatively quick to arrive at the conclusion that traditional Western thought, particularly anthropocentrism, can be identified as the philosophical background that determined the behaviour leading to the death of Nature, that is, "the ideological culprit." Literally, the term anthropocentrism refers to human-centred thinking and is based on the principle that Man is the only being that has moral value or intrinsic value, all other entities having only instrumental value, that is, it defines other entities as merely things [1].

Philosophers agree that the human being of moral status has intrinsic value. While environmental ethics also accepts this thesis, it also argues that Man is not the only being in possession of such value, other natural entities (animals, species, ecosystems) possess intrinsic value as well, thus secular environmental ethics is quick to occupy a non-anthropocentric position [2].

Some critics of traditional anthropocentrism wish to eliminate anthropocentrism from human thinking altogether (e.g. Lynn White, Jr.); while others consider anthropocentrism necessary, but in a renewed version and not in the traditional sense [3-4]. The latter call their views "anthropocentrism" in a weak sense or relative anthropocentrism (e.g. Jan Deckers). Both non-anthropocentric authors and those wishing to renew anthropocentrism generally agree that the traditional Western anthropocentrism is anthropocentrism in a strong sense.

II. The three main founts of anthropocentrism according to critics

This system of ideas goes back to three roots: classical Greek philosophy (Socrates, Plato, Aristotle), the Judeo-Christian tradition, and Modernism which started with Descartes [5] (p. 215).

(i) The anthropocentrism of classical Greek philosophy is most concisely summarized by Aristotle in his *Politics* (1256 b): "If therefore nature makes nothing without purpose or in vain, it follows that nature has made all the animals for the sake of men." [6]

(ii) The emblematic Biblical quote on Judeo-Christian anthropocentrism appears in the 28th verse of the first Chapter in the book of Genesis (Gen. 1:28), in which God commands the first couple:

"Be fruitful and increase in number; fill the earth and subdue it. Rule over the fish in the sea and the birds in the sky and over every living creature that moves on the ground."

(iii) The ontological basis for Cartesian anthropocentrism is Descartes' assumption that there are two kinds of substances in the created world. One is *res cogitans:* consciousness unfolding through doubt, whose main attribute is the ability to think, to have free will, that is, the ability to have self-determination. This is the essence of the ego that the philosopher identifies with the immortal, eternal soul. The other substance is *res extensa*, the extended body the attributes which are nothing like the attributes of the conscious substance and that is just helplessly obeying the laws of mechanics. The *res cogitans*, possessing value in its own, exists only in Man, everything exterior to the human being is only *res extensa.*

In mainstream secular environment philosophy, these three intellectual sources share the attribution of exclusive importance to Man, thus leaving no moral space for the adequate consideration of non-human entities. In this view, Western anthropocentrism is a homogenous ideology that assigns intrinsic value exclusively to man and considers everything else to have only instrumental value, thus leading to the arrogance of today's Western consumer society towards the environment.

III. A proposal for a different interpretation

As the conclusion of my research, I prove that these anthropocentric systems show significant differences.

(i) While analysing the anthropocentrism of classical Greek philosophy, there is no way to avoid the term *kalokagathia.* The anthropocentrism of classical Greek philosophy is incompatible with the consumerism prevailing today and the related anthropocentrism.

(ii) While the theoreticians of environmental ethics who are critical of the anthropocentrism of the Judeo-Christian tradition refer to the Book of Genesis 1:28, they ignore the fact the original Hebrew word used in the Bible is וּרְדוּ , meaning not only "to rule over," but also "to cultivate with care." Man has been created in the image of God (in Latin: *imago Dei*), who must assimilate God wanting the beauty and harmony of the created world. The duty of being imago Dei, as formulated in the Old Testament, is expanded in the New Testament with the duty of following Jesus Christ (in Latin: *sequela Christi*), that is, to walk in Christ's footsteps, to follow Jesus Christ.

(iii) The three types of anthropocentrism are actually only two. Even though there are substantial differences between classical Greek and Judeo-Christian worldviews, from the perspective of environmental ethics I put them both in the same group and I separate Cartesian anthropocentrism from the previous two concepts. To justify the classification, I argue that Cartesian anthropocentrism can be criticized based on arguments proposed by environmental ethics, but the critique is not applicable in the case of the other two concepts. I think that the environment philosophers who criticize anthropocentrism essentially misinterpret Christian anthropocentrism along with the classical Greek interpretation.

Simultaneously, I admit the possible causes for this misinterpretation, and I conclude that the environmental-ethical anthropocentrism-critique does have practical significance. These works reveal the practical errors that come from a distorted interpretation of Christianity. (Although non anthropocentric environmental philosophers think that these errors originate from Christianity itself, and not its distorted interpretation.) They accuse anthropocentrism of being responsible for ecologic unsustainability, which may motivate Christians to cleanse their understanding regarding religious doctrines, this way enabling them to separate ecologically harmful misinterpretations from the message of universal love that is the essence of Christianity and that can serve as a possible and efficient ethical foundation for ecologic sustainability.

IV. Concretes

The general outline of the problem is followed by the summary of the analysis of two specific trends.

A. Panexperientialism

Panexperientialism starts out from the process-philosophy developed by Alfred North Whitehead and Charles Hartshorne [7-8]. They consider anthropocentrism to be derived from a dualist view, but if such a dualistic interpretation of the world can be transcended, anthropocentric thinking can be transcended as well. The pursuit is present in the panexperientialist criticism of Christianity.

Panexperientialism repudiates the inherent value surplus in Man that is so evident in Christian thought, the foundation for the argument being that experience is the principle guiding line in the process (or the history) of the world. Experience is produced by living entities at different levels, but these differences do not generate any value-based hierarchy regarding these entities. The inherent values of each entity are present in all the other entities, the intrinsic values of entities form a common network of values. This mutual dependence and mutual experience-transit disregards not only the value-based differences among entities, but also the difference between the immanent and the transcendent. Therefore, within the world's experience network, neither Man's value surplus in the traditional sense proposed by Western ethics, nor its privileged position within the created world are acceptable [8] (p. 22, p. 215).

In my examination, I point out that the ontology of Christianity is not dualistic, therefore the privileged role of Man cannot originate in dualism either. The ontological non-duality of Christianity is corroborated by the doctrine of existence analogy that dominates Christian ontology. Thomas Aquinas distinguishes four levels of existence in the *De veritate.* [9] The first is the level of inorganic matter, the second is the level of vegetative organisms, while the third is the level of the sensitive animals. The fourth level of the spiritual/intellectual level, the level of Man. Christianity would be dualistic if the fourth level were separated from the other three in the same way as the Cartesian *res cogitans* is separated from *res extensa.* However, these levels of existence are not so rigidly delimited, they mutually presuppose and pervade each other. The intrinsic value of entities is to be interpreted based on this principle of gradation as well. Thus, the ontological context of Christianity cannot integrate any kind of ethics that would allow Man to ignore his duties to treat non-human entities respectfully.

B. Malthusian thinkers

Malthusian thinkers (Ehrlich, Hardin, Lovelock) combine their anti-democratic views – intended to convincingly introduce the necessity to reduce global population – with catastrophic eco-pessimism. These thinkers consider that welfare states need to deny food aid and they also need to refuse to take climate refugees. Lovelock argues that due to the environmental crisis, the survival now takes priority over ethical justice [10-11].

In Lovelock's view, the solution to the demographic crisis can only come from the transformation of mankind into a social structure similar to the State of termites. He deems the arrival of a new Churchill in possession of total control and the appearance of the Earth First Battalion to maintain order as necessary [12]. Lovelock also deals with the embarrassing question of whom and how many people will inhabit the global, in a termites-state-like human society that he envisages. In the future, mankind will be reduced to a maximum of half a billion people who must all serve Gaia [13] (p. 270).

In Christian demographic thinking, there is no need for such adjustment of Man. Christian researchers point to declining population growth and to the fact that Europe and East Asia are affected by what is called natural population decrease – decrease without migration –, therefore the solution they suggest is conscious and responsible family planning. In their view, Earth has enormous capacities and resources, therefore the question that we need to elucidate is why our goods are so unfairly distributed within our global society – unfair distribution that allows the population of the United States to use 100 times more resources (ecological, biologic) than the population of the poorest continent of the South, Africa [14] (p. 16).

In my opinion, the principal cause of environmental pollution is not global overpopulation, but the profit-centred selfishness of some, and the continuous spreading of the worldview serving this profit-centred selfishness. If the self-realization of too many individuals leads to the accumulation and possession of material goods, the process will fundamentally undermine the cause of environment protection.

References

[1] A. Brennan, Y. Sze Lo, "Environmental Ethics," *Stanford Encyclopedia of Philosophy,* http://plato.stanford.edu/entries/ethics-environmental/ Accessed: January 13, 2018.

[2] I. J. Tóth, "Karteziánizmus és ökofilozófia," [Cartesianism and Ecosophia] http://acta.bibl.u-szeged.hu/9455/1/esz_005_129-135.pdf. Accessed: February 13th, 2018.
[3] Jr. Lynn White, "The Historical Roots of Our Ecologic Crisis;" *Science,* vol. 155, no. 3767, pp. 1205–1206, Mar. 1967.
[4] J. Deckers, "Christianity and Ecological Ethics: The Significance of Process Thought and a Panexperientialist Critique of Strong Anthropocentrism," *Ecotheology*, vol. 9, no. 3, pp. 359-387. Dec. 2004.
[5] P. W Taylor, "The Ethics of Respect for Nature," *Environmental Ethics*, M. E. Zimmermann, ed., University of California Press, Berkeley, 1981.
[6] Aristotle, "Politics" http://www.perseus.tufts.edu/hopper/text?doc=Perseus%3Atext%3A1999.01.0058%3Abook%3D1%3Asection%3D1256b. Accessed: February 13th, 2018.
[7] C. Hartshorne, "The Compound Individual" *Whitehead's Philosophy: Selected Essays, 1935–1970,* Charles Hartshorne, ed., University of Nebraska Press, Lincoln, pp. 41-46, 1972.
[8] A. N. Whitehead, *Science and the Modern World.* London, Free Association Books, 1985. Taylor, Paul W.: *The Ethics of Respect for Nature* in *Environmental Ethics* (ed. Zimmermann, Michael E., University of California Press, Berkeley 1981).
[9] Thomas Aquinas, "De veritate" q.1.a.9 in *CORPUS THOMISTICUM, Sancti Thomae de Aquino ,Quaestiones disputatae de veritate* http://www.corpusthomisticum.org/qdv01.html. Accessed: February 13th, 2018.
[10] R. Revelle, "Paul Erlich: New Priest of Ecocatastrophe," *Family Planning Perspective,* vol. 3, no. 2, pp. 66-70, Apr. 1971.
[11] G. Hardin, "Lifeboat Ethics: the Case Against Helping the Poor." http://www.garretthardinsociety.org/articles/art_lifeboat_ethics_case_against_helping_poor.html. Accessed: July 1st, 2018.
[12] J. Godel, "James Lovelock, the Prophet," *RollingStone,* https://www.rollingstone.com/politics/news/james-lovelock-the-prophet-20071101 Accessed: July 1st, 2018.
[13] J. Lovelock, *The Vanishing Face of Gaia: A Final Warning.* New York, Basic Books, pp. 270, 2009.
[14] H. Boulad, *A túlknépesedés mítosza* [The Myth of Overpopulation], Kecskemét, Korda Kiadó, 2012.

Ecocentrism or the Attempt to Leave Antropocentricity

Dejan Donev

Institute of Philosophy, Faculty of Philosophy,
University "St. Cyril and Methodius", Skopje, R. Macedonia

Abstract

With the emergence of industrialization as a social phenomenon, an event identifying the onset of dysfunctionality of the biotic community emerges: the issue of environmental sustainability. This question carries with it the need for expanding the views of traditional ethics that rests on anthropocentric theses. One of these approaches is ecocentrism, where the issue of the rights of individuals is crucial (whether it is just humans or people along with other living beings), stating that ecocentrism takes into account all participants in the biotic community as integral parts of the system. Logically, the mutual relations of the individuals of such a community and the consequences arising from their actions, which are the subject of research in this essay, are the main subjects of this position.

Keywords

Ecocentrism; bioethics; anthropocentrism; Aldo Leopold; land ethics.

I. Introduction

Ethics, as a theory of morality, has appeared and been considered as anthropocentric for most of its history. Human behaviour as an object of ethics has been explored strictly in relation to other people. Starting from Aristotle's theory of virtue, to the deontological ethics of Kant and conventionalism in Bentham and Mill, all ethical systems that are imposed as classical in contemporary philosophy were almost entirely based on anthropocentrism.

But with the advent of industrialization as social phenomenon, an event that particularly marks the beginning of the dysfunctionality of the biotic community, the issue of environmental sustainability has emerged. At the same time, the question of our attitude towards it is raised, because the reality shows that man is increasingly destroying "the bed of his life," thus threatening his survival. These questions carry the need to extend the views of traditional ethics that rests on anthropocentric theses for our relationship to nature and life.

This is why the conception of nature (especially from the perspective of the new century's natural sciences, whose paradigms, with its anthropocentrism, are considered the basis of the contemporary industrial civilization), treats nature as an object of human activity only. Moreover, for an instrumentalized nature which is "pure lumber - organic or non-organic material" [1] (p. 67), it is a mean to achieve the goals. Namely, from Bacon onwards, the implications are strengthened on the basis of the claim that knowledge is power, and science, as the highest degree of knowledge, is only a tool for man's control over nature, which is increasingly considered hostile and alien.

Namely, the mathematized, quantitative, and experimentally-oriented science has removed all of its qualities from its subject matter and thereby resolved the nature of its internal values. The quality of Aristotle in the new age science has been replaced by quantity as a priority task of the new sciences, which in the mathematical expression and formulas received the most perfect expression. Thus, modern science until today, along with modern technique as its practical application, is shown as a manifestation of the changed attitude of man towards nature established since the beginning of the new century.

Hence, this conception of nature is considered to be the most responsible for situations of general ecological crisis, disturbance of balance, and dying as a consequence of human intervention. The most powerful weapon of man, i.e. his product, projects nature as a means, material, and not as "a kind of life force" [2] (p. 151) which has value in itself.

II. The necessity of new solutions, new approaches, and positions

Modern civilization is anthropocentric, characterized by an infinite belief in human reason by the means of which man raises above all other living forms (and even as the master of nature). This subjectivism, since the new age philosophy, in symbiosis with the understanding of nature as presupposed by modern science, is a spiritual basis of the contemporary civilization. It is a civilization dominated by the subjective mind, while wisdom is expelled from nature and exists only in man. Nevertheless, because of the negative consequences of actions towards nature, it slowly becomes aware of the urgency of changing the approach, otherwise it will go extinct.

This need is a logical consequence of the admixture of the living community to which the man belongs. As such, all the consequences that this community suffers from, directly or indirectly, are reflected in the man himself. Hence, in certain (bio)ethical circles in 20th, and especially the 21st centuries, several new positions expand the boundaries of anthropocentric ethics in order to cope with the problems that are neglected in the traditional approach.[1]

Such positions which extend the moral objects of the traditional anthropocentric ethics (where man, as an intelligent being, is central and represents the axis of everything around him) are the only object of moral behaviour:

- *Pathocentrism*; is an attempt to expand the subject of all sensitive living beings. These are the positions most clearly expressed through the views of Peter Singer, according to which animals that have developed a sophisticated sensory system tend to avoid situations that cause them pain.Namely, this attempt hierarchically selects highly-developed animal species (primarily mammalian) as moral objects to which the conduct of the moral subject should be the same as in a situation where the moral object is the man himself. In other words, every living being in which we can, by behavioural observation, establish analogous behaviour as of that in man, has the potential to be a moral object, that is, the object of reviewing our behaviour.
- *Biocentrism*, expands the ethical horizons to even greater distances, the moral object determining here a step further and wider than pathocentrism (where it is determined by behaviour), i.e. by biological path. Namely, everything that is living is an object of moral behaviour or, starting from plants and insects, to the most

[1] That is why the ecological crisis, because of this necessity and urgency, shattering the faith in the correctness of the key assumptions of our civilization, is shown as a general crisis or as the basis of the general crisis.

sophisticated living beings, every living individual should be taken into account when making a decision that includes that living individual. This is therefore an individualistic and egalitarian position which argues that living beings are individuals who should possess and should be respected according to their respective rights. This is an important feature which separates this position from that of ecocentrism.

- *Ecocentrism.* While for all the other ethical theories the question of the rights of individuals is crucial (whether it is only humans or humans along with other living beings), ecocentrism is a position which considers all participants in the biotic community as integral to the system. It argues that this is a holistic system in which the mutual relations of individuals of communities and the consequences arising from their actions are the main subject of review.

III. Pro et contra ecocentrism

Every ethical position has tangible implications. Every ethical attitude is essentially pragmatic and manifests itself through certain intentional behaviour. In the case of eco-centrism,[2] its procedures take into account the harmonious state of the ecosystem. This fundamentally distinguishes it from the other two positions, in particular from biocentrism. It is so because of the following reasons [3] (p. 298):

- First, eco-centrism has a comprehensive perspective. This is because procedures based on the eco-centric position seem to be able to solve a number of contemporary environmental problems, but also problems that the environment faces in general. Although by modern positions of various animal communities, this concept of living beings as resources is rigorously and deeply condemned, it is justified by the eco-centric theory because the use of a member

[2] In social theory, ecocentrism, which has an ecological sign, is often equated with the so-called "deep ecology." although it may seem more appropriate to consider it as a kind of stream within ecocentrism. Compare with Arne Naess, *Ecology, Community, Lifestyle: An Outline of an Ecosophy* (Cambridge: Cambridge University Press, 2001), 37; Mathew Humphery, *Preservation Versus the People: Nature, Humanity and Political Philosophy* (New York: Oxford University Press, 2002).

(prey) in a community can represent a natural resource for another member (predator), as long as it does not adversely affect the natural equilibrium of the ecosystem.[3]

- Second, the eco-centric position avoids counterintuitive conclusions that are represented by other ethical positions, above all biocentrism, which advocate the defence of animal rights as individuals.[4]
- Third, eco-centrism is a completely non-anthropocentric position. Unlike other ethical systems (anthropocentrism, pathocentrism, biocentrism), where man is the only moral subject, or man besides other animals enjoys privileged rights as an individual, man is considered in eco-centrism as a simple member of the biotic community. Therefore this position, through its holistic views, gives moral dignity to natural objects and the ecosystem as a whole. With this, it is in accordance with the teachings of ecology.

A deeper analysis of these views, as well as the requirement to find a soundly-based answer to the question of why it is reasonable to adopt ethical holism in relation to the ecological community, necessarily refers to a nominal review of one of the most relevant advocates of ecocentrism: Aldo Leopold and his key essay "Land Ethic." This is the first systematic presentation of ecocentric ethics, published in the classical text of the ecological movement, the collection of essays entitled *A Sand County Almanac* in 1949. The ethics of the land, the idea on which Leopold built his ethical holism expressed in three philosophical pivotal moments, states: "a certain procedure is correct when it strives to preserve the wholeness, stability and beauty of a

---

[3] Of course, this assertion cannot and must not be generalized, which means that ecocentrism does not justify the mass destruction of entire ecosystems by systematically cutting out forest areas or industrially growing large quantities of goods, which are scientifically proven to be one of the most important factors for the negative consequences on all ecosystems, such as the issue of global warming.

[4] For example, killing an animal by a farmer or hunter, or cutting wood in the yard for fire are fully justified by ecocentrism as these activities are aimed at securing personal existence and satisfying biological needs, while they do not harm the overall functionality of the biotic community, and are in the interest of some members of the same community, i.e. they are part of the natural equilibrium.

particular biotic community. It is incorrect, however, when it tends to a different state" [4] (p. 262).

Leopold **first** [5] (p. 299) writes a study on the practical cause of the adoption of ethical holism. He argues that it will be seen when[5] we will allow ourselves to "think like a mountain" [6] (p. 137). This is alluding to our limited opinion on individual situations that do not allow us to understand the interdependent mechanisms of the entire ecosystem. In this sense, when "we think like a mountain," when we identify with the mountain as epistemological subjects, we are able to see the interdependencies of a functioning living community. This is of particular importance, because unlike other ethical systems that have an individualistic approach, the ethics of the land, with its holistic vision, provide positive practical consequences not on individuals, but on the entire biotic community as a system.

Secondly, Leopold explores the foundations of the epistemological holism that allows the existence of its ethical system, i.e. it is a cognitive basis for the existence of the ethics of the land. Therefore, he believes that it is necessary to identify and catalogue the roles of all members within an ecosystem, so that we can safely derive the correct ethical implications. Namely, it is necessary to know each other's role in order to claim that everything which is being done is done in order not to allow the hyperpopulation of one kind of animals in relation to others. The same issue concerns maintaining the balance of the biotic community.

Lastly, Leopold developed the metaphysical holism as the necessary presupposition to justify the existence of ethical holism. Namely, metaphysical holism is an extension of the concept of the "living being.". Thus the biotic community is seen as a system built by individual organs or parts which not only include people, animals, and plants, but also certain parts of the planet itself, such as mountains, rivers, the atmosphere, etc. With the inclusion of these elements in the biotic system, traditionally regarded as inanimate nature, Leopold gives an entirely new dimension to the term "living being." As he explicitly claims: "philosophy expresses a certain reason why we cannot destroy the land, without being punished; it is because the 'dead' nature is an

[5] Although expressed as a syntagm with metaphorical meaning.

organism that possesses a certain kind or degree of life, which we as such intuitively respect" [7] (p. 47-48).

Hence, ethical holism is impossible without assuming that the land has a certain ***livelihood***. Moreover, the land can no longer be treated as a pure object, a dead element that can be used and shaped to the extent of human desire.[6] Land should be understood as a living organism that can be healthy or ill, hurt, or killed. "Land, therefore, is not only ground; it is a source of energy that flows through the circuit of the ground, plants and animals" [4] (p. 253).

Several notes on the ethics of the land

However, despite the seductiveness of the ecocentric positions, especially the notion of the ethics of the land, when they are more deeply and critically analyzed, they are not immune to systematic criticism. Two of these criticisms are of fundamental philosophical significance.

The **first** critique addresses concerns regarding the extended metaphysics of Leopold (i.e. that the categories "health", "death" and the like, which can be attributed to any single living organism) cannot be attributed to the whole environment as a system. This way, Leopold himself, as well as the advocates of this position, must and should admit that the "vitality" of his ecosystem is merely metaphorical, and not literal.

This is because the ethical implications of such learning rise doubt in the realization of a natural equilibrium advocated by the ethics of the land. More specifically, doubt appears at the impossibility of a concrete answer to the question of what exactly the "integrity, stability and beauty" of the ecosystem relate to. Despite the systematic research of

[6] This new perspective, which can be called "environmental awareness," has been developing for decades, and was fully presented in the late 1940s in the form of *Ethics of the Land*. This is a turning point from the understanding of nature that possesses only an instrumental value, towards the understanding that recognizes the intrinsic value in natural systems. This is why our short-sightedness bothers more than it contributes to natural balance. Until we learn to "think like a mountain" and understand nature from wider and long-term perspectives, we are doomed to inefficiently manage ecosystems. More detail on this in the essay "Thinking Like a Mountain," in *A Sand County Almanac*.

existing ecosystems (i.e. finding the appropriate place for each member in the biotic community, as well as their roles as interrelationships on which the eco-system is built), and despite perceiving the consequences which would arise from each particular manipulation with each individual member of the community, the possibility of the full awareness of the procedure which would bring us closer to establishing the "wholeness and stability" of the ecosystem is questionable. In this way, we cannot even come up to the way these facts are related to this valuable conclusion.[7]

Secondly, justifying the sacrifice of individual animals and/or plants as resources, as long as they are in the interest of the ecosystem, puts the land's ethics and ecocentrism to a fierce and serious test. Within this test, the land's ethics is referred to as "totalitarian ethics" by Marty Kiel [9] (p. 135-147), a learning that undermines and destroys respect for the individual [10] (p. 241-256), thus it is also called "ecological fascism." As Tom Reagan sums up in his work "The Case of Animal Rights" [10] (p. 241-256): "the weight and implications for the development of rights based on ecological ethics imply the consent of the individualistic nature of moral rights with the holistic understanding of nature... Aldo Leopold is an example of this tendency ... The implications of this understanding contain a clear perspective that claims that an individual can be sacrificed in the interest of the great biotic good, in the name of 'the wholeness, stability and beauty of the biotic community.' It is difficult to see how the notion of the right of the individual can find a place in an understanding that can rightly be called ecological fascism."

This is why the criticism of the two premises (i.e. that the sacrifice of individual animals and/or plants as resources is justified as long as it is in the interest of the ecosystem and the second, that humans, together with other living beings are only members of the living community) leads to the conclusion that certain sacrifices are in the interest of maintaining the "wholeness, stability and beauty" of the eco system. Furthermore, this holistic ecocentrism and the ethics of the land are unable to respond to this criticism, which raises the question of

---

[7] For a more detailed defence of this view, see: Laura Westra, *An Environmental Proposal for Ethics: The Principle of Integrity* (Rowman & Littlefield, Lanham Md., 1994).

whether such an ethical approach is appropriate and whether the answer should be sought elsewhere.

IV. Conclusion

At the forefront, it is noticeable that these are undefined positions of holistic ecocentrism and the theory of the land because they completely ignore the rights of individual members in the interest of the biotic community as a system. However, when considering the remarks of their criticisms, the possible compromise offered by this position should also be considered.

Namely, a systematic research of the existing ecosystems (i.e. finding the appropriate place for each member in the biotic community, as well as their roles and interrelationships on which the eco system is built), would allow us to know what consequences could arise in each particular manipulation with each individual member of the community. Then we may know which procedure could bring us closer to establishing the "wholeness and stability" in the ecosystem, and which would distract us from such a state. Thus, the ecocentric approach, if properly understood, could give positive results which would primarily respect the balance of the animal community, while also respecting the rights of individual members of the system.

Moreover, the very existence of ecocentrism as an ethical position, although nominally, yet unequivocally, allows for progress in ethics. This is why any opposition to and refutation of the arguments of this theory could only bring us closer to finding such an ethical system that would allow the harmonious functioning of all members of the living community without disturbing individual, single rights.

The justification of this view could be found in the claim that ecocentrism[8] is actually trying to question and change our conventional understanding of the environment. The "main point of overlap between all authors who have been advocating this approach is actually a critique of civilization conditioned conception of nature as an element which primarily appear as supplies for production of our humanity" [11] (p. 197).

[8] As one of the many streams inside the ecological movement that appeared in the mid-70s of the last century.

Hence, instead of seeing nature as a relatively simple phenomenon, the prism should be turned to nature and be regarded as an almost infinitely complex system. This, in turn, leads to the conclusion that it is only about blurring our core notch networks with Prometheus watching nature. So everything that is in the domain of "natural" is susceptible to exploitation that can be exclusively limited by the logic of the work itself. This gives the right to Naess and Goldsmith to insist on an additional effort to revise the ontological status of nature itself, not just to improve laws and internal rules of production.

References

[1] H. Marcuse, *Counterrevolution and Revolt.* Belgrade: Grafos, 1979.

[2] H. Marcuse, *A Man of One Dimension.* Sarajevo: Veselin Masleša, 1968.

[3] J. R. DesJardins, *Ecological Ethics - Introduction to Ecological Philosophy.* Fourth edition. Belgrade: Official Gazette, 2006.

[4] A. Leopold, "Land Ethic,"*A Sand County Almanac.* New York: Ballantine, 1970.

[5] J.R. DesJardins, *Ecological Ethics - Introduction to Ecological Philosophy.* Fourth edition. Belgrade: Official Gazette, 2006.

[6] A. Leopold, "Thinking Like a Mountain,"*A Sand County Almanac,* New York: Ballantine, 1970.

[7] A. Leopold, "Some Fundaments of Conservation in the Southwest,"*The River of the Mother of god and Other Essays,* Susan Flader and Baird Callicott eds., Madison: University of Wisconsin Press, pp. 86-97, 1991.

[8] A. Leopold, "Land Ethic,"*A Sand County Almanac.* New York: Ballantine, 1970.

[9] M. Kheel, "The Liberation of Nature: A Circular Affair," *Environmental Ethics*, vol. 7, pp. 135-149, Summer 1985.

[10] E. Katz, "Organicism, Community, and the Substitution Problem," *Environmental Ethics*, vol. 7, pp. 241-256, Fall 1985.

[11] S. Prodanović, "Ecocentric and anthropocentric understanding of human needs in ecology," *Nature, ethics, politics - Ecological (pre)occupation and (pre)orientation*, Ana Birišev et al., eds., Belgrade: Institute of Philosophy and Social Theory, University of Belgrade, pp. 195-211, 2012.

Debating Public Policy: Ethics, Politics and Economics of Wildlife Management in Southern Africa

Matthew Crippen,[1] John Salevurakis[2]

[1] Grand Valley State University, USA
and Humboldt University of Berlin, Germany

[2] American University in Cairo, Egypt

Abstract

Based on field research in Africa, this essay explores three claims: first, that sport hunting places economic value on wildlife and habitats; second, that this motivates conservation practices in the interest of sustaining revenue sources; and, third, that this benefits human populations. If true, then sport hunting may sometimes be justifiable on utilitarian grounds. While not dismissing objections from the likes of Singer and Regan, we suggest their views – if converted into policy in desperately impoverished places – would destroy animals and the habitats on which they depend. There are empirical verifications of this, which we discuss.

Keywords

Ecology; economics; environmental ethics; hunting; policy; wildlife management; utilitarianism.

I. Introduction

This essay is a précis for a larger project underway on the ethics and economics of sport hunting. More specifically, it is based on field research, and aims at testing the claims:

1. That sport hunting places economic value on wildlife and habitats.
2. That this motivates conservation practices in order to maintain business enterprises, however distasteful sport hunting may be.

3. That this, in turn, benefits local human populations.

The field research – presented in a number of case studies – specifically focuses on regions of Namibia, South Africa, and Zimbabwe. While some of the findings suggest that sport hunting can have the benefits just outlined, political opposition in the name of conservation thwarts such practices. So without necessarily challenging philosophical outlooks that would forbid sport hunting such as those advanced by Jeremy Bentham [1], Peter Singer [12], and Tom Regan [9], the case studies presented suggest that these positions, if enacted in policy, can have the counterintuitive effect of destroying that which they seek to protect, in this case, animals and the habitats on which they depend.

II. Case Study 1: Zimbabwe's Zambezi Valley

The first case study focuses on northern Zimbabwe's Zambezi Valley, specifically the Mbire District in the Dande region, an area bordering Zambia and Mozambique. This region has long suffered from poaching, which employs indiscriminate methods such as snares and poison, stripping the land of game. Porous and unpatrolled boarders make matters worse. At the same time, rampant killing of animals is bad for the hunting business, and those running legal operations have a vested interest in preventing it.

In line with this, the sport hunting operation we observed – Charlton McCallum Safaris – has gone to lengths to thwart poaching. Measures include funding anti-poaching patrols, and establishing reward systems for captured snares, poachers, hunting equipment and so forth. These measures appear effective, so that during our brief research visit, we witnessed a leopard poacher arrested and another convicted for killing a juvenile elephant. The anti-poaching patrol we travelled with also removed numerous snares during our stay, with approximately 6.500 seized between 2014 and 2016.

It is important to note that overhunting is bad for business. Consequently, the sport hunting carried out here is not the mass slaughters that we picture when thinking of colonial adventurers – the sort detailed in books such as Theodore Roosevelt's *African Game Trails* [11]. Quotas restrict the numbers killed. So do prices that clients pay for hunts, sometimes exceeding $110.000. In the 457.000 acre area in which Charlton McCallum Safaris operate, the number legally hunted accordingly pales in comparison to poaching levels prior to the introduction of regulated trophy hunting. This is enough so, for example, that with anti-poaching measures the total amount of elephants killed has dropped. Thus in 2010 – the year Charlton McCallum Safaris established the Dande Anti-Poaching Unit or DAPU – 40 poached elephant carcasses were found from the previous season. By 2013, the number was down to four, with amounts fluctuating in subsequent years, but remaining below 10. The number hunted on legal safaris, by contrast, averaged 11.25 per year from 2013 onward. This means that in even the most egregious years, the number of animals killed was roughly half of that occurring in the year prior to the introduction of DAPA.

Species accordingly appear to benefit, and advantages also accrue to human populations, enough so that the local government pays half the salaries of DAPU scouts, with Charlton McCallum Safaris covering the remainder, plus food and equipment. This makes economic sense since hunting realizes 90% of revenues in the Mbire District, with locals prospering from jobs that would otherwise be absent, along with lease payments. Just as importantly, future generations should yield benefits from the continued existence of wildlife and habitats that would otherwise be destroyed.

III. Case Study 2: Namibia Ranch Country

The second case study differs from the first insofar as land is not leased from local indigenous populations, but owned by ranchers in northwestern Namibia. In at least the short term, ranchers' interests tend to conflict with wildlife conservation. However, the introduction of sport hunting generates enough revenue to compensate for losses caused by wildlife, and in this way motivates a conservation spirit.

One problem ranchers face is elephants breaking through fences, allowing breeding stock to mix. Elephants also destroy watering holes. Another common challenge is predators eating stock. Hence both have historically been shot. The introduction of sport hunting, however, markedly alters the situation, assuring profits exceeding damages, even for ranchers not directly involved in safari operations. This, in turn, dissuades them from shooting so called "problem animals" in the 650.000 acre Loxodonta Africana Conservancy jointly established by them for regulated sport hunting on their lands. As in the Zambezi case, moreover, only small numbers – for example, one or two elephants every five years – are hunted.

The situation is of course optimal when ranching is mostly abandoned in favour of using lands for sport hunting, with ecosystems benefiting from the removal of cattle, which strip vegetation, thus harming wildlife. Such is the case in the 37.000 acre parcel run by Westfalen Hunting Safaris that we observed. Moreover, it is not just the removal of cattle that yields benefits. Operators also cultivate wildlife – including giraffe, oryx, springbok, warthog, zebra and other species hunted – by putting out saltlicks and alfalfa pellets for animals and by maintaining waterholes, which are essential, more so during increasingly common droughts. The co-owner of Westfalen Hunting Safaris, John Westhuizen, further notes that sport hunting – which he incidentally does not enjoy – is more profitable than raising cattle, particularly in times of drought. Notice, moreover, that animals will be killed even if owners opt exclusively for ranching, only in this case it will be cattle raised for slaughter, plus wildlife extinguished because of stripped vegetation, making the total number higher.

Although most of the financial benefits accrue to affluent landowners, who happen to be overwhelmingly white, this does not diminish the fact that wildlife and habitats are preserved. Financial rewards, moreover, would still go disproportionately to land owners if they engaged in ranching instead of sport hunting, just as it goes disproportionately to those who already have much in every society. That this is so, in any case, does not obviate the fact that poorer black populations nonetheless benefit. This is not so much because more jobs are created, although this happens in times of expansion. Rather, it is because jobs are higher skilled and hence better paying. In addition to this, uniforms, shelter and meat are supplied to workers and local schools, with about four tons donated annually by Westfalen Safaris alone. Because of the environmentally destructive nature of ranching, and for other reasons stated, changing to sport hunting – in addition to benefiting locals – conserves habitats and wildlife for future generations.

IV. Case Study 3: Makuya Lands

The third case of sport hunting – this time in Makuya lands in South Africa along the Mozambique and Zimbabwe boarders – appears less successful in terms of achieving conservation goals. Here tribespeople control both the hunting rights and the land abutting Kruger National Park, a protected area from which game wanders onto their territory.

A problem here is that one group in common controls the land, and the same group has access to it, with little incentive against illegal hunting since individuals are rarely held responsible, and since, from their perspective, there is almost unlimited game passing onto their territory from Kruger. As if to exemplify this, our hosts boasted in front of tribal leaders that they served us poached buffalo after drinks had loosened tongues. When touring the land, we observed lower densities and less variety of animals than when in Namibia, which is shocking given the proximity to Kruger Park, although more thorough surveys over longer periods would be needed to determine the extent to which this is actually so. Another problem with the Makuya operation is that revenues earned from hunting appear to be funneled disproportionately to local elites, though this is hard to ascertain for certain. To some extent, this is the case everywhere, but the difficulty appears worse here.

V. Political Obstacles

Though sport hunting is not always beneficial, there are cases in which it is, as in the Zambezi and Namibia examples. Be this as it may, political agendas tend to thwart even these beneficial arrangements.

As an illustrating example, one might consider the 2014 Dallas Safari Club auction of the right to hunt a black rhinoceros bull from a national park in Namibia. A wealthy Texan named Corey Knowlton won the bid, paying $350.000 to hunt this critically-endangered species. Interestingly, after the auction, Knowlton declared that he was unhappy with the result. He anticipated (and wished for) a much higher price. However, weeks of negative publicity had scared off potential bidders. Perhaps more interestingly, Knowlton wondered why individuals or organizations so repulsed by the potential killing had not bothered to bid and then simply opted not to shoot the animal. He suggested that this – with the broader anti-hunting venom – detracted from conservation aims by lowering the value of the animals and hence local motivation and capacity to protect them.

A parallel example involves land within Kruger National Park which after much political wrangling, was restored to the Makuleke, with rights granted to any commercial ventures in the area as long as they coincided with the conservation efforts of the park itself. The most obvious prohibited activities included homebuilding, mining or farming within the "contractual park" [13]. Notably, commercial trophy hunting was permitted and took place in an effort to raise revenues, in turn used for community development projects and to pursue an eco-tourism arrangement, eventually carried out without hunting. The projections were an estimated 150 jobs, paying $400 per month [7], a very good wage by South African standards. While unpalatable to many, it should be made clear that hunting revenue allowed the construction of several lodges within the area which serve today as a source of funds for the community [10]. The trophy hunting, of course, received abundant criticism, likely more so because the activity was taking place within the crown jewel of African national parks – an area, however, subject to occasional culls due to overpopulation that is not only unsustainable but also stressing local vegetation [4]. In circumstances similar to the Knowlton incident, an uproar ensued, and consequently only a handful of hunts occurred, nonetheless yielding roughly $500.000 revenue [2, 6, 9].

It is important to note community involvement nonetheless protected the area to an extent that Kruger National Park elected to relocate rhinos into it. More recently, however, the Makuleke have come to see politically fuelled conservation pressures as a hindrance. This is directly related to the reality that revenue has dropped as hunting has become too politically unpopular as a management option. Further, the imposition of less profitable luxury lodges is not without ecological impact. Without sufficient revenue, one wonders about the long-term prospects for local wildlife [3], though in this case matters should at least be mitigated by the fact that these lands lie within Kruger.

VI. Ethics and Policy

One premise advanced throughout is that human populations stress non-human habitats. Land is generally used one way or another. Cattle farming, for example, is extremely harmful to habitats and the entire planet insofar as it contributes to climate change, water pollution and depletion, vegetation destruction and more. Growing crops also damages wildlife habitats and contaminates water with pesticides and fertilizers, among much else. Sport hunting obviously kills wildlife too. In all cases, accordingly, animals die.

Anti-hunters, moreover, cannot protect animals without a sufficient revenue source, and will inevitably "sell" some animals and let them die, even if merely opting to protect some over others. Sport hunting, as carried out in the first two case studies, is less destructive to habitats and wildlife in comparison to farming and perhaps in comparison to not doing anything since unoccupied land is more vulnerable to poaching. Conversely, wildlife plummets have been documented following hunting bans, as in Kenya from the late 1970s onwards [8]. Reasons are complicated, some independent of hunting or its absence, as with increases in human population and consequent strains on habitats. However, bans can relate directly to wildlife destruction, as when environmental stresses follow because people, whose livelihood depends on doing something on available land, opt to raise livestock or grow crops, thereby destroying habitats. In comparison to livestock, moreover, sport hunting yields a kind of double-value per animal. This is because safari operators receive trophy fees and other remunerations while also keeping the meat, either consuming or donating it, whereas the economic value of slaughtered cattle is solely in money exchanged for meat and skins. Predators and other animals such as baboons, which trophy hunters sometimes shoot, are exceptions and not consumed. In some cases, jobs are also supplied that would otherwise be absent, as with the Makuleke and Zambezi examples.

As intimated at the outset, it is not our aim to repudiate Regan's inherent value argument [9] or Singer's anti-speciesist position [12], both suggesting that sport hunting is morally indefensible. However, these ideas – like the fortress conservation practices they would imply – are ineffectual in regions where people are struggling just to get by. In fact, insofar as these positions contribute to the political unpalatability of sport hunting, they may reduce conservation outcomes by depleting the economic value of wildlife, while simultaneously harming desperately poor people. For example, ivory import bans that the United States government issued against Zimbabwe due to rampant mismanagement and lack of enforcement lowers the price of elephant hunts, with revenues precipitously dropping by half in the case of Charlton McCallum Safaris. The price of black-market ivory, however, remains unaffected, and the decreased funding from legal hunts impedes anti-poaching initiatives and the economic motivations to mount them. The ban on rhinoceros horn trade is similarly irrational. In this case, it is irrational because horns regrow and can be harvested without harm from farmed animals. Increasing the legal supply, in turn, would lower black-market prices and hence incentives to poach.

The idea in our research is to critically weigh the costs and benefits of hunting within a utilitarian framework, and within that framework it appears that sport hunting is defensible in certain circumstances, namely, those in which it benefits habitats, wildlife and human beings. We do not maintain utilitarianism is the only framework and are not even necessarily committed to it philosophically, but we do think it is a good one if the aim is saving animals and habitats, while improving living standards of impoverished people. Leaving philosophical standpoints aside, we emphatically believe it is time to ask whether hunters and environmentalists dislike wildlife and habitat destruction more than they dislike one another, and whether they can accordingly find common ground that promotes all their interests, along with those of non-human and human populations in effected regions.

References

[1] J. Bentham, *An Introduction to the Principles of Morals and Leglislation*. New Edition.Oxford: Clarendon Press, 1823.

[2] G. Collins, Personal Communication, 2005.

[3] T. Kepe, "Land Restitution and Biodiversity Conservation in South Africa: The case of Mkambati, Eastern Cape Province," *Canadian Journal of African Studies*, vol. 38, no. 3, pp. 688-704, 2004.

[4] K. Lange, "Desperate Measures," *National Geographic* vol. 214, no. 3, pp. 64-69, 2008.

[5] D. Lewis, P. Alpert, "Trophy Hunting and Wildlife Conservation in Zambia," *Conservation Biology,* vol. 11, no. 1, pp. 59-68, 1997.

[6] H. Magome, *Sharing South African National Parks: Community Land and Conservation in a Democratic South Africa.* Pretoria: South African National Parks, 2002.

[7] L. Makuleke, "The Makuleke Story" [2008], *Earthlore Communications Web*, http://www.earthlore.ca/clients/WPC/English/grfx/sessions/PDFs/session_1/Maluleke.pdf. Accessed: Jan 9, 2010.

[8] J. O. Ogutu, H-P Piepho, M.Y. Said, G.O. Ojwang, L.W. Njino, S.C. Kifugo, P.W. Wargute, "Extreme Wildlife Declines and Concurrent Increase in Livestock Numbers in Kenya: What Are the Causes?" *PLoS ONE*, vol. 11, no. 9, e0163249, 2016.
[9] T. Regan, *Case for Animal Rights*. Berkeley: University of California Press, 2004.
[10] H. Reid, "Co-Management of Contractual National Parks in South Africa: Lessons from Australia," *Conservation and Society*, vol. 2, no. 2, 377-409, pp. 2004.
[11] T. Roosevelt, *African Game Trails: An Account of the African Wanderings of an American Hunter-Naturalist.* New York: Charles Scribner's Sons, 1910.
[12] P. Singer, *Animal Liberation.* New York: Harpers Collins, 2002.
[13] M. Wells, T. McShane, "Integrating Protected Area Management with Local Needs and Aspirations," *Ambio*, vol. 33, no. 8, pp. 513-519, 2004.

The Role of the Codex Alimentarius Commission in the Controversy over Genetically-Modified Food

Ivica Kelam

Faculty of Education/Centre for Integrative Bioethics,
Josip Juraj Strossmayer University of Osijek, Croatia

Abstract

The Codex Alimentarius Commission (Codex) is the main international body responsible for adopting food safety rules. A particular problem is the issue of genetically-modified food since this topic has been in the public attention from the first instances of sowing genetically-modified crops in the mid-1990s. Because of the importance of the recommendations in the Codex, corporations are particularly interested in adopting the recommendations which cannot affect their profits – this is particularly evident in their effort to prevent the mandatory labelling of genetically-modified foods. In this chapter, through a brief historical overview of the process of labelling genetically-modified food, we will point to the complexity of this issue, which is manifested within a twenty-year process of issuing rules on labelling genetically-modified food.

Keywords

Codex Alimentarius Commission; genetically-modified food; labelling; bioethics; corporations.

I. Introduction

The exponential increase in food safety incidents across the globe in the past decades has resulted in mushrooming regulatory initiatives, including new standards and requirements from national governments, international organizations, and private actors. Such regulatory initiatives, both public and private, have primarily emerged to address the rapidly decaying public trust in modern global food chains. This is further complicated by several factors including the globalization of economic activities, advancements in food science and transportation technology, the multinationalization of the food industry, and the advent of the World Trade Organization (WTO) in 1995. Given the significantly transformed production, transportation, and consumption of food [1], recent food safety incidents, with their intensified scope, severity, frequency, and impacts have become extremely challenging to cope with. A World Health Organization (WHO) report indicates that food safety problems contribute to 1.5 billion cases of diarrhoea in children and over three million premature deaths annually, both in developed and developing countries [2]. For example, yearly, approximately 125 000 children die in developing countries due to foodborne diseases caused by contaminated food and water [3]. The United States Centers for Disease Control and Prevention (CDC) estimates that 48 million illnesses, 128,000 hospitalizations, and 3,000 deaths result from foodborne diseases each year in the United States [4]. Despite the unprecedented numbers of food safety incidents, no multilateral treaty exists to monitor or regulate global food safety, other than the Agreement on the Application of Sanitary and Phytosanitary Measures (SPS Agreement) [5], which lightly touches upon the side issues of harmonization and scientification from an international-trade-law (facilitation of food trade) perspective. In addition, as contaminated food outbreaks do not stay within national boundaries, unilateral measures adopted by national governments, even with a certain level of extraterritorial effects, fail to provide an effective or efficient approach to addressing the global food safety problem.

II. History of the Codex Alimentarius Commission

The origins of the Codex can be traced back to the beginning of the 20th Century. In 1911, Austria recognized the private notes of experts on the evaluation of food items as the "Codex Alimentarius Austriacus" [6] (p. 19). "Codex Alimentarius" is a Latin term meaning "Food Code," and the current Codex derived its name from a title of the food standards adopted by the Austro-Hungarian Empire [7]. In 1958, The European Codex Commission was set up to effectively promote international co-operation in setting a common "Food Code." The European Codex Commission promoted the idea that the objective of achieving an international food code should be taken under the auspices of the United Nations [8] (p. 43). In 1959, several Latin American countries also created their own Code, the Código Latino-Americano de Alimentos [9]. In overcoming the fragmentation of this system and better addressing the challenges of international trade, the FAO and WHO launched the joint FAO/WHO food standards programme, in 1962 in Geneva [10] (p. 12-13). One year later, the Codex was born as a subsidiary body of FAO and WHO, with the mandate of implementing the programme. The Codex was also subsumed and continued the work of the Codex Europaeus. Currently, the Codex has 189 Codex Members comprised of 188 Member Countries and one Member Organization (The European Union) [11], as well as 226 Codex Observers: 56 international organizations, 154 non-governmental organizations, and 16 UN organizations [12]. The member states of the Codex represent 99% of the world's population.

The Codex is a multilayered international body, composed by a multitude of committees dealing with both horizontal and vertical issues: next to an Executive Committee, there are 6 active commodity committees (e.g. Codex Committee on Fresh Fruits and Vegetables, and Codex Committee on Spices and Culinary Herbs) and 10 active General subject committees (e.g. Codex Committee on Contaminants in Foods, Codex Committee on Food Labelling, and the Codex Committee on Food Hygiene) [13]. Each committee is administered, organized, and financed by a Codex Member (a 'host country'); these are most commonly countries that have a specific interest in a certain Codex issue, who then become the host of that specific committee (e.g. Canada is the host country for the Codex Committee on Food Labelling). The scientific meta-analysis used in Codex decision-making processes are produced by one of the several expert committees, which have been jointly established by the WHO and FAO and provide the studies that are further used to adopt standards. The decision-making process of the Codex consists of an eight-step procedure, starting with the proposal of a standard by a Committee or a Member, followed by several evaluations and discussions (step 2 to 7) and ending with the submission of the standards for adoption by the Codex [14] (p. 31). On step 2, a draft text and an assessment by an expert body (e.g. JECFA) are prepared, followed by recommendations of Acceptable Daily Intakes (ADIs) and/or the preparation of Maximum Residue Levels

(MRLs). This is arguably one of the most important phases of the decision-making process because expert recommendations are typically endorsed in the final standards. According to Article 1 of the Statutes of the Codex, the purpose of the Codex is as follows:

> The Codex Alimentarius Commission shall, subject to Article 5 below, be responsible for making proposals to, and shall be consulted by, the Directors-General of the Food and Agriculture Organization (FAO) and the World Health Organization (WHO) on all matters pertaining to the implementation of the Joint FAO/WHO Food Standards Programme, the purpose of which is:
> (a) protecting the health of the consumers and ensuring fair practices in the food trade;
> (b) promoting coordination of all food standards work undertaken by international governmental and non-governmental organizations;
> (c) determining priorities and initiating and guiding the preparation of draft standards through and with the aid of appropriate organizations;
> (d) finalizing standards elaborated under (c) above and, after acceptance by governments, publishing them in a Codex Alimentarius either as regional or worldwide standards, together with international standards already finalized by other bodies under (b) above, wherever this is practicable;
> (e) amending published standards, after an appropriate survey in the light of developments [15].

The purpose of the Codex is to provide international standards to ensure the health and safety of consumers worldwide and to guide member countries in adopting harmonious food standards and thereby promote international food trade. This purpose is also reiterated on the Codex website as follows:

> Codex Alimentarius follows the principle that consumers have a right to expect their food to be safe, of good quality and suitable for consumption. In this regard, the safety and essential quality of internationally traded food is of paramount importance. Codex has set a number of

> standards and codes on foods for vulnerable groups such as infants and young children, to provide adequate nutrition while protecting them from foodborne risks and to reduce infant mortality and morbidity worldwide. Codex also aims at protecting consumers against deceptive practices. Codex work in food labelling contributes to providing consumers with accurate and useful information to guide their choice of food. Codex assists in the harmonisation of national food legislation and regulation of countries which want to use Codex texts as a benchmark. International harmonisation of standards facilitates food trade and sustainable economic development. Codex plays an important role particularly for developing countries that may lack the necessary infrastructure and expertise to put in place adequate standards, food safety controls, and management systems [15].

Although the adoption of the Codex texts is voluntary, many countries choose to adopt these guidelines because policy makers see the benefits for consumers and international food trade. By providing a harmonious standard that could be adopted worldwide, Codex texts serve a dual purpose of protecting consumer welfare whilst promoting international food trade. Codex texts are based on scientific principles; hence, countries that do not have sufficient infrastructure and expertise consider Codex texts as valid standards of food safety and control over international food trade.

The guidelines/standards established by the Codex are of a recommendatory nature and are not a substitute for national legislation. Countries may adopt these guidelines in their national legislation and make provisions suitable to their specific requirements. This is reiterated in the General Principles of the Codex Alimentarius as follows: "Codex standards and related texts are not a substitute for or an alternative to national legislation. Every country's laws and administrative procedures contain provisions with which it is essential to comply. Codex standards and related texts contain requirements for food aimed at ensuring for the consumer a safe, wholesome food product free from adulteration, correctly labelled and presented [...]" [17].

After establishing the WTO, according to some authors, the voluntary nature of the standards and the small size of meetings to reach agreements were radically changed. The principal legal framework that has enabled this transformation is the WTO Sanitary and Phitosanitary Agreement (SPS Agreement), which provides that Members shall either base their measures on international standards or on science and risk assessments. Article 3.1 and the 3.2 SPS Agreement provide that

> to harmonize sanitary and phytosanitary measures [...] Members shall base their sanitary or phytosanitary measures on international standards, guidelines or recommendations, where they exist; and that (2) Sanitary or phytosanitary measures which conform to international standards, guidelines or recommendations shall be [...] presumed to be consistent with the relevant provisions of this Agreement and of GATT 1994 [18].

It is clear that conformity with Codex standards implies almost invariable conformity with WTO law since the Codex is listed among the international standardization bodies. According to Van den Bossche under the WTO rules, disputes and the adoption of rulings are quasi-automatic. In case of non-compliance, complaining Members are authorized to retaliate (suspension of concessions) [19]. The WTO enforcement system is accordingly one of the most powerful in the international legal arena. For this reason, according to Veggeland and Borgen, after the establishment of the WTO, the nature of the Codex standards has been characterized as quasi-binding [20]. By virtue of the Technical Barriers to Trade Agreement (TBT Agreement), the Codex is also relevant for technical barriers to trade, such as labelling. While this Agreement does not explicitly refer to Codex, it does refer to international standards as a benchmark for legality of regulatory measures that may adversely impact trade, and it is highly likely that Codex standards qualify as international standards.

As of today, virtually no measures disputed at the WTO and diverging from Codex standards have been found to be WTO compatible, which may be indicative of the new role played by the post-SPS Codex standards. The words of a European Commission representative, commenting on the new regime, well capture this transition: "In the past, if we disagreed with Codex Standards [...] we could ignore and take our own legislation. Now we can't" [20] (p. 675). The reason for this is explained by Ching-Fu Lin when he says,

> the Codex, despite being an international organization jointly established by the WHO and FAO, often becomes an extended WTO battlefield of public health versus international trade. Therefore, the Codex has gone far beyond its original purpose as a mere scientific reference point, and become the most controversial "lawmaker" in the international food safety regulation arena. Moreover, as the controversies over beef growth hormones, genetically modified organisms (GMOs), and ractopamine demonstrate, the Codex has become politicized [21] (p. 21).

In the opinion of Elizabeth Smythe, the politicization of the Codex is particularly evident in the case of genetically modified foods, as exporting countries and food importing countries play a key role in the development of Codex standards [22] (p. 99-102). There is another problem that contributes to the Codex politicization. Since the Codex is a highly technical, it needs a specialized scientific body in charge of collecting and processing, specific knowledge that needs to be conducted in a large number of tests, researches, and analysis. An increasing problem is the lack of appropriate funding and technical capacity to assess various health risks. According to Huller and Maier,

> the FAO/WHO experts' committees perform these tasks by reviewing the evidence contained in existing studies, mainly conducted by the food manufacturers' scientists, rather than by conducting first-hand empirical studies of their own. The advice to the Codex which results from risk assessment frequently comes in the form of specific maximum residue levels (MRLs)— i.e., values for the amount of a given substance that can safely be consumed with the food in which it is contained. Alternatively, the conclusion that such values need not be specified because the substance in question is safe (if used according to standard agricultural or industrial practice) can also be a result of the risk assessment [23] (p. 284-285).

In a crucial evaluation of the Codex's work in 2002, a major problem was highlighted:

> the work of expert groups is the least open and transparent part of the international policy-making process, reflecting certain old-fashioned attitudes about how science should be applied to policy. Most expert body meetings are closed to the public. Reports by the expert bodies that support Codex committees are often not available for months or years while recommendations (concerning standards) might be issued promptly, the basis for those positions might not be public for a long time. This lack of transparency has led to public distrust of certain conclusions by expert bodies on controversial topics [24] (p. 7).

Thus it is not always clear what is actually independent knowledge, i.e. knowledge without a hidden agenda.

Corporate employees try to improve their authority and legitimacy over controversial issues regarding food safety or other health issues through the creation of what Buse and Lee called "Institutionalised, non-profit, industry-established and -funded, scientific networks with an interest in health issues" [25]. This is evident in the International Life Sciences Institute (ILSI), a non-profit international organization, claiming that their mission is

> to provide science that improves human health and well-being and safeguards the environment. All ILSI scientific activities have a primary public purpose and benefit. Scientists from geographically diverse regions of the world can best address complex science and health issues by sharing their unique skills, insights, and perspectives. ILSI believes scientists from industry, government, and academia and other sectors of society can *and should* work together to identify and address topics of common interest [26].

However, according to Susan Sell, this organization was founded by various companies in the food industry, including Coca-Cola and Pepsi Cola, and has had a long-standing relationship with the tobacco industry that only came to light after protracted legal proceedings [27]. ILSI has strong links to the FAO and is very active in lobbying the

Codex, particularly in the food labelling committee. According to its own acknowledgment, the ILSI Biotechnological Food Board aims

> to support the development and harmonisation of science based regulations around the world for biotechnology-derived food products and to disseminate science-based information regarding the safety assessment of these products to governments, industry, academia, and other interested groups around the world [28].

ILSI strongly opposes the mandatory labelling of genetically modified food, which is unsurprising especially if we see that the organization's Board of Members consists of one representative from each member company, most of which are global firms with a significant footprint in agrifood, food additive, and food chemical markets like Cargill, Monsanto, Mars, Coca-Cola, Nestle, PepsiCo, and Unilever [29]. Discussing ILSI's mission in the face of increasing food regulation in the late 1980s, Alex Malaspina from Coca-Cola declared that "ILSI is prepared to meet new challenges by continuing to generate scientific data to resolve pending issues, providing relevant scientific data to state agencies, and working to harmonize food regulations and facilitate international trade" [30]. ILSI enjoys close relationships with regulatory bodies including the Joint WHO/FAO Expert Committee on Food Additives [31] and the European Food Safety Authority [32], which rely on the substantial resources controlled by the ILSI. The organization not only influences Codex standards through direct cooperation with national and international regulators but also indirectly through the networks it develops.

Despite the accusations made by critics of secrecy and non-transparency, the Codex is, contrasting with other international organizations such as the WTO, more open to representing the various interests of all bodies involved in food production, from producers, processors to consumer rights associations. This relative openness has strengthened corporate influence in the food safety decision-making process of the Codex. According to the Codex Principles Concerning the Participation of International Non-governmental Organizations (INGOs) elaborated in 1999, any organization with an international structure and scope of activity (in three or more countries) can be accorded status and can participate as observer, receive all documents, and speak at the discretion of the chair (normally after interested country delegations have occurred). In addition, many national delegations include representatives of various NGOs or industry and corporate associations in their delegations for full commission and committee meetings. As many critics point out, large corporate agribusiness have a vested interest in uniform standards and global market access, reflected in their participation in the Codex work, and they have identified Codex as a critical target to influence, either through lobbying the body itself or by having their interests represented by government delegations. The official governmental delegations sent to the Codex meetings include many industry representatives – 40% of the delegates in the case of the USA, for instance. In a report to the president of the European Commission in 1999, one observer of the Codex committee meetings discussing standards on the cattle growth hormone rbST stated that "the ways and means by which rbST was re-evaluated last year strengthens our belief that powerful politico-economic interests and multinational companies exercise improper influence and control in the work of Codex Alimentarius and its scientific committees whose supposed primary task is to protect human health" [33] (p. 212).

A report by National Food Alliance indicates that of the total of 2,578 participants on the various Codex committees, over 660 represented industry [34] (p. 32). The commission in 1993 included 105 national delegations, while 140 corporations were also represented. This trend has continued. The meetings of the different Codex Committees in 2000 and 2002 included about 150 of these NGOs, of these about 70 percent represented industry [35] (p. 159-174). By 2018, the number of NGOs had reached 154 [36]. The number of observers has, in fact, increased more rapidly than state membership [23] (pp. 277).

III. Codex Committee on food labelling

The Codex Committee on Food Labeling (CCFL) is the key body in the process of adopting international food labelling standards, with Canada as its host country. According to the long-time chairman Anna MacKenzie, the CCFL

> examines international food labeling issues; drafts labeling provisions that are applicable to all foods; and endorses labeling provisions prepared by Codex Committees charged

> with drafting standards, codes of practice, and guidelines. The high controversy generated by many issues before the CCFL demonstrates the importance given by member countries to the development of international labeling standards [37] (p. 204).

The Codex agreed on the need to address the issues of biotechnology and GM foods in 1991, and the CCFL agreed that work on labelling aspects of biotechnology should begin. It should be noted that this happened before the commencement of commercial sowing of genetically modified crops in 1996. In April 1993, the United States, as the leading country in the industry, was asked to prepare a discussion paper which was initially discussed in the October 1994 session. Debate centred around whether labelling should be required only when there were health and safety concerns, and whether labelling should be required if the foods in question did not differ substantively from traditional equivalents. However, for many country delegates, including those of the EU, the issues were premature since they were still reviewing and developing domestic regulations. Consumer groups – in this case, Consumers International (CI) – favoured a system of comprehensive labelling based on the consumers' "right to know." Others also argued in favour of labelling that indicated how food was produced in order to permit consumers to make choices based on values other than just those of health and safety [22] (pp. 103). Others have used this argument to mandatorily label genetically modified food, enabling consumers to make informed decisions, not only on the basis of health and safety issues but inclusive of other values such as ethical or religious ones. Due to the lack of a clear consensus among member states, the issue of labelling genetically modified foods has returned to the CCFL for redrafting. The issue was a constant source of conflict for nearly 20 years between the producers of genetically modified crops, led by the US, Canada and Argentina and the opposite bloc led by the EU and developing countries which were seeking mandatory labelling. The final agreement was reached on the 39th session of the CCFL, held in Quebec City from 9 to 13 May 2011, and was attended by delegates from 60 countries [38]. After almost twenty years of discussions on genetically modified food labelling, the CCFL finalized a document composed of various food labelling texts applied to genetically modified foods. Although the discussion of the document itself was also a disputed and long-lasting one, the CCFL members eventually agreed on the text. It is interesting to note that in the 39th session, the title of the proposed document was changed to *Proposed Draft Compilation of Codex Texts Relevant to Labelling of Foods Derived from Modern Biotechnology* [38]. The Committee agreed with the proposal of the facilitated session in that the purpose should read as follows: "The purpose of this document is only to recall and assemble in a single document some important elements of guidance from Codex texts, which are relevant to labelling of foods derived from modern biotechnology" [38]. The document listed the following relevant circumstances:

> Different approaches regarding labelling of foods derived from modern biotechnology are used. Any approach implemented by Codex members should be consistent with already adopted Codex provisions. This document is not intended to suggest or imply that foods derived from modern biotechnology are necessarily different from other foods simply due to their method of production" [38]. At its meeting in Geneva, from the 4th to 9th July 2011, the Codex adopted the proposed text of the CCFL. In the days and weeks following this move, a number of press releases and articles were written describing the importance of the labeling guidance. One article described the outcome as "Codex has capitulated on the GM labeling issue after a battle spanning approximately 20 years, stating that it will allow countries to label GMOs and the WTO will not legally challenge them for it [39].

Consumer groups and the biotech industry are issuing different interpretations of a new Codex agreement on biotech labelling. Consumers International, in a press release, highlights that,

> the new Codex agreement means that any country wishing to adopt GM food labeling will no longer face the threat of a legal challenge from the World Trade Organization (WTO). This is because national measures based on Codex guidance or standards cannot be challenged as a barrier to trade [40].

However, a July 5th, 2011 article in *The Hagstrom Report* noted a disagreement among organizations as to the meaning of the new guidance. The report quoted the spokesperson of a Biotechnology Industry Organization stating that

> the agreement is totally consistent with the U.S. position, which we support since it says no new guidelines are needed, because the guidelines for other foods apply to biotech foods as well. The agreement is just a compilation of existing texts with a consideration statement that says foods derived from biotech are no different from other foods based on method of productions. It also encourages companies to be consistent with Codex guidelines [41].

IV. Conclusion

That these guidelines on labelling genetically modified foods are a compromise made to satisfy both sides, opponents, and supporters of genetically modified food. It should be noted that the GM labelling guidance document was adopted after 20 years from the moment when the Codex was inquired about how to handle the food produced by modern biotechnology. Although GM labelling guidance did not recommend mandatory labelling of genetically modified food, there was no ban on labelling. Nevertheless, it should be noted that Codex's labelling guidelines are a step forward in recognizing the rights of the countries which require mandatory labelling, and these guidelines nonetheless respected the right of consumers to understand what decisions they were making in terms of eating. It should be further highlighted that the application of these guidelines is left to the decision of the member states, which have an autonomous right to introduce mandatory labelling without fear of lawsuits. As for the Codex, we can conclude that it is not a sinister mysterious society, but the international body that aims to enable the development and promotion of standards that guarantee food safety. A special risk for the Codex lies in the overwhelming power and lobbying of corporations, which, although formally have no right to vote, act as observers and have increasing influence on the decision-making process of the Codex, as evidenced by the case of adopting guidelines for the labelling of genetically modified foods. Corporations used the transparency and openness of the Codex to covertly promote its only goal: profit. From a bioethical point of view, the work of the Codex on issues of labelling genetically modified food is especially interesting since the issue of genetically modified food is extremely controversial in the public since the very beginnings of sowing genetically modified crops.

References

[1] World Economic Forum, "Global Risk 2008: A Global Risk Network Report," http://www.weforum.org/pdf/globalrisk/report2008.pdf.; Y. Motarjemi et al., "Future Challenges in Global Harmonization of Food Safety Legislation", *Food Control*, 1, pp. 339-346, 2001.; F. K. Käferstein et al., "Foodborne Disease Control: A Transnational Challenge," *Emerging Infectious Diseases*, vol. 3, no. 4, pp. 503-510, Oct-Dec 1997.

[2] WHO Food Safety Programme, *Food safety: an essential public health issue for the new millennium*, Geneva : World Health Organization, 1999.

[3] WHO, "WHO's first ever global estimates of foodborne diseases find children under 5 account for almost one third of deaths," https://www.who.int/news-room/detail/03-12-2015-who-s-first-ever-global-estimates-of-foodborne-diseases-find-children-under-5-account-for-almost-one-third-of-deaths. Accesed: May 10, 2018.

[4] CDC, "CDC Estimates of Foodborne Illness in the United States," http://cdc.gov/foodborneburden/2011-foodborne-estimates.html. Accessed: May 10, 2018.

[5] WTO, "The WTO Agreement on the Application of Sanitary and Phytosanitary Measures (SPS Agreement)" https://www.wto.org/english/tratop_E/sps_e/spsagr_e.htm#top. Accessed: May 10, 2018.

[6] FAO, "ENVIRONMENTAL AND SOCIAL STANDARDS, CERTIFICATION AND LABELLING FOR CASH CROPS," http://www.fao.org/3/a-y5136e.pdf. Accessed: May 10, 2018.

[7] FAO, "Understanding The Codex Alimentarius," http://www.fao.org/3/a-i5667e.pdf. Accessed: May 10, 2018.

[8] B. M.J. van der Meulen, „Development of Food Legislation Around the World," *Ensuring Global Food Safety: Exploring Global Harmonization*, C. Boisrobert, A. Stjepanovic, S. Oh, and H. Lelieveld, eds., London: Elsevier, pp. 5-69., 2010.

[9] C. Guajardo, *Código alimentario argentino: su valoración juridica*, Mendoza: Ediciones Jurídicas Cuyo, 1998.

[10] M. D. Masson-Matthee, *The Codex Alimentarius Commission and Its Standards*. The Hague: T.M.C. Asser Instituut, 2007.
[11] FAO, "Members," http://www.fao.org/fao-who-codexalimentarius/about-codex/members/en/, Accessed: May 10, 2018.
[12] FAO, "Codex Observers," http://www.fao.org/fao-who-codexalimentarius/about-codex/observers/en/. Accessed: May 10, 2018.
[13] FAO, "List of Codex Committees: Active," http://www.fao.org/fao-who-codexalimentarius/committees/en/. Accessed: May 10, 2018.
[14] Codex Alimentarius Commission, "Procedural Manual", Rome: WORLD HEALTH ORGANIZATION FOOD AND AGRICULTURE ORGANIZATION OF THE UNITED NATIONS, 2013.
[15] FAO, "Statutes of the Codex Alimentarius Commission (adopted in 1961 by FAO and in 1963 by WHO)," http://www.fao.org/DOCREP/005/Y2200E/Y2200E00.HTM. Accessed: May 10, 2018.
[16] Codex Alimentarius Commission, "FAQ's-Purpose of Codex Alimentarius," http://www.codexalimentarius.net/web/faq_gen.jsp. Accessed: May 10, 2018.
[17] FAO, "General Principles of the Codex Alimentarius," http://www.fao.org/DOCREP/005/Y2200E/y2200e05.htm. Accessed: May 10, 2018.
[18] WTO, "The WTO Agreement on the Application of Sanitary and Phytosanitary Measures (SPS Agreement)", https://www.wto.org/english/tratop_e/sps_e/spsagr_e.htm. Accessed: May 10, 2018.
[19] P. van den Bossche, W. Zdouc, *The Law and Policy of the World Trade Organization*. Cambridge: Cambridge University Press, 2013.
[20] F. Veggeland, S. O. Borgen, "Negotiating International Food Standards: The World Trade Organisation's Impact on the Codex Alimentarius Commission," *Governance*, vol. 18, no. 4, pp. 675-708., 2005.
[21] C. F. Lin, "Scientification of Politics or Politicization of Science: Reassessing the Limits of International Food Safety Lawmaking," *The Columbia Science & Technology Law Review*, Vol. 15, pp. 1-40., 2013.
[22] E. Smythe, "In Whose Interests? Transparency and Accountability in the Global Governance of Food: Agribusiness, the Codex Alimentarius, and the World Trade Organization," *Corporate Power in Global Agrifood Governance*, J. Clapp and D. Fuchs eds., Cambridge: MIT Press, pp. 93-123., 2009.
[23] T. Huller, M. L. Maier, "Fixing the Codex? Global Food Safety Governance under Review," *Constitutionalism, Multilevel Trade Governance and Social Regulation*, C. Joerges, E. U. Petersmann eds., Portland: Hart Publishing, pp. 267-299., 2006.

[24] S. Suppan, "Codex Standards and Consumer Rights," *Consumer Policy Review*, vol. 16, no. 1, pp. 5-13, 2006.
[25] K. Buse, K. Lee, "Business and Global Health Governance," http://www.odi.org.uk/sites/odi.org.uk/files/odi-assets/publications-opinion-files/1997.pdf. Accessed: May 10, 2018.
[26] International Life Sciences Institute (ILSI), "Mission & Operating Principles," http://ilsi.org/about/mission/. Accessed: May 10, 2018.
[27] S. Sell, "Lobbying Strategies of Multinational Corporations in Biotechnology," http://citation.allacademic.com/meta/p_mla_apa_research_citation/1/8/0/8/0/pages180805/p180805-1.php. Accessed: May 10, 2018.
[28] International Life Sciences Institute (ILSI), "International Food Biotechnology Committee," http://web.archive.org/web/20080106130251/http://www.ilsi.org/About ILSI/IFBIC/. Accessed: May 10, 2018.
[29] Board of Directors, "INT'L LIFE SCI. INST.," http://www.ilsi.org/Europe/Pages/Board-of-Directors.aspx (Accessed: May 10, 2018); Leadership and Supporting Companies, "INT'L LIFE SCI. INST.," http://www.ilsi.org/Pages/Leadership.aspx. Accessed: May 10, 2018.
[30] Minutes of the Joint Meeting of the ILSI and ILSI-NF Boards of Members 1 (Jan. 21, 1989), available at http://legacy.library.ucsf.edu/tid/pyu56a00/pdf. Accessed: May 10, 2018.
[31] R. Verkerk, "The International Food Code of Codex Alimentarius, ETHICAL CONSUMER," http://www.ethicalconsumer.org/commentanalysis/factsvgreenwash/code xalimentarius.aspx. Accessed: May 10, 2018.
[32] EFSA, "EUR. Food Safety Auth., EFSA Meeting Summary Report: EFSA/WHO International Conference With Support Of ILSI Europe On Risk Assessment Of Compounds That Are Both Genotoxic And Cytotoxic," http://www.efsa.europa.eu/en/ home/publication/ efsawho2006.pdf. Accessed: May 10, 2018.
[33] R. Peet, *Unholy Trinity: The IMF, World Bank and WTO*. London: Zed Books, 2009.
[34] N. Avery, M. Drake, T. Lang, *Cracking the Codex: An Analysis of Who Sets World Food Standards*. London: National Food Alliance, 1993.
[35] L. Sklair, "The Transnational Capitalist Class and Global Politics: Deconstructing the Corporate-State Connection," *International Political Science Review*, vol. 23, no. 2, pp. 159–174., 2002.
[36] FAO, "Codex Observers," http://www.fao.org/fao-who-codexalimentarius/about-codex/observers/en/. Accessed: May 10, 2018.

[37] A. MacKenzie, "The Process of Developing Labeling Standards for GM Foods in the Codex Alimentarius," *AgBioForum*, vol. 3, no. 4, pp. 203-208, 2000.

[38] Joint FAO/WHO Food Standards Programme, Codex Alimentarius Commission, "Report of the thirty ninth session of the Codex Commission on Food Labelling," http://www.fao.org/fao-who-codexalimentarius/shproxy/en/?lnk=1&url=https%253A%252F%252Fworkspace.fao.org%252Fsites%252Fcodex%252FMeetings%252FCX-714-39%252FREP11_FLe.pdf. Accessed: May 10, 2018.

[39] B. Peterson, "Codex Commission – Voluntary GMO Labeling Okay with WTO?," http://farmwars.info/?p=6408. Accessed: May 10, 2018.

[40] Consumers Union, "Consumer Rights Victory as US Ends Opposition to GM Labelling Guidelines," http://consumersunion.org/news/us-ends-opposition-to-gm-labeling-guidelines/. Accessed: May 10, 2018.

[41] The Hagstrom Report, "Biotech labelling Interpretations differ," http://www.hagstromreport.com/news_files/070511_biotech.html. Accessed: May 10, 2018.

PART IV

Ethical and Legal Challenges in Medicine and Research

UDBHR:
An Interpretation in the Indian Medico-legal and Bioethical Context

Anamika Krishnan
Mahatma Gandhi University, India

Abstract
The UDBHR is the most advanced set of guidelines ever set forth in global bioethics. Despite its ambiguity and issues in the drafting process, it provides guidance to developing and developed countries to uphold bioethical principles. India is a country with rich potential in health care, which is unfortunately distorted by bioethical issues like non-protection of surrogate mothers, pharmaceutical scams, complex questions of autonomy and consent etc. The study is made to critically appraise the UDBHR, and discover the scope and prospects of its applicability in Indian medico-legal and bioethical scenario.

Keywords
UDBHR; bioethics; Indian medico-legal scenario; autonomy and consent; pharmaceutical and reproductive ethics; de-stigmatisation; amendments of medical laws

I. Introduction

> "Ethics and equity and principles of
> justice do not change with the calendar."
> *D. H. Lawrence*

Of all the areas in applied and professional ethics, bioethics is the most diverse and it contains debates that are more often philosophically sophisticated. Many of the particular concerns of contemporary bioethics- such as abortion, prenatal diagnosis, cloning, informed consent and voluntary euthanasia commonly raise deep issues of perennial importance, about when life begins, the value of parenthood, the notion of respect for others and what it means to be human. The Universal Declaration of Bioethics and Human Rights (UDBHR) came into existence as an instrument to guide the legal frameworks, health care providers and professionals to attempt to strike a balance between public health and human rights. The fact that law and morality rests on two different levels pose both as a question and a roadblock towards integrating health, human rights and bioethics under a common paradigm. The search for common rules for different communities in the field of bio-law is arduous. The governments of various countries have endeavoured to bring the bioethical guidelines into the practical realm of law, seeking exhaustive study of international documents like UDBHR and ongoing and emerging situations in the medical arena. UDBHR sets an example on how to embrace co-operation, responsibility and self-reflection and the principles enshrined in it seems to serve as a possible solution for most of the medico-legal and ethical problems that exist. But, there is absence of solid evidence to prove that its ideals are universal in nature. UDBHR also generated many controversies while it was being drafted. There is also widespread opinion that it fails to identify the dynamic nature of socio-economic inequalities, environmental resources and other factors that affect health and bioethics.

Among the poor and developing nations, India occupies an enviable position of possessing a health services system which has an extensive infrastructure, an army of trained personnel and well-equipped institutions for research and education. An overview of the system, however, reveals that despite its professed commitment to serve all- especially the poor- it is riddled with unevenness [1]. There is sharp inequality in terms of resource distribution, participation and proper practical knowledge in bioethics. Since India has no agency till date to track medico-legal cases and bioethical concerns, most of the events go unreported. India needs to take a leap to create a comprehensive legal framework to combat bioethical issues related to pharmaceutical scams, vagueness of informed consent, organ rackets, and non-protection of surrogate mothers and impersonal approach of doctors towards patients. Thus, the researcher modestly attempts to critically analyse the UDBHR, assess the reasons for its failure in implementation, and identify ways to inculcate, most possibly through public policy, the universally acceptable principles in the Indian legal context in order to ameliorate factors that obstruct law and the society from embracing bioethical principles.

II. Literature Survey

A. Bioethics

Bioethics can be defined as the humanistic knowledge of scientific methodology to handle living organisms in an acceptable survival or a sustainable society within a healthy ecosystem or as a branch of ethics, philosophy and social commentary that discusses the life sciences and their potential impact on our society [2]. A widely used ethical framework for health care consists of four ethical principles: the principles of autonomy, privacy, beneficence and justice [3].

B. Bioethical issues

A significant factor in the emergence of bioethics as a discipline was the critique by many philosophers of paternalistic practices in medicine. This led to a new emphasis on the importance of patient autonomy in medical decision making [4]. Some of the earliest writing in bioethics focused on the ethics of abortion and questions of foetal and moral status of human embryos. Peter has famously argued that mere membership of the human species does not itself confer any special moral status [5]. But, Harris has surveyed a range of ethical objections to human cloning, and argues that the instinctive aversion which many feel towards reproductive cloning lacks any solid ethical foundation. He also argued that current prohibitions on human cloning restrict our interest in procreative autonomy without sufficient reason, and so may themselves violate human dignity [6]. Another topic is reproductive ethics. A recurrent concern in the literature of commercial surrogacy is that such arrangements exploit the surrogate mother. Wilkinson argues that there is good reason to believe a commercial surrogate mother is exploited by the commissioning couple, in the sense that they take unfair advantage of her [7]. Ethical issues about the end of life have been prominent. Because of developments in medical technology and disease control, patients can be kept alive longer and this has prompted many health professionals and bioethicists to question whether this is always appropriate. Helga Kuhse criticises the claim, on which advocates of the sanctity of life doctrine heavily rely, that there is an asymmetry between killing a patient and letting them die because only the former involves the intentional cause of death [8]. Some bioethicists saw the prevalence of unjustifiable medical paternalism as symptomatic of the insularity of professional role-based ethical standards themselves, and so argued that doctors should reject a professionally generated ethic altogether and rely for guidance solely on broad-based ethical theories such as Kantianism or Utilitarianism [9]. Early discussions of research ethics focused on the importance of informed consent and international statements of the rights of researched subjects such as the Nuremberg Code and the Declaration of Helsinki. Even after clinical trials have shown drugs to be safe and effective, there are further ethical concerns that have been raised about the pharmaceutical industry [10].

C. Bioethics and Public Policy

The most practical side of bioethics is perhaps, its engagement with public policy. The development of democratic public policy on these issues of life and death, reproduction, genetics and biotechnology highlights the tensions between populist and deliberative conceptions of democracy. Populist approaches hold that the will of the majority serves as the appropriate reference point for law-making, because uneducated people are as capable of using their reason to judge what is right and wrong as are educated people [11].

D. The future of bioethics

Bioethics examines and challenges existing and likely future practices in health care, reproduction, genetics, biotechnology and biomedical research. At a broad level, there is a distinction emerging between bioethical issues facing current health professionals and bioethical issues involving future applications of various biotechnologies, some of which remain fairly speculative. There will continue to be ethical debates about more familiar issues in clinical practice along with discussions about the pros and cons of moves towards greater transparency and issues involving stem cell research, embryo selection decisions in pre-implantation genetic diagnosis and so on. Whichever directions eventuate, bioethics has now become an established field of enquiry and clearly has an exciting future ahead [12].

III. Reading the UDBHR

A. Background of Global Bioethics

The term "bioethics" was coined in 1926 by Fritz Jahr in an article about a "bioethical imperative" regarding the use of animals and plants in scientific research. In 1970, the American Biochemist Van Rensselaer Potter used the term to describe the relation between the biosphere and a growing human population. The scientific advancements of the last few decades have opened up doors in many different fields like reproduction, organ transplants, sexuality, surrogacy etc. This has had a significant social consequence: facts have turned into choices. Human beings are now left with the option of choosing life and death. The scientists are even waging a war against human aging. Millions of venture capital dollars are being dumped into longevity research and there are ongoing debates about the 'Right to life v. Right to die'. In such a situation, we can essentially apprehend the intrusions into right to privacy and dignity of an individual. Law has not been able to keep up the pace with science. The international conscience demanded universal ethical rules after the disastrous events of human experimentation during World War II without consent, resulting in death, permanent impairment, trauma and disfigurement of the subjects. Subsequently, Nuremberg Code, Helsinki Declaration and UDHR came into existence to ensure that the worst past doesn't repeat. In *Newcrest Mining (WA) Ltd v. The Commonwealth* [(1997) 190 CLR 513], it was held that; "The central ideas of the UDHR were incontestably influential in the subsequent development of the United Nations treaty law; the emergence of customary international law; and the incorporation of variations on the UDHR language in national constitutions, legislations and common law." UN continued to strive towards upholding international humanitarian law on many fronts and the chief of it has been the United Nations Economic, Scientific and Cultural Organisation (UNESCO). It is the think tank of the UN. Initially, the discussions on any international standards related to bioethics primarily focused on health care issues. But, the advent of biotechnology, had presented the UNESCO with the opportunity to delve deeper into the Bioethical concerns. To fulfil the perceived responsibilities of UNESCO in the field of bioethics, an International Bioethics Committee (IBC) was formed in 1993 and it was asked to examine the bioethical questions of international concern. The Universal Declaration on the Human Genome and Human Rights was adopted unanimously at UNESCO's 29th General Conference on 11 November, 1997. The Intergovernmental Bioethics Committee (IGBC) was created in 1998, under Article 11 of the Statutes of the IBC. The International Declaration on Human Genetic Data was adopted in 2003. At the 32nd session, the General Conference considered that it was opportune and desirable to set universal standards in the field of bioethics and invited the Director-General to continue preparatory work on a declaration on universal norms on bioethics, by holding consultation with member states, the other international organisations concerned and relevant national bodies and to submit a draft declaration at its 33rd session. The drafting of the text took place between April 2004 and January 2005. On October 19, 2005, at the UNESCO General Conference, the member states by resolution adopted the Universal Declaration on Bioethics and Human Rights (UDBHR) unanimously.

B. What does the UDBHR promise?

UNESCO's UDBHR has laudable goals in its ambit. It is the first document to set global standards in biological and medical ethics. UDBHR respects human dignity, human rights and fundamental freedoms (Article 3). The word "dignity" comes to us, via the Latin *dignus* and *dignitas*, from Greek and Roman antiquity, in whose literature it means something like "worthiness for honour esteem." Human dignity forms the basis of human rights in the 1948 UDHR as it is considered central to human identity. Thus, human dignity is the recognition that human beings possess a special value intrinsic to their humanity and as such are worthy of respect simply because they are human beings. While the prior documents only accorded respect to autonomous persons, the UDBHR considered and inculcated the plight of all those who could not exercise autonomy. After covering the existing persons, UDBHR moved forward and embraced the respect and dignity for future persons, in the era of genetics and assisted reproduction. Integrating the human rights framework into bioethics is meaningful because most basic human prerogatives are relevant to biomedical activities, for instance the rights to life, physical integrity, privacy and access to basic health care [13]. The UDBHR has tried to cover the basic fundamental freedoms and rights that are desirable to all human beings- all persons must be treated with due respect (Article 3, 5, 8 and 9), be treated justly and equitably (Article 10), to enjoy the highest attainable standard of health (Article 14) and not to be subjected to discrimination and stigmatisation (Article 11). UDBHR also includes considerations about the environment, biosphere and other living beings (Article 17) because of the fact that it is no longer possible to advance science and technology without reflecting on the impact of our actions on our environment and other living beings. Thus, by focusing on common principles and shared values, the UDBHR has established global standards that could be used by both developing and developed countries. UNESCO understands that not all of its member states have adequate infrastructure to uphold bioethical values and therefore, sets benchmarks which are universal in nature to address such gaps.

C. The polarisation of promises and results: a critical view

There has been several criticism and controversies regarding the draft process and sharp conflicts in terms of opinions about the contents have made the world wonder if the UDBHR really would achieve its primary intended objectives. The researcher does not go into the problems of process as I understand that the declaration was unanimously adopted in 2005 at the general conference. The focus is on rooting out the differences in promises and results. The main obstruction is the usage of abstract language throughout the declaration. In order to set forth universal guidelines for the member states whose resources vary, simple language should have been used as in UDHR. The UDBHR lacks conceptual clarity, for what is implicit in abstract language? Abstract terms and clauses will be spoilt over by numerous interpretations and subsequent conflicts will question the synchronicity of principles and objectives. Bioethicists and scholars have rightly pointed out some examples- Can the statement on the freedom of scientific research and the need to respect human dignity be read apart from recent debates on stem cell research?, Can the injunctions on human solidarity in Article 13 or the statement on protecting the environment or Article 29 on the assessment and management of risk be read apart from the vast inequities of wealth and power? These doubts will certainly shed light on the conceptual errors wherein the UNESCO and the IBC specifically failed to consider the social, economic and cultural disparities among its member States. How can we formulate universal principles by ignoring the context in which they are going to be applied? How can one call such principles universal if it lacks universal applicability? So, there has been a serious mistake of presuming universality disregarding the structural inequalities. Unfortunately, the Bioethics Declaration simply relies on concepts of universality, rights, persons, and equality that derive from the logic of fraternity and Enlightenment philosophies of man's common sense. As feminist philosophers have demonstrated, these concepts are not innocent and inevitably reflect a certain history of power [14]. Rawlinson and Donchin rightly opines that the formidable task of articulating 'universal principles' requires a critical recognition of the implication of Enlightenment concepts of persons, rights and equality in the hegemony of particular racial, cultural, and sexual identities [15]. The interests of the individual should have been placed higher to societal interests. The reality of human rights and bioethics being placed on different levels cannot be ignored either. Richard Ashcroft, Head of the Medical Ethics Unit at Imperial College, comments, "Human rights statements are positive declarations of what is obligatory, whereas bioethics documents are frequently exploratory or speculative in nature, arguing about what may or may not be permissible or necessary." To be frank, the expectations of the international community about bioethical guidelines were not met. Already, being a declaration, having no legal binding in terms of legal status, UDBHR will not prevent any ethical lapses from being committed. So, it would have been better if brisk, yet powerful language was used to write down the declaration. The guidelines should have been such that it will influence the governments to incorporate them in their policies, by making them realise the dire need for bioethics. But, it also does not mean that UDBHR has no scope of applicability through public policy. The scope of implementing the guidelines in the Indian context will be discussed in coming sections. The researcher very well accepts the truth that declarations cannot cover all possibilities. UDBHR has however succeeded to call attention of the countries to bioethics. It may lack rigour and accuracy, but the principles pave way for the States to formulate public policies and programmes related to bioethics. Hence, the Bioethics Declaration remains an important advance for the international community.

IV. Bird's eye view of the Indian medico-legal and bioethical scenario

A. The Constitutional framework of medical law and bioethics

In a developing society as in India, steeped in unbridgeable and ever widening gaps of inequality in status and opportunity, law is catalyst and rubicon to the poor to reach the ladder of social justice. The Constitution of India which was adopted by the Constituent Assembly on 26th November, 1949, came into force on 26th January, 1950. It is the longest and most detailed Constitution in the world. Even though health as a sector does not appear in many places of the Indian Constitution, there are indirect and tacit references to health of people and the role of State. As held in *Paramanand Katara* v. *Union of India* (AIR 1989 SC 2039), "the human life is more valuable and it must be preserved at all costs and members of the medical profession are ethically bound to provide necessary medical care to the patients. There can be no second opinion that preservation of human life is of paramount importance. This is so on account of the fact that once the life is lost, the *status quo ante* cannot be restored as resurrection is beyond the capacity of man". In India, the right to health care and protection has been recognised since early times. India is a founder member of the United Nations and has ratified various International Conventions promising to secure health care rights of individuals in the society. Part III of the Indian Constitution guarantees fundamental rights to persons against the State. The preamble to the Constitution of India, which strives to provide for a welfare state with socialistic patterns of society under Article 21, guarantees the right to life and personal liberty. It states that: "No one shall be deprived of his right to life and personal liberty except according to the procedure established by law."

Because of the vibrant interpretation given by the Apex Court to Article 21, it has become the thrust and throb of the Constitution itself. It is a well accepted fact that the foundation of bioethics is the protection of human dignity. The framers of the Constitution were well aware of the concept of dignity and they incorporated the term in the preamble. According to Kelsen's pure law theory, the Constitution is the basic norm. He was aware that in some legal orders, it can be difficult to draw a line between constitutional and other norms. So, in 1964, he relativised the idea of "Constitution", holding that each relatively higher legal norm is a 'constitution' in relation to any norm created under its authority. The Constitution of India takes care of every section of the society. The liberal approach of the Apex Court in the interpretation of Constitution has resulted in derivation of various rights from Article 21, the most important of which is the right to live with human dignity. In *Kharak Singh* v. *State of Uttar Pradesh* (AIR 1963 SC 1295), the Supreme Court held that, "By the term 'life', is meant more than mere animal existence and the inhibition against its deprivation extend to all those limbs and faculties by which life is enjoyed."

Article 14 ensures that there can be no valid law which arbitrarily denies to persons the right to health. Article 38(1) lays down the basis for human rights. It summarises that in order to secure socio-economic justice to the people and to promote their welfare, the State is under a constitutional duty to take effective steps to provide proper health care and to ensure a life of dignity to them. Article 39 enumerates that the State shall, in particular, direct its policy towards securing, inter alia, that the health and strength of workers, men and women, and the tender age of children are not abused and that citizens are not forced by economic necessity to enter avocations unsuited to their age or strength and that children are given opportunities and facilities to develop in a healthy manner and in conditions of freedom and dignity. Article 43 enjoins upon the State to make endeavour to ensure a decent standard of life and full enjoyment of leisure and social and cultural opportunities. Article 47 places responsibility on the State to raise the standard of living of its people and improvement of public health by ensuring proper health care and health facilities as among its primary duties. Article 48-A further provides that the "State shall endeavour to protect and improve the environment and to safeguard the forests and wildlife of the country."

Thus, it is evident that the right to public health is considered at the top of Indian Constitutional agenda and the State is duty bound to ensure it to all in all circumstances. The Apex court has beautifully explained this aspect of Constitution in *State of HP* v. *Umed Ram Sharma* (AIR 1986 SC 847), wherein it held; "the jurisprudence of personhood or philosophy of the right to life envisaged under Article 21, enlarges its sweep to encompass human personality to its full blossom with invigorated health. Thus, the right to life with human dignity covers within its fold, some of the finer facets of human civilisation which makes life worth living. It includes the quality of life as understood in its richness and fullness by the ambit of the Constitution."

B. Human rights and Bioethics in India

The United Nations has explained that "the denial of human rights and fundamental freedoms not only is an individual and personal tragedy, but also creates conditions of social and political unrest, sowing the seeds of violence and conflict within and between societies and nations." The concept of human rights, to the core embraces certain universe of values significant to the protection of human dignity which apply beyond the national borders and the cultural divides [16]. Locke opined that man is born with a title to perfect freedom and an uncontrolled enjoyment of all the rights and privileges of the Law of Nature and by nature he has the power to preserve his property-life, liberty, and estate, against the injuries and attempts of other men. The United Nations since 1945 has tried to secure the promotion and protection of the Human Rights [17]. The International Bill of Human Rights consists of the Universal Declaration of Human Rights, the International Covenant on Economic, Social and Cultural Rights, and the International Covenant on Civil and Political Rights and its two Optional Protocols.

Article 12-35 of the Indian Constitution ensures various human rights guaranteed by the UDHR in its most sacrosanct manner. Though there is no explicit implication to the declaration in any of the provisions is the Constitution, yet the Supreme Court now follows an approach to evaluate fundamental rights in the light of UDHR. In *Chairman, Railway Board* v. *Chandrima Das* (AIR 2000 SC 988), the Apex Court had made reference to the provisions of the International Bill of Rights and observed that "the applicability of the universal declaration of human rights and the principles thereof may have to be read, if need be, into the domestic jurisprudence." Section 2(d) of the Protection of

Human Rights Act, 1993 has defined the human rights "to mean the rights relating to life, liberty, equality and dignity of the individual guaranteed by the Constitution or embodies in the International Covenants and enforceable by Courts in India." Thus, the judiciary applies all such international instruments to uphold the human values and rights. Now, the prime focus is on strengthening the legal mechanism in consonance with the international standards.

Unlike in the West, where the medical professional emerged largely as a part of an entrepreneurial profession, which was only later brought into a nationalised service, in India, modern medicine was "built up" by both the colonial and the independent Indian State as the vehicle of modernity and welfare. Considering the recent advances in life-sciences, bioethics has many roles to play, namely, to raise the questions about non-ethical practices among health care institutions, to compete with the new bioethical problems and to challenge the public health issues and the deliverance of health care facilities among economically underdeveloped parts of the world [18]. The Indian Council for Medical Research (ICMR) formulates, coordinates and promotes biomedical research in India. In 1980, they formulated the first national ethical guidelines. They also have designed curriculum for implementing bioethics in medical schools. The ethical principles that guide bio-research in India are essentiality, voluntariness, non-exploitation, privacy and confidentiality, precaution and risk minimization, professional competence, accountability and transparency, maximisation of the public interest, totality of responsibility and compliance [19]. A number of laws have been enacted to protect the health interests of the people including the Fatal Accidents Act, 1855, the Indian Medical Degrees Act, 1916, Dangerous Drugs Act, 1930, Drugs and Cosmetics Act, 1940, Pharmacy Council of India Regulations 1952, the Indian Medical Council Rules 1957, the Medical Termination of Pregnancy Act, 1975, the Dentist Code of Ethics and Regulations 1976, Narcotic Drugs and Psychotropic Substances Act and Rules, 1985, The Pre-Natal Diagnostics Techniques Act, 1994, the Transplantation of Human Organs and Tissues Act(THOTA), 1994 etc. Moreover, the ICMR publishes ethical guidelines for the health providers and medical professionals periodically. Further, there are various NGOs working towards unearthing pharmaceutical scams, protesting against unethical practices of medical professionals in clinical medicine and trial and demanding IPRs for tribal and rural people who have protected various medicinal herbs and plants, which are continuously being tried to be stolen and experimented upon, without their permission.

C. Bioethical issues in India that demand urgent attention

It is well known that bioethics evolved in the West as a result of a series of scandals and public outcry over the misuse of authority and control by the medical profession. There are several bioethical concerns in India, and if each of it has to be gone through in depth, it will necessitate writing of separate and elaborate research papers. Considering the constraints, the researcher focuses on prime bioethical issues that demand immediate attention, unless which the public health care system in India may collapse in few years.

- *The forgotten 'Informed Consent' in clinical trials*: In any field affecting man, research is necessary as it provides data to improve the health of human beings and this appreciable objective demands encouragement and motivation at all levels of participation. "Informed consent" is a concept that is required at clinical trials as well as treatment of individuals. Informed consent implies that a person from whom consent is sought should have adequate reasoning ability, a clear understanding of the facts, a good idea of the choices available, an appreciation of the implications and comprehension of the consequences, and he/she should be able to voluntarily choose a particular option. Despite significant economic advances and strides in education, India still remains a feudal society with a great differential in power between doctors and patients. Informed consent is just looked upon as a necessary burden of legal documentation procedure and its ethical side is ignored. Further, human experimentation in India is seen to take place without the "informed" consent of individuals and big pharmaceuticals test their drugs on rural patients who are mostly illiterate and untaught about the basic medicines given to them for even common illness. As reported on 25th April, 2013, the Central government informed the Supreme Court of India that as many as 2644 people, called subjects, died during the clinical trials of 475 new drugs on human beings in the period of 2005-2012 (7 years) and only 17 of the medicines were approved for marketing in

India. Clinical trials of two drugs- Bayer's Rivaroxaban and Novartis's Aliskiren vs. Enalapril- accounted for maximum number of deaths [20]. The former was first used for human trials in 2008 resulting in death of 21 of which it claimed that only five were related to clinical trial but it has till paid compensation to kin of only two. Two years later, the same drug was again put on human trial and this time 125 deaths were reported.

- *Is there patient or subject Autonomy? Laws overriding the decisions in surrogacy and organ transplants*: The age-old paternalistic model in clinical medicine has changed into a contractual relationship between physician and patient. The established philosophical tradition argued that knowledge is always good in itself and that ignorance deprives people of their choice and consequently, their autonomy. Western philosophy and individualistic societies set great store by personal autonomy. They often assign less value to health than to autonomy. On the other hand, many in India tend to value health more than personal autonomy [21]. One disease which captured the attention of the medical professionals all over the world in early 1900s and throughout the first half of the 20th century was solid organ failure. With the advent of advanced medical technology in transplant procedures, an end stage organ failure patient got a ray of hope to lead a healthy and prolonged life by undergoing organ transplant. With the revolutionary changes brought about in social, political and scientific fields of human activity, human organ transplant as a curative medical technique gained public acceptance in India too. While the technology expanded possibilities of organ transplants, the number of organ donors does not increase correspondingly and this gap in demand and supply of transplantable organs results in illegal trade in organs. To make the regulation of organ transplant procedure with all the incidental paraphernalia meaningfully effective, periodic amendments have been made to the THOTA, 1994. We know that we all have a duty to act at all times in the best interests of human kind. This includes respecting the autonomous choices made by persons regarding the treatment of their body after their death. Autonomy is a kind of deliberated self-rule which makes us to act according to our own wishes on the basis of informed choices, thought and decision, freely and independently [22] (p. 158). THOTA 1994 provides that a person can make a wish to donate his organs/tissues or can also register his objection for such donation after his death. As per the Act, even if the deceased had consented to donation of organs during his life-time, the wishes of his near ones are given priority over the deceased's wishes and organ retrieval cannot be done without the consent of the near relatives. In such a situation, autonomy of the deceased is not respected. In India, religious beliefs generally discourage practices such as stockpiling and collecting organs from cadavers [23]. Though all religions permit donation, people still are unwilling to donate organs. Reproductive tourism, an important subset of the medical tourism industry, has become a USD 500 million industry in India [24]. In light of the booming and unregulated nature of the surrogacy business, its ethical aspects need to be evaluated. The Western Liberal feminist perspective of considering surrogacy as an exercise of reproductive choice and of women's economic autonomy do not apply to Indian Surrogates, owing to the socio-economic realities which determine their choice. In India, surrogates mostly come from poor families whose husbands might have died or absconded. Further, the female literacy rate in India is also low. The principle of "Equal pay for equal work" has not yet been entirely established in the country and the economic prospects of surrogacy push them to this business. But, it is pertinent to note that commercial surrogacy is culturally stigmatized by the conservative Indian attitude towards sex and procreation. Surrogate mothers are usually thrown out of their homes and communities. In such a scenario, the existence of voluntary choice or autonomy on the part of these women is contentious. It is debatable whether women are choosing freely to become surrogates, or that their will is socially and economically constructed [25]. A conspicuous paradox manifests itself when one compares reproductive rights and reproductive health facilities available to a common woman in India when compared to a woman who has volunteered to be a surrogate. The World Health Organisation (WHO) gives a working definition of the term sexual and reproductive rights as the "highest attainable standard of sexual health, including access to sexual and reproductive health care services." UNICEF has observed that due to social and economic inequalities and shortages in primary health facilities, India's fight to lower maternal mortality rates is failing. For every maternal death in India, an estimated 30 more women suffer injury, infection and pregnancy-related disabilities [26]. In Jaitun v. Maternity Home, MCD, Jangpura & Ors (W.P. No. 10700/2009), the Delhi High Court emphasized that there has been a complete failure of public health care system and a failure in implementation of government schemes designed to reduce maternal and neonatal mortality by encouraging institutional delivery for poor pregnant women. Medical services are denied to poor women because they are unable to demonstrate their BPL Status for medical services. The Court observed that the Indian government should be facilitating their access to these essential services and therefore, it is stressed that 'no woman, more so a pregnant woman, should be denied the facility of treatment at any stage irrespective of her social and economic background.' The judiciary is sensitive to the plight of the poor and destitute pregnant women, but the implementation at the ground level is lacking as highlighted by the courts themselves.

- *Absence of adherence to Pharmaceutical ethics*: Another alarming bioethical issue is- 'Ethical violations by the Pharmaceutical industry in India'. According to the World Health Organisation (WHO), 35 percent of the fake drugs sold all over the world comes from India and it occupies the counterfeit drug market of nearly Rs. 4,000 crore and 20 percent of drugs sold in India are fake. Subsequent to a research study by the National Institute of Biologicals (NIB) conducted during 2014-2016; it was revealed that more drugs of poor standards were found in government hospitals as compared to the pharmacies of the market and most of such fake drugs are made in extremely polluted atmosphere [27]. The Standards created by the Pharmacy Council of India are regularly violated. It is saddening that the Government hospitals which are the only resort of the Indian poor to access health care fails to provide the common masses with reliable and authentic drugs. In a shocking disclosure, some drugs companies were caught red-handed writing scientific recommendations of their own products and submitting them to the Drug Controller General of India (DCGI) after getting them endorsed by top doctors for a quicker marketing approval. Usually, scientific recommendations are submitted by experts after they have studied a drug's content [28].

V. Significance of the study: the dire need to analyse the situation

Law and medicine join a common pilgrimage towards all pervasive welfare of human life. Because bioethics and law are so strongly connected, even a bare reading of the UDBHR requires basic understanding of law. There have been so many discussions and criticism against the draft process and contents of the UDBHR, but its relevance cannot be overlooked. It has set global benchmarks for all and it is for the countries to adopt the principles into the domestic law and customise it as per their legal and ethical conditions. UDBHR has undoubtedly broadened the scope of bioethics. India is a developing country and it has been a laboratory of various landmark legal and jurisprudential experiments. It has got almost all of the bioethical issues covered under separate laws. But, the recent development of civil society and widespread awareness regarding the medico-legal and bioethical problems has unearthed several scams, corruption, blatant violations of ethics and laws. There is an asymmetry in the existence and implementation of laws. India has sufficient space to incorporate the universal principles of bioethics in its public policy. Thus, this paper gains importance considering the fact that the principles laid down by the UDBHR if correctly revised and modified and applied in the Indian context can create wonderful and enviable results.

VI. Data collection and methodology

The current study is descriptive and analytical in nature. The researcher has made use of both primary and secondary resources. The study is a unique combination of empirical, non-doctrinal mode of research based on field work by way of personal interviews of the bioethicists, legal experts and health care professionals as well as secondary source based analysis of Indian medico-legal context, the existing and emerging bioethical concerns and the interpretation of the UDBHR. To make this work original and authentic, attempts were made to gather views of eminent legal experts regarding the scope of applicability of the

UDBHR through public policy in India. Sincere efforts have been made to collect secondary data in the form of medical and law journals, resource books, research studies, ICMR Guidelines and independent media reports.

VII. Delimitations and limitations of the study

By delimitations and limitations, the intention is to bring out those characteristics that limit the scope, i.e. to define the boundaries of the enquiry as determined at the time of conscious exclusionary and inclusionary decisions that were made throughout the development of its proposal and execution. The first limiting step was the choice of the problem itself; while formulating the proposal basic focus; many aspects were thought of like collecting data from patients of different hospitals and research centres and documenting separate testimonies. Implicit are the problems that could have been chosen, but were screened off from practical bound feasibility point of view.

Lack of prior research studies on the scope of application of UDBHR in India was another limitation. In fact this is the first time a research is being done on interpreting UDBHR in the Indian context. In fact, this limitation has always been the motivational factor to explore the unexplored area. Absence of an agency to track medico-legal and bioethical issues in India posed another roadblock to extract out the real position of medical laws and their implementation. Smaller sample size is a usually a limitation, but not in this kind of qualitative research.

VIII. Findings: scope of applicability of UDBHR in Indian context

A. Is the UDBHR applicable to India?

As far as India is concerned, there is no contradiction between the individual and the social good and natural rights and development. India recognised that there are certain unalienable rights that are inherent to the concept of the dignity of man. India firmly believes that a country's overall performance and its resolve to translate into reality the enjoyment of rights by its people is of paramount importance [29] (p. 34). The Bioethics Declaration provides legal and political guidelines which proposes to have universal applicability. The plural perspectives and multidisciplinary approach of the integrative bioethics construct a perfect methodological framework for a theoretical justification and an in-depth explication of the most important and the subtlest recommendations of the declaration. Although, the UDBHR technically has no legal authority, it is not unusual for such statements to become incorporated in national legislation and court rulings. There are 3 broad reasons why UDBHR becomes applicable to the Indian legal context. Firstly, India is an active member of the UNESCO and any declarations, treaties or resolutions passed by it shall become binding on India. Further, India also has an international responsibility to follow bioethical guidelines, considering its scientific advancements moving ahead in the quickness of light years. Secondly, India is a founder member of the United Nations, and has ratified various International conventions promising to secure health care rights of individuals. The UN adopts large number of declarations, resolutions and other statements that are not treaties; they do not have parties to them, they are not ratified and their legal effect is less certain. However, as they are products of the UN system, they are considered to be highly influential and there is an argument that compliance is a necessary consequence of membership of the UN. In this context, Article 51 of the Indian Constitution needs to be looked at. Article 51 states that the "State shall endeavour (a) to promote international peace and security, (b) maintain just and honourable relations between nations, (c) foster respect for the international law and treaty obligations in the dealings of organised people with one another and (d) encourage settlement of international disputes by arbitration." Thirdly, Section 2(d) of the Protection of Human Rights Act, 1993 recognises that the human rights include the rights relating to life, liberty, equality and dignity of the individual guaranteed by the Constitution or embodies in the International Covenants and enforceable by courts in India. Thus, India has no bar from adopting the guidelines put forth by the UDBHR and in fact, must take the responsibility to implement it in the domestic law.

B. Translating the UDBHR into public policy: the tug of war between law and morality

Ethics has claimed a right to criticise legal arrangements and recommend changes to them. Many debates about law, when they are not merely debates about how legal precedent mechanically applies in a particular situation, are also ethical debates. The work of ethicists is in many ways both oriented towards the law and influenced by law. Conversely, many legal concepts like the right to privacy, dominate moral discussions. The perceptions of ethics (moral philosophy) vary from a society to another. The relationship between bioethics and public policy has become a rather broad subject that asks a rather simple question; namely, which moral imperatives that arise out of the study and consideration of bioethical issues should be reflected in public policies that govern us all. Being social animals, it is unavoidable for us to establish rules of conduct that are thought to serve the collective interests of individuals. These rules must, however, be constantly reviewed and perhaps revised and updated. As Jhering rightly opines, "The relation of law to morals is the cape horn of jurisprudence. The juristic navigator who would overcome its perils ran no little risk of fatal shipwreck." In a given community at a specific time, there is a continuous interaction among the various types of norms signified by the terms law, morals and ethics. Members of a community are in a position to change the relationship between law and ethics. These norms are not clearly distinguishable from each other. None of the regimes can maintain a complete normative monopoly [30]. Despite the enormous achievements made in science and technology, this century has produced a very high level of apprehension about the future unlike the past ones. It is true that science has given us a gigantic set of opportunities, but, it also has raised the levels of moral responsibility that falls on our shoulders. The same phenomenon demanded the set of bioethical guidelines for the Countries to adopt and follow.

The best and optimal way to achieve the feats of bioethics in any country is to implement the principles through the public policy. The overall impact of public policy on the shape and scope of the scientific agenda has become so widespread that, we barely notice it in special sense. Public policy influences science through the problems it deems worth solving, the methods it finds acceptable and the resources it makes available for the task. The Bioethics Declaration gives us abstract concepts which are for the respective States to interpret, break down into understandable chunks of rules and draft public policies in accordance to them. It is a strenuous, yet worthy process. In India, most of the bioethical issues arise due to the patriarchal set up of the society and the religious influences on an individual's decisions. It can be seen in instances of people denying becoming organ donors, stigmas on surrogate mothers etc. We are in a chaos wherein we do not know how to place morality and law on the same levels. Law seems to be compulsive, so the only alternative is to inculcate updated moral concepts into the grass root levels of society, which public policy alone can do. Thus, one of the greatest responsibilities placed on us is to consider the social and human repercussions of our rapidly accumulating new knowledge and the appropriate stance of public policies with respect to these matters. Indeed, the more dependent we become on new science and technology, the more essential it is to generate renaissance after renaissance in moral philosophy.

C. Solutions for bioethical issues in India

In our own time, of course, we must both celebrate and contend with the fact that science and technology are advancing at a rapid pace. We may see the science undermining our faith in existing arrangements. India is in such a situation. Organ transplants, surrogacy, and global pharmaceutical companies coming into India has changed the medical field and further raised several complex questions in the context of ethics and absence of proper laws. This is a serious issue. But, UDBHR sets forth certain principles that serve as the best solution to most of these bioethical concerns worrying the Indian State.

The principles of autonomy and informed consent: Paternalism cannot be considered as morally wrong in entirety. But, if certain acts attempt to maximise welfare by disregarding autonomy, it will be wrong. The concept of consent arises from the ethical principle of patient autonomy. Under the THOTA, 1994, the provision of free will of the deceased is nothing but a misnomer because in spite of his opting for the organ donation, his wish becomes secondary to the wishes of his family. Despite the open declaration of his consent and wish to donate organs after his death, it is the approval of family which is needed for such a donation. The ethical and religious constraints also pose a challenge. The grief-stricken family opposes organ donation as they apprehend that the corpse may be mutilated. Again, surrogacy contracts pose a sufficiently great number of risks to personal autonomy to justify their impermissibility like surrogate's loss of control over her body and daily activities during pregnancy. Non-protection of surrogate mothers is a big issue. Currently, there is no specific legislation to deal with surrogacy and India relies on rules and guidelines issued by ICMR. The Surrogacy (Regulation) Bill of 2016 has reinforced the legendary constitutional historian Sir Ivor Jennings' comment: "The truth is that the rule of law is apt to be rather an unruly horse...if analysis is attempted, it is found that the ideas includes notions which are essentially imprecise [31]. Also, there are several flaws in the bill like anti-liberal and unscientific regulation, exclusion of live-in couples and homosexuals, banning of commercial surrogacy, and camouflaged regulatory designs. Thus, the organ transplants and surrogacy are being discouraged by the State itself, which is a bad socio-medical indicator. Other bioethical issues like researches done on human subjects and clinical treatments shed light on the complex questions of informed consent. There is inadequacy in human subject protection in India. The State also has failed to impart shared decision making, i.e. to direct the doctors to decide the extent of information to be conveyed only after carefully assessing the individual's beliefs and choices. Most of the doctors see informed consent's legal aspects alone. Thus, the challenge is to change the prevailing culture within medicine and health care. Article 5 of the UDBHR considers the importance of autonomy and individual responsibility and also covers the persons who are unable to exercise autonomy. Article 6 specifically elaborates the ethical and legal aspects of consent. Both these articles serve as the yardstick to measure and evaluate the Indian laws in the topics of informed consent and autonomy. They also enable the State to draft and amend the laws accordingly adhering to bioethical principles.

- *Weighing the aspects of benefit and harm.* Preventing harm is present in the Hippocratic Oath. So, issues regarding the minimising of possible harm and maximising of potential benefits to patients is not of recent origin. India has a well-established pharmaceuticals industry, but its ethical aspects give us a jerk. Since last century, there has been a rise of human experimentation in Asia, and specifically in India. There has been widespread abuse and exploitation of the subjects, and causing harm and even death as a result of adverse drug reactions for which no compensation is usually paid. It is thus opined that the local populations are used as guinea pigs with double standards in operation [32]. Inspections into the laboratories are unorganised and fragmented. Severe violations are reported ranging from low hygiene to inadequate control systems to store data. Despite elaborate rules regarding acceptance of a new drug, they are very easily violated and remain unquestioned. There is an imbalance between benefits and harm. Article 4 of the UDBHR needs to be followed not only in ethical paradigm, but also in legislations and their implementation. Only then shall the utilitarian ethics of maximising benefits and minimising harm be achieved.

- *Social responsibility of the State to promote health: Reproductive rights*: The Indian medical tourism industry promotes itself by the slogan, "First world treatment at Third world price." However, the health care provided to India's poor continues to be of third world standards. The last few decades witnessed policy makers pre-occupying with pregnancy and contraception-related services and failing to recognise and address the fundamental problems. Recently, WHO published some disturbing facts about the state of pregnant women in India; it revealed that at least one woman every five minutes, dies during pregnancy and child birth and this can be attributed to the absence of proper emergency obstetric care, missing trained mid-wives, lack of management capacity in health system, and lack of comprehensive maternal care services [33]. Unethical nature of trials of contraceptives on women is also frequently reported. Reproductive rights are the fundamental rights of every woman. Article 14 of the UDBHR provides the best guideline for India to ensure reproductive health of women, wherein it holds that the Governments should advance access to quality health care and essential medicines, especially for the health of women and children, because health is essential to life itself and must be considered to be a social and human good. India should abide by that principle in order to regulate reproduction-related issues and ensure healthy environment and adequate resources to pregnant women.
- *De-stigmatisation*: Article 11 of the UDBHR states, "No individual or group should be discriminated against or stigmatized on any grounds, in violation of human dignity, human rights and fundamental freedoms." Globalisation and changes in the forms of governments in various countries have resulted in unfamiliar

and unappreciable phenomena, such as homophobia, xenophobia, ethno-centrism etc. The health sector also subsequently suffered from inhumane and impersonal approaches of clinical professionals, researchers towards their patients and subjects respectively. India is not free from stigma and discrimination. Stigma prevails when one is made to feel inferior to another and considered to be below other fellow beings. Dignity is an intrinsic human quality and thus, cannot be alienated. So, stigma in broader sense is the slaughter of human dignity. This interpretation stems from Kantian philosophy, according to which respecting the dignity of others is not making it just a means; all rational beings are subject to the law that each of them should treat him/herself, and treat everyone else, never as a mere means, but always at the same time as an end in itself. Kant opines that this is not just a relative value or a price, but rather an internal value that is, dignity [34]. When a woman does not have the right to bodily integrity in reality, when her body is invaded against her will, when her choices are judged by the prejudiced society, it is unlikely that she can play an active role in decision making. Commercial surrogacy has been attacked in many ways like harm to surrogates, their children; the commoditisation of surrogates and women in general; and the exploitation of women, especially those who are poor or vulnerable in some other way [35]. Further, a survey conducted in 2012 by the Delhi Women's group, Sama revealed that Indian couples were prepared to pay more for a light-skinned surrogate from the right caste [36]. The practice of surrogacy is rising among the Indians, but many of the communities still consider it unacceptable. Studies of surrogate mothers have reported that they are often stigmatized because surrogacy is widely believed to involve sexual intercourse [37]. Further, stigma also exists in the context of organ donations. In India, the consent of nearest relative is required for a hospital to obtain organs from a body and it is a major roadblock as relatives are often uncomfortable about organs being harvested [38]. All these issues can be put to an end if the Principle of Non-discrimination and Non-stigmatization is weaved in through the laws and programmes in India. It will ensure that surrogates and organ donors are protected from societal ostracisation. The State and the civil society can initiate national-level awareness drives about the myths and realities of organ donation and surrogacy. The prospects of organ donation should be included in as a subject right from the school curriculum as it is the right way to shape a new generation who strongly rely on reasoning and logic rather than myths and stigma.

IX. Recommendations/ Suggestions

A. An efficient public health care system: the need of the hour

India's health care system is already in a crisis. The poverty induces the masses to approach government hospitals which provide free health care to them. But, unfortunately and alarmingly, there is just one government allopathic doctor for every 10,189 people. Adding to it, there are dead and mutilated animals, rotting corpses and infections in the hospitals and laboratories. It then makes the poor rush to private hospitals which drain their pockets and households and render them ever poorer. Further, the system is becoming more and more unequal. There is no agency to track medico-legal and bioethical issues in India. To uphold bioethics, an efficient public health care system becomes the dire need of the hour.

B. Amending the laws

The Indian laws discussed in the previous sections need to be seriously and meticulously reviewed and amended. The State should ensure that they adhere to the bioethical principles. The UDBHR provides the accurate path to amend the laws, especially issues involving consent, benefit and harm, autonomy etc.

C. Bringing law and morality under the same roof

The final suggestion, though not easily achievable, is to restructure the perceptions about morality and bring them under the ambit of law. Bad impressions and prejudices towards organ donors and surrogates points out the unacceptable imbalance between law and morality in India. It is impossible to equate law and morality. However, the images of morality can be modified from grass root levels, by inculcating the acceptable values amongst the children at the schools. It is easier to build a new reasonable generation than running behind self-deafened society and norms. Optimism lies in seeing the initiatives of various NGOs to protect and promote bioethical values in the citizens. It is pertinent to note that law and morality can also not be separated. As Jeremy Bentham quoted, "Law has the same centre with morals, but it has not the same circumference." The challenge and prospect lies in the task of deciding that central point.

X. Conclusion

All national and international efforts in achieving health care must be supported by positive action of the forward looking legislature, dynamic executive and independent judiciary. The harmonious blend of ethics and law is needed to reach the International bioethical standards, by the turn of this century. Socially committed bioethics defends diversity, not allowing it to be suffocated by the hegemonic perspective. The UDBHR remains to be a significant contribution to global bioethics. The last but the most important question remains as to who will take the initiative? The emerging trends of citizens' protests against ethical violations and the rising contribution of Indian scholars and students to the area of bioethics give hope. It is vital to remember that it is the poor, the illiterate and the common people of India who took it to the streets with protests whenever their democratic rights were threatened. The media is also in its full swing in making the people aware about the bioethical issues in India. Considering the extraordinary complexity of the Indian medical scenario and texture of ethics and culture, the researcher concludes that the principles enshrined in the Bioethics Declaration has the potential to lead India in achieving its constitutional goals of health and international responsibility towards ensuring equity, liberty and justice.

References

[1] Q. Imrana. *Public Health in India.* New Delhi: Daanish, 2011.

[2] M.K. Sateesh, *Bioethics and Bio-safety.* New Delhi: I K International, 2008.

[3] Tom Beauchamp and James Childress, *Principles of Biomedical Ethics.* Oxford: OUP, 2009.

[4] G. Dworkin, "Autonomy and Behavior Control," *Hastings Center Report,* vol. 6, no. 1, pp. 23-28, 1976.
[5] P. Singer, *Practical Ethics.* Cambridge: Cambridge University Press, 1993.
[6] J. Harris, "Goodbye Dolly? The ethics of human cloning?" *J. Medical Ethics*, vol. 23, no. 6, pp. 353-360, 1997.
[7] S. Wilkinson, "The exploitation argument against commercial surrogacy," *Bioethics*, vol. 17, no. 2, pp. 169-187, 2003.
[8] H. Kuhse, "A Modern Myth that Letting Die is not the Intentional Causation of Death: Some Reflections on the Trial and Acquittal of Dr. Leonard Arthur," *Journal of Applied Philosophy*, vol. 1, no. 1, pp. 21-38, 1984.
[9] R. M. Veatch, *A Theory of Medical Ethics.* New York: Basic Books, 1981.
[10] T. W. Pogge, "Human Rights and Global Health: A Research Program,",*Metaphilosophy*, vol. 36, no. 2, p. 182-209, 2005.
[11] P. Devlin, "Law, Democracy and Morality," *University of Pennsylvania Law Review*, vol. 110, no. 5, pp. 635-349, 1962.
[12] J. Oakley, *Bioethics.* Ashgate: Routledge, 2009.
[13]R. Magnus, "The Universality of the UNESCO Universal Declaration on Bioethics and Human Rights," *Basic Medical Key*, https://basicmedicalkey.com/the-universality-of-the-unesco-universal-declaration-on-bioethics-and-human-rights/ (accessed: November 2018).
[14] L. Irigaray, *A Two Subject Culture: In democracy begins between two.* London: Athlone Press, 2000.
[15] M. C. Rawlinson, Anne Donchin, "The Quest for Universality: Reflections on the Universal Draft Declaration on Bioethics and Human Right," *Developing World Bioethics*, vol. 5, no. 3, pp. 258-266, 2005.
[16] H. Anaya Lillich, D. Sheton, *International Human Rights, problems of law, Policy and practice.* Aspen: Aspen, 006.
[17] K. M. Smith Rhona, *Textbook on International Human Rights.* Oxford: OUP, 2012.
[18] R. Dal-Re, "Addressing exploitation of poor clinical trial participants in North America and the European Union," *Eur J Intern. Med*, vol. 34, no, 1, pp. 37-38, 2016.
[19] M. Qurram, "Principles of bioethics in India: A Review," *Journal of Medical and Health Sciences*, vol. 5, no. 3, pp. 22, 2016.
[20] D. Mahapatra, "2644 died during clinical trial of drugs in 7 years: Govt to SC,"*The Times of India.* https://timesofindia.indiatimes.com/india/2644-died-during-clinical-trial-of-drugs-in-7-years-Govt-to-SC/articleshow/19719175.cms (accessed: November 2018)
[21] K.S. Rajan Jacob, "Informed consent in India," *Br J Psychiatry*, vol. 1, no. 1, pp. 158-576, 1991.
[22] A. Vali Tikoo, "Transplantation of Human Organs: The Indian Scenario," *ILI Law Review*, Summer Issue, vol. 1, no. 1, 2017.

[23] M. Moore, J. Anderson, "Kidney Rackets Riles Indians," *The Washington Post*, https://www.washingtonpost.com/archive/politics/1995/04/30/kidney-racket-riles-indians/4c51d8b0-429a-4294-afd6-5a991cd29ea0/?utm_term=.a4dceba1c5cd (accessed: November 2018).

[24] K. D. Kraweic, "Altruism and intermediation in the market for babies," *Washington and Lee Law Review*, vol. 65, no. 1, pp. 203-257, 2009.

[25] J. A. Gupta, "Towards transnational feminisms: Some reflections and concerns in relation to the globalization of reproductive technologies," *European Journal of Women's studies*, vol. 13, no. 1, pp. 23-38, 2006.

[26] NFHS, National Family Health Survey-3, International Institute for Population Services, http://www.nfhsindia.org/nfhs3.shtml, (accessed: April 2006).

[27] "Majority of the drugs found in India are either fake or ineffective," *Think Change India*, https://yourstory.com/2017/06/india-fake-drugs/ (accessed: November 2017).

[28] K. Sinhal, "Drugs scam: Pharmaceutical companies wrote letters signed by doctors," *The Times of India*, https://timesofindia.indiatimes.com/india/Drugs-scam-Pharmaceutical-companies-wrote-letters-signed-by-doctors/articleshow/13072709.cms (accessed: November 2017).

[29] P. Shah, "International Human Rights: A Perspective from India," *Fordham International Law Journal*, vol. 21, no. 1, 1997.

[30] G. C. Hazard Jr., "Law, morals and Ethics," *Faculty Scholarship Series*, vol. 19, no. 1, 1995.

[31] A. Singhvi, A. Dayal, "Her reproductive autonomy," *The Times of India*, https://blogs.timesofindia.indiatimes.com/candid-corner/her-reproductive-autonomy-surrogacy-regulations-must-respect-diversity-of-personal-choices/ (accessed: 2016).

[32] R. Macklin, *Double Standards in Medical Research in Developing Countries*. Cambridge: Cambridge University Press, 2004.

[33] S. Kalpana, "A pregnant woman is dying every five minutes in India," The *Times of India*, https://timesofindia.indiatimes.com/life-style/health-fitness/health-news/A-pregnant-woman-is-dying-every-5-minutes-in-India-Heres-why/articleshow/52866404.cms (accessed: 2016).

[34] I. Kant, *Fundamental Principles of the Metaphysics of Morals*. London: Dover Classics, 2004.

[35] E. Anderson, "Is Women's labour a commodity?" *Philosophy and Public Affairs*, vol. 19, no. 1, pp. 71-92, 1990.

[36] A. Dhillon, "Womb service: How surrogacy still carries stigma in India," *South China Morning Post*, http://www.scmp.com/lifestyle/family-

education/article/1296130/womb-service-how-surrogacy-still-carries-stigma-india (accessed: 2015).
[37] A. Pande, "Not an angel, not a whore: Surrogates as dirty workers in India," *Indian J Gend Stud.*, vol. 16, no. 2, pp. 141-173, 2009.
[38] A. Srivatsa, "Drive India to Aware People About Organ Donation," NDTV, https://sites.ndtv.com/moretogive/drive-india-aware-people-organ-donation-inspire-sign-donors-2016/ (accessed: 2017).

An Evaluation of Faith-Based Perspectives on the Withdrawal of Life-Sustaining Treatment from Patients in a Permanent Vegetative State

Kartina A. Choong,[1] Mahmood Chandia[2]

[1] Lancashire Law School, University of Central Lancashire, UK

[2] School of Humanities & Social Sciences,
University of Central Lancashire, UK

Abstract

In England and Wales, when patients are diagnosed as being in a permanent vegetative state (PVS), continuation of life-sustaining treatment is considered futile. Consequently, it is legally deemed to be in their best interests for this to be withdrawn, alongside clinically assisted nutrition and hydration. This article investigates whether English Law's stipulations on the matter align with the religious doctrines of the Abrahamic Faith traditions. Based on the findings, it highlights a number of important issues relating to the development of a religiously-sensitive decision-making framework which merits public debate.

Keywords

Permanent vegetative state; best interests; medical futility; life-sustaining treatment; clinically assisted nutrition and hydration; Abrahamic faith traditions; end of life.

I. Introduction

For the last quarter of a century, English law has consistently countenanced the withdrawal of life-sustaining treatment (LST) from patients diagnosed as being in a permanent vegetative state (PVS). Continuing with interventions that are not conferring any medical benefit is considered not to be in those patients' best interests, thus doctors are released from any further duty to provide medical care, including clinically assisted nutrition and hydration (CANH). In 2013, we published an article which examined the compatibility between, on the one hand, the conceptual matrix and value base of the secular legal framework that medical professionals in England and Wales are guided by, and, on the other hand, the Islamic worldview [1]. This article is an extension of that previous mono-faith effort. It is the summative critical analysis of a multi-faith symposium that we hosted at our institution, the University of Central Lancashire, on 29 April 2015.

In essence, a key question of this work is to explore how secular legal understandings challenge Abrahamic faith teachings, and what collective or individual epistemology the Abrahamic Faiths offer alongside faith-based legal strategems or understandings (e.g. Talmudic Law, Canon Law, Sharia Law) to balance patient care and religious values, whereby medical practicalities are reasonably considered and religious values are also honoured. The discussion is contextualised, in the next section, with an exploration of the medical understandings of PVS and its legal implications. This is followed, in the subsequent section, by an analysis of the viewpoints of the three Abrahamic faiths regarding the management of this patient group from a religious perspective. Through a comparative analysis of the three Abrahamic faith traditions, namely Judaism, Christianity and Islam, this section aims to identify the similarities, and particularly the differences and challenges that the secular legal approach holds for patients from these respective faiths. The article concludes with reflections on some of the questions which need to be considered in negotiating religious sensitivities within the existing legal framework.

II. The Permanent Vegetative State

When patients are diagnosed with serious brain injury, intervention by healthcare professionals is most often on the side of life preservation. In England and Wales, patients are afforded the benefit of sophisticated medical technology which can save their lives [2]. If the patient is subsequently diagnosed as brain-stem dead, the LST will be removed, as patients with this diagnosis are recognised as dead from medical and legal perspectives [3]. If the brain-stem is still alive but wakefulness and awareness are compromised, the patient would be diagnosed as having a "disorder of consciousness" [4]. One of the main conditions which fall under this umbrella term is the vegetative state.

Patients in a vegetative state may show signs of being awake. For example, they may open their eyes, have sleep-wake cycles and blink [5] (p. 163-164). They are also able to breathe and retain a heartbeat without mechanical assistance [4], although some may require ventilation. However, they do not show signs of environmental awareness or purposeful movement. For example, they do not respond to instructions and voice, do not follow objects with their eyes, nor display any emotional responses [4, 6]. Furthermore, they do not demonstrate any flinching reaction to pain stimuli, thus prompting the generally accepted medical viewpoint that they do not experience pain [7] (p. 96). Notwithstanding this, a number of studies have now challenged this understanding by claiming that such patients do retain some residual awareness, and are therefore able to perceive and experience painful stimuli [8] (p. 336; p. 3).

Vegetative patients who have been in this condition for over 6 months following anoxic or other metabolic brain injury, or over one year after traumatic brain injury, would be classified as being in a permanent vegetative state [9] (p. 10). Inasmuch as they are medically and legally recognised as still being alive, they are also not believed to be imminently dying [10] (p. 328). If LST is provided, they could live for many years but doctors are generally of the opinion that recovery of consciousness is "extremely unlikely" [4]. Thus, is its continuation in the patients' best interests in view of the poor medical prognosis?

It is necessary to highlight that English Law has taken a consistent approach since 1993 [11]. A diagnosis of this condition has led, and still leads, to the automatic conclusion that the continuation of LST is not in the patients' best interests on the grounds that it is futile [11]. Consequently, it is lawful for it to be withdrawn and the patients be allowed to die. This includes CANH, which is classified as medical treatment. It is pertinent to observe that whilst suspension of CANH should always be preceded by court approval, the Supreme Court ruled in July 2018 that this is no longer necessary in situations where there is agreement between the medical team and the patients' family that this would be in the best interests of the patients [12].

The retraction of LST and CANH is considered as an omission rather than an action, and doctors would not be in breach of their duty of care to the patients as it is deemed not to be in those patients' best interests to be in receipt of those interventions [13] (p. 14-18).

Table 1: Medical terminology and definitions

Terminology	Medical Definition
Vegetative State	A state of wakefulness without awareness where there is preserved capacity for spontaneous or stimulus-induced arousal, evidenced by sleep-wake cycles and a range of reflexive and spontaneous behaviours. It is characterised by complete absence of behavioural evidence for self- or environmental awareness [9] (p. 3).
Permanent Vegetative State	When a vegetative state has persisted for over 6 months following anoxic or other metabolic brain injury; or more than 12 months following traumatic brain injury [9] (p. 10).
Brain-stem Death	Irreversible loss of the capacity for consciousness, combined with irreversible loss of the capacity to breathe, as produced by the irreversible cessation of the integrative function of the brain-stem [14] (p. 11).
Life-Sustaining Treatment	Treatment that replaces or supports ailing bodily functions (e.g. ventilation, cardiopulmonary resuscitation, antibiotics and dialysis) [15].
Clinically Assisted Nutrition and Hydration	All forms of tube-feeding (e.g. via nasograstric tube, percutaneous endoscopic gastronomy (PEG) or parenteral nutrition) [16] (p. 6).

In the next section, we turn to the important question of whether the secular legal framework and its underpinning concepts are congruent with religious values. For this purpose, we will explore whether the Abrahamic religions have at their disposal conceptual mechanisms that help their adherents determine how such patients should be managed. In other words, do religious values allow all medical interventions, including CANH, to be withdrawn and subsequently withheld from such patients when they are not expected to bring any medical benefit?

III. The perspectives of the Abrahamic Faiths on the management of PVS patients

The religious doctrines of the three Abrahamic faiths of Judaism, Christianity and Islam pre-date the development of modern life-support technology by many centuries. These doctrines are stipulated in their religious texts. Each faith community has its independent set of values, albeit some values are quite common, which influence how it approaches this phase of life. The following discussion offers an analysis of the variant positions they take on the notion of "futility."

A. Judaism

Our discussion begins with Judaism, the oldest of the three Abrahamic religions. In Jewish Law as in English law, a PVS patient is considered alive. The notion of life is defined by an ability to breathe based on the Book of Genesis: "Then the LORD formed a man from the dust of the ground and breathed into his nostrils the *breath of life,* and the man became a living being [17]." The word "breath" is understood as "soul" in this context [18] (p. 694). Therefore, the lack of an ability to breathe is considered to be tantamount to the removal of the soul, given that the removal of the soul is understood to be the definition of death [19] (p. 19-22). In light of this Judaic scriptural understanding, the question remains: if life is based on the ability to function, i.e. to breathe; then if a person's ability to breathe independently or to feed themselves are compromised, to what extent is there a need to provide such a person with treatment and sustenance?

In order to address this question, it is important to note that Orthodox Judaic legal thinking, which this article confines itself to, does not support the concept of "medical futility" nor letting the PVS patient die as being lawful [20] (p. 121). It argues that as long as the brain-stem is still alive, treatment should not be considered entirely futile [21] (p. 1267-1268). This understanding is underpinned by scriptural understandings. The starting scriptural reference point is the Book of Exodus, where it is recorded that killing any individual is forbidden [22]. Further, in accordance with the Book of Genesis: "Whoever sheds human blood, by humans shall their blood be shed; for in the image of God has God made mankind [23]." In light of this verse, the Judaic understanding is that human life is sacred since man is in the image of God on the grounds that human life is the closest reflection of God [24] (p. 94). Given that human life is sacred, violating something sacred would be contra to Judaic teachings and will have severe consequences.

To the extent that Orthodox Judaism does not support the concept of "medical futility" nor the letting die of PVS patients as being lawful, it also maintains this position in the event of a competition for the same resources. In fact, Jewish Law states that no life is worth more than another [25, 29] (p. 1, p. 278). Therefore, if there is another person who requires the LST equipment, allocation would be based on a first come first served formula. The second person and subsequent people would be considered "pursuers" with not necessarily being primary claimants over resources, in the event of a constraint over resources [26] (p. 203-205). The implication of this is that an individual with a better chance of resumption of consciousness or survival could theoretically be overlooked in treatment, with preference given to an individual with a lesser chance of consciousness recovery or survival. The Jewish law understanding is that preservation of life is paramount, and anyone attempting to remove aid from a PVS patient would be considered as a pursuer and murderer. Indeed, there is a Jewish concept used in civil law, which can also apply here. Namely, if a person can prove ownership of something that is in the possession of someone else, then it can be removed from the second party and lawfully returned to the rightful owner [27] (p. 71). This argument is based on the premise that a court will recognise the right of a person to appropriate property unless conclusive evidence is presented to the court to show that it is the property of the disputant [28]. This analogy can also be applied in this case and the person receiving treatment would be considered as having lawful current possession and holding primary stake over resources.

Notwithstanding the above, one is not allowed to cause an ill person discomfort or pain [29] (p. 279). Therefore, it may also be necessary to provide any person with hydration and nutrition [30] (p. 5). However, if this causes them distress or pain and is only prolonging the dying process, it would be deemed forbidden. No doubt levels of pain would need to be considered with regard to this. The level of discomfort that is required for stopping CANH is one that has the potential to cause a damaging amount of pain [31] (p. 47). This dilemma must be resolved by the application of an informed judgement undertaken by clinicians, applying their knowledge and experience, then coming to a professional judgement on behalf of the patient. However, clinicians would need to approach such decisions on a case by case scenario and in consultation with Jewish religious leaders [32] (p. 48). The latter equally cannot make the decision without clinical knowledge or consultation with clinicians [32] (p. 48). It would be anticipated from a Rabbi or a decider that a judgement is made after taking into consideration logical thinking, the minutiae of previous rulings (analogical thinking) and from meta-halakhic rulings [33]. It is anticipated that this cognitive clinical process is also creative but firmly underpinned by relevant principles that are related to the meta-halakhic rulings. Further, as mentioned previously, it is forbidden to carry out an intervention that would cause pain or suffering.

It should be noted, though, that there exists a plurality of opinions amongst Jewish legal deciders. There are strands of Judaism that may disagree with the Orthodox view on these issues. The change in opinion is normally governed by variations in extenuating circumstances i.e. there is a text and context relationship.

In essence, the Judaic epistemology is based upon the application of scriptural wisdom and human reasoning. It has a set of higher aims, scriptural references and legal or halakhic processes of legal deduction. It is a structured process that aims to balance medical guidance with faith obligations. There are three key stakeholders in this process: the patient or his family, the team of clinicians (medical experts), and the Rabbi (Jewish law expert).

B. Christianity

Like Judaism, there are many Christian understandings or opinions on the topic. This study only comprises the view of the Church of England. The central point upon which its understanding is based, is how the intrinsic worth of every individual human being is understood. This is highly intertwined with the concept that each human being is created in the image of God in the context of the here and now, and eternity [34]. By extension, all human life is therefore sacred from birth to death [35]. Further, the human body itself is regarded as a sacred temple of the Spirit of God i.e. God's breath is breathed into the human being [36] (p. 99). Given this understanding, since PVS patients are still breathing whether naturally or with technological assistance, it is accepted that they are alive and are to be treated accordingly [37].

This perspective is further underpinned by the understanding that the aim of all human relations centres around the Christian commandment to love God and to love one's neighbour as oneself [38] (p. 50). However, this can attract a diversity of approaches in pastoral care depending on how those involved interpret the call to love in all circumstances. The pastor may support a decision to allow a patient to die in limited cases, whereas in other situations the pastor may support continued preservation of life, given the possibility of variant understandings of the "call to love" [39] (p. 4). Nevertheless, the service provided to PVS patients and their relatives will be customised to their needs while the general principle (the "call to love") will be the same. The adherents of the Church of England believe that there is a life after this one and an individual needs to hope that there is a better life beyond the suffering and physical incapacity [40]. The merciful or "call to love" approach in such a circumstance would be not to prolong the patient's suffering beyond necessity, as any meaningful life has seemingly ended [41]. Further, at all times, the wishes of the patient would be respected in compliance with the law. The Church of England would support such patients by prioritising the need to respect their dignity. Since the merciful approach would be *not* to prolong patient suffering; in such circumstances all medical, legal and religious considerations would become more patient-centred and be considered for what is in the best interests of the patient [42]. The understanding is that it is better for these to coincide rather than conflict with one another.

The Church of England is therefore comfortable with how the law determines the best interests of the PVS patient [42]. It further argues that it is right and proper to allow the patient to die. As human beings, we have no right to hold on to such patients [42]. However, there is no broad line opinion on what is right and what is wrong in this type of situation [42]. Equally, there is no generic perspective that can be applied for the decision-making process other than the "call to love" complemented with patient-related understanding of the notion of suffering.

This Christian denomination (like other denominations) holds the view that suffering is part of a soul-enrichment process [43] (p. 14).

Humans have to live, it argues, in a challenging and uncertain environment where little is assured, and the enhancement and denial of life, individual prosperity and catastrophe happen almost randomly [43] (p. 11). So, suffering is seen as a test of faith and an opportunity for spiritual growth. The sufferer is not alone. Rather, he is supported by God whilst being in the process of being drawn to Him [44] (p. 145-146).

In the belief that life is fast-approaching an end, appropriate pastoral support is more focused on the holiness of death. Death is viewed as the final healing from the suffering and sorrows of this world [45]. Further, the teachings of Jesus Christ, it is argued, indicate that it is the Will of God that humans take responsibility for their own lives and acknowledge that God gives life and God receives life [46]. The Church of England also considers that as humans, relatives and doctors may not be sure whether the right thing was administered but can be assured that they have made the best decision possible and assign the rest to the Grace of God, with an emphasis on redeeming love rather than Divine Judgement [47]. The belief is that judgement has already been passed and Christians are not judged twice [48]. Given this, Christians are prepared for death as the beginning of a new eternal life. In the final extension, for the Christian in PVS, death leads inevitably to life; eternal life [45].

The post-death situation is approached in a similar manner. Prayers following the patient's death will focus on reconciliation and the long-term peace of the patient in affirming a Godly grace that is already in operation [49] (p. 6-14). Consequently, the patient can move on to the next life and the relatives can move on with the rest of their earthly lives. Therefore, the withdrawing of all medical support from a patient is in line with Christian values or at least Church of England values to allow the person to die.

The Church of England Christian perspective is based upon the Christian commandment of "call to love." The application of this notion can manifest a diversity of approaches to individual cases since the application of this commandment can be customised on a case by case scenario. The outcome of this application is subject to the condition and prognosis of the patient. The approach seems patient-orientated and comprises an asset of values: any suffering should not be prolonged; the wishes of the patient should be respected and illness is an opportunity for spiritual growth. There are three prominent stakeholders in this process: the patient or his family; the team of clinicians (medical experts), and the pastor (who assists in the application of the "call to love").

C. Islam

In Islam, a key notion that underpins all human life actions is that the main purpose of human creation is the worship of God [50]. Although PVS patients are physically unable to carry out any acts of devotional worship, and also do not even possess the ability to smile, which is in itself an act of virtue [51] (p. 1383), it is believed that their state of being ill is itself a source of Divine mercy and being patient upon this is meritorious and worthy of reward in the Hereafter [52] (p. 395). In light of this belief, such patients are considered alive in Islam and subject to its commands according to their mental and physical states [53].

In Islam, similar to Judaism and Christianity, life *per se* is valuable [54] (p. 95). Therefore, there should always be an ardent attempt to seek a treatment given that the primary goal of medical intervention is the preservation of life [55] (p. 432). Nevertheless, to what extent the sanctity is maintained through the indefinite provision of life support, is a debated issue. This is approached in different ways. A range of issues (e.g. status of seeking medical treatment in Islam, potential outcomes of treatment, condition of the patient, etc.), and a set of variables are considered. The more embracive question with regard to a PVS patient is whether administration of medical treatment is mandatory or less in Islam. This would be adjudicated by Islamic legal consults. In Islamic law, the opinions of the classical Sunni scholars are considered to be authentic and followed across the globe on the proviso that they are not compromising fundamental Islamic principles (i.e. those that contravene the primary scriptures: the Quran and the main Hadith collections) [56] (p. 3-6). There are a number of viewpoints from the classical Sunni schools of Islamic law that range from the "literal" opinion that advocates total reliance on the Will of God and contra behaviour potentially being reprehensible [57] (p. 213), to the majority view of the four established schools of legal thought (Hanafi, Hanbali, Maliki and Shafi'i) who collectively deem it permissible [58] (p. 238), whilst some scholars from the latter three schools advocate treatment as being preferable [1]. This structured approach and thinking is at the disposal of the family, patient and the scholar to care for their PVS relative. It is very much related to the clinical condition of the PVS patient and takes into consideration practicalities. Each Sunni school of legal thought has an informed position subject to its mode of engagement with the primary sources [59] (p. 1-35; p. 1-8).

The preservation and sustenance of life is the *key* notion for treatment being mandatory [60] (p. 267; p. 965). Where this outcome is not anticipated, the case of medical treatment would be graded as preferable or permissible, subject to the presumption of cure [61] (p. 433; p. 163; p. 530). Members of the Sunni Muslim community of different regions follow different schools of legal thought. Each respects the view of the other schools [62] (p. 2). Whilst it could be difficult for a medic to navigate through the maze of Islamic legal formulations, without a doubt, collaboration between clinicians, scholars, and the family would enrich patient care and comfort.

The generic underpinning is that even when medical intervention is considered futile or even harmful for the body, CANH must be continued, regardless of the futility of the medical treatment [63] (p. 203). This would be maintained up to the point that the harm outweighs the benefit. However, medical intervention can be stopped if the treatment is considered futile for a PVS patient. Omission then becomes necessary and the provision of medical intervention would be understood as significantly compromising the efficient use of resources [64]. This factor would need to be determined by analogical deduction by taking into account past experiences in similar cases, with a view to preserve medical resources for a patient with a better prognosis [65].

Another consideration of the decision-making process is the severity of the illness and the certitude or the probability of intervention efficacy. If the ailment is tolerable with patience and forbearance, then it would not be mandatory to seek treatment. However, if the illness is considered life-threatening or even intolerable, then treatment would be graded accordingly [66] (p. 98), for example:

- Mandatory: if treatment will definitely provide a cure
- Encouraged: if treatment is likely to provide a cure
- Permissible but not encouraged: if treatment may or may not provide a cure
- Necessary: if treatment is considered futile, then omission is necessary

There is nevertheless an argument that treatment and care should never be considered futile [60], i.e. that it may be definitively unknown what the outcome of treatment could be but there should still be an attempt to seek a treatment irrespective of the prognosis. Further, treatment is not considered mandatory for a conscious Muslim patient with diminished capacity who cannot regain accountability. However, this does not negate the permissibility of providing treatment [67] (p. 4).

Notwithstanding the above range of understandings, a non-cognisant person can still gain reward for the afterlife albeit in a vegetative state, as he is still a living individual [68] (p. 180). However, it is argued that reward is the providence of Divine Mercy, it can be manifold and not just a source of recompense [69]. In Islam, death is seen as a bridge to the Hereafter, and a natural journey to the eternal next life. The time of death has been Divinely fixed [70]. In this light, medical intervention or the lack of it should be understood as the Will of God. In essence, the opinions consider notions of *Taklif* (i.e. accountability) or *Hurma* (i.e. sanctity) [70]. The basis of the decision would be on the efficacy of treatment and severity of illness, and by extension perhaps on putting a value on life in regards to PVS.

In Islam, any discussion with regard to the care of the PVS patient must consider a set of concepts as part of a decision-making process. This includes a consideration of the concept of sanctity of human life, the concept of seeking treatment for illness, the concept of the human as a trustee of his body, and the definition of death and its determinants. The consideration of these values can greatly assist to determine the nature of treatment and care a PVS patient could be administered. The nature of such concepts implies that any determination of treatment would be considered on a case by case basis where medical views and Islamic scholarship views would need to be balanced. Further, it should be a collaborative exercise in order to minimise error [71] (p. 187-188). Ultimately, the wish of the patient and the family is that the patient dies honouring the faith. Islamic law has several precepts in any given scenario to facilitate this. It does not acknowledge that a PVS patient has clinically died but does have a range of tools to balance clinical medical care and costs, and patient welfare and spiritual care.

IV. Conclusion

The discussions regarding patients in a PVS are inextricably linked to the following question: is the patient to be considered alive or not? The three Abrahamic faiths share with English law the view that PVS patients are still alive. There is nevertheless, as explored above, a variation in the approach religions advocate for the withdrawal of LST. Further, even in situations where they believe that LST could or should be withdrawn, they are in agreement that CANH should be continued, and not stopped unless this is causing pain and suffering. Abrahamic Faith scholarship and followership represent understandings and practices that manifest a diverse range of scriptural interpretations and devotional practices that are all based on human engagement with revelation and post-revelation literature across time and space. It represents an effort to be loyal to faith whilst dying and being on the journey of meeting one's Creator. This diversity also demonstrates an effort to be loyal to scripture and to respective schools of interpretation. These two diachronic and synchronic principles represent the epistemology of the Abrahamic Faiths in determining a decision. They are applied to varying degrees across the Abrahamic Faiths to balance the spiritual and the temporal aspects with a view to provide religious-specific spiritual comfort for the patient. Abrahamic Faith scholarship has in broad terms considered faith values, medical assessment, and resources. They equally hold common perspectives and concerns which although diverge, are still subject to the patient's condition.

The implications of such a broad common perspective, albeit in principle rather than in detail, are that with a sizeable proportion of the UK population belonging to these faith groups [72], it is important that religious values on this issue be accommodated, as part of the medico-legal decision-making process. But as healthcare in the UK is publicly funded, the findings give rise to two other questions which merit further reflection. One, how could these faith-based values be synthesised and accommodated within the secular legal framework? Two, how should the potentially disproportionate use of resources such an accommodation may entail be addressed, seeing as public health resources are finite? Should it still be funded by the National Health Service (NHS) in furtherance of the equality and diversity agenda [73]? Or should it be financed by the faith communities themselves, the patient or the patient's family? Or will the advancement of technology have an impact on faith understandings? These questions require public debate.

References

[1] K. A. Choong, M. Chandia, "Technology at the End of Life: 'Medical Futility' and the Muslim PVS Patient," *International Review of Law,* vol. 9, 2013, available at http://dx.doi.org/10.5339/irl.2013.9. Accessed: August 15, 2018.
[2] Department of Health, "The NHS Constitution for England", 27 July 2015.
[3] *R v. Malcherek*; *R v. Steel* [1981] 1 *Weekly Law Reports* 690.
[4] NHS, "Disorders of Consciousness", available at http://www.nhs.uk/conditions/disorders-of-consciousness/. Accessed: June 7, 2018.
[5] L. Bonfiglio, M. C. Carboncini, P. Bongioanni, P. Andre, F. Minichilli, M. Forni, B. Rossi, "Spontaneous Blinking Behaviour in Persistent Vegetative and Minimally Conscious States: Relationships with Evolution and Outcome," *Brain Research Bulletin,* vol. 68, pp. 163-170, 2005.
[6] Parliamentary Office of Science and Technology, "Vegetative and Minimally Conscious States," *POSTnote*, vol. 489, 2015.
[7] A. Markl, T. Yu, D. Vogel, F. Müller, B. Kotchoubey, S. Lang, "Brain Processing of Pain in Patients with Unresponsive Wakefulness Syndrome," *Brain and Behavior*, vol. 3, no. 2, pp. 95-103, 2013.
[8] A. Demertzi, C. Schnakers, D. Ledoux, C. Chatelle, M. A. Bruno, A. Vanhaudenhuyse, M. Boly, G. Moonen, S. Laureys, "Different Beliefs About Pain Perception in the Vegetative and Minimally Conscious States: A European Survey of Medical and Paramedical Professionals," *Progress in Brain Research*, vol. 177, pp. 329-338, 2009. F. Riganello et. al., "Pain Perception in Unresponsive Wakefulness Syndrome May Challenge the Interruption of Artificial Nutrition and Hydration: Neuroethics in Action," *Frontiers in Neurology*, vol. 7, doi 10.3389/fneur.2016.00202, 2016. Accessed: January 31, 2019.
[9] Royal College of Physicians, "Prolonged Disorders of Consciousness: National Clinical Guidelines," 2013.
[10] J. Keown, *The Law and Ethics of Medicine: Essays on the Inviolability of Human Life.* Oxford: Oxford University Press, 2012.
[11] See the House of Lords' ruling in *Airedale NHS Trust v. Bland* [1993] 1 *All England Law Reports* 821.
[12] *An NHS Trust and Others v Y (by his Litigation Friend, the Official Solicitor) and Another* [2018] *UKSC* 46.

[13] R.W.M. Law, K.A. Choong, "Disorders of Consciousness: Is a Dichotomous Legal Approach Justified?" *European Journal of Health Law*, vol. 25, pp. 1-22, 2018.

[14] Academy of Medical Royal Colleges, "A Code of Practice for the Diagnosis and Confirmation of Death," 2008.

[15] NHS, "End of Life Care: Advanced Decision (Living Will)", available at https://www.nhs.uk/conditions/end-of-life-care/advance-decision-to-refuse-treatment/. Accessed: January 31, 2019.

[16] Royal College of Physicians and British Medical Association, "Clinically-Assisted Nutrition and Hydration (CANH) and Adults Who Lack the Capacity to Consent: Guidance for Decision-Making in England and Wales," 2018.

[17] Genesis 2:7 (our emphasis).

[18] G. Dreifuss, "Experience of the Self in a Lifetime," *Journal of Analytical Psychology*, vol. 46, no. 4, pp. 689-696, 2001.

[19] F. Rosner, M.D. Tendler, "Definition of Death in Judaism," *Journal of Halakha and Contemporary Society*, vol. 17, pp. 14-31, 1989.

[20] B.M. Kinzbrunner, "The Terri Schiavo Case – From the Viewpoint of Jewish Law', *B'Or Ha' Torah*, vol. 18, pp. 117-133, (5768/2008).

[21] D. Inwald, I. Jakobovitz, A. Petros, "Brain Stem Death: Managing Care When Accepted Medical Guidelines and Religious Beliefs are in Conflict," *British Medical Journal*, vol. 320(7244), pp. 1266-1268, 2000.

[22] Exodus 20:13.

[23] Genesis 9:6.

[24] J. Bloomberg, *The Jewish World in the Modern Age*. New Jersey: KTAV Publishing House, 2004.

[25] J.Z. Abrams, "Introduction," *Jewish Perspectives on Theology and the Human Experience of Disability*, J.Z. Abrams and W.C. Gaventa, eds., New York: Routledge, pp. 1-4, 2006.

[26] D.B. Sinclair, *Jewish Biomedical Law: Legal and Extra-Legal Dimensions*. New York: Oxford University Press, 2003.

[27] A. Cohen, *An Introduction to Jewish Civil Law*. Jerusalem: Feldheim Publishers, 1991.

[28] Choshen Mishpat, 139:1.

[29] E.N. Dorff, "Applying Traditional Jewish Law to PVS," *NeuroRehabilitation*, vol. 19, no. 4, pp. 277-284, 2004.

[30] M.N. Staitman, "Withdrawing or Withholding Nutrition, Hydration or Oxygen from Patients," *Death and Euthanasia in Jewish Law: Essays and Responsa*, W. Jacob and M. Zemer, eds., Pittsburgh: Rodef Shalom Press, pp. 1-10, 1995.

[31] A.J. Rosin, M. Sonnenblick, "Autonomy and Paternalism in Geriatric Medicine: The Jewish Ethical Approach to Issues of Feeding Terminally Ill

Patients, and to Cardiopulmonary Resuscitation," *Journal of Medical Ethics*, vol. 24, pp. 44-48, 1998.

[32] S.N. Ali, "Religious Minorities and Medicine: The Collision of Health Care and Faith," *Stigma and Prejudice: Touchstones in Understanding Diversity in Healthcare*, R. Parekh and E.W. Childs, eds., Switzerland: Human Press, pp. 41-66, 2016.

[33] For a discussion on the variety of Jewish-legal considerations regarding PVS, see F. Friedman, "The Chronic Vegetative Patient: A Torah Perspective" *Journal of Halacha and Contemporary Society*, vol. 24, pp. 88-109, 1993; M.R. Gillick, "Artificial Nutrition and Hydration in the Patient with Advanced Dementia: Is Withholding Treatment Compatible with Traditional Judaism?" *Journal of Medical Ethics*, vol. 27, no. 1, pp. 12-15, 2001.

[34] Holy Bible, Genesis 1:26.

[35] Holy Bible, Genesis 2:7.

[36] D.P. Gushee, *The Sacredness of Human Life: Why an Ancient Biblical Vision is Key to the World's Future*. Cambridge: William B. Eerdmans Publishing Co., 2013.

[37] S.B. Rae, "Views of Human Nature at the Edges of Life: Personhood and Medical Ethics," *Christian Perspectives on Being Human: A Multidisciplinary Approach to Integration*, J.P. Moreland and D.M. Ciocchi, eds., Oregon: Wipf & Stock Publishers, pp. 235-262, 2015.

[38] S.J. Jones, *The Evangelistic Love of God and Neighbor: A Theology of Witness and Discipleship*. Nashville: Abingdon Press, 2003.

[39] B. McCarthy, "Medical Ethics: A Christian Perspective," available at https://www.churchofengland.org/sites/default/files/2017-11/Medical%20Ethics%20a%20Christian%20Perspective.pdf, pp. 1-5. (Accessed: June 17, 2018).

[40] Church of England, "What We Believe," available at https://www.churchofengland.org/our-faith/what-we-believe. (Accessed: June 17, 2018).

[41] G. Tulloch, *Euthanasia, Choice and Death*. Edinburgh: Edinburgh University Press, 2005.

[42] A position confirmed during the symposium by Reverend Dr Brendan McCarthy, the Church of England's National Adviser on Medical Ethics and Health and Social Care Policy.

[43] J. Taylor, "Introduction," *Suffering and the Sovereignty of God*, J. Piper and J. Taylor, eds., Illinois: Crossway Books, pp. 1-14, 2006.

[44] D. Powlison, "God's Grace and Your Suffering" in J. Piper and J. Taylor, eds., *ibid.*, pp. 145-174.

[45] Church of England, "Ministry at the Time of Death", available at https://www.churchofengland.org/prayer-and-worship/worship-texts-

and-resources/common-worship/death-and-dying/funeral#mm113. Accessed: June 18, 2018.

[46] Holy Bible, Job 1:21; Deuteronomy 32:39.

[47] Hebrew, 9:27; "Bishop of Durham Highlights the Importance of Chaplaincy and A Whole Person Approach to End of Life Care", available at https://churchinparliament.org/2017/03/14/bishop-of-durham-highlights-the-importance-of-chaplaincy-and-a-whole-person-approach-to-end-of-life-care/. (Accessed: June 17, 2018).

[48] 2 Corinthians, 5: 8-11; C.B. Cohen, et. al., *Faithful Living, Faithful Dying: Anglican Reflections on End of Life Case.* Pennsylvania: Morehouse Publishing, chapter 1, 2000.

[49] A. Chapple, C. Swift, S. Zieland, "The Role of Spirituality and Religion for those Bereaved Due to a Traumatic Death", *Mortality,* vol. 16, no. 1, pp. 1-19, 2011.

[50] Quran: Surah Al-Dhariyat, verse 56.

[51] M. Ibn Al-Hajjaj, *Sahih Muslim* (translated by A.H. Siddiqi), Lahore: Pakistan, vol. 4, Hadith 6359, 1972.

[52] I. M. Al-Bukhari, *The Translation of Meanings of Sahih al-Bukhari* (translated by M.M. Khan), Beirut: Dar al-Fikr vol. 8, Hadith 582, 1391/1971.

[53] Quran: Sura al-Baqarah, verse 286.

[54] According to verse 5:32 of the Quran, "taking a single life is tantamount to killing all of mankind". See also A. Shah and M. Chandia, "The Relationship Between Suicide and Islam: A Cross-National Study," *Journal of Injury and Violence Research*, vol. 2, no. 2, pp. 93-97, 2010.

[55] M.Z. Al-Shahri, A. Al-Khenaizan, "Palliative Care for Muslim Patients," *Journal of Supportive Oncology,* vol. 3, pp. 432-436, 2005.

[56] Quran: Sura al-Nisa, verse 59; Sura al-Nahl, verse 43 and Sura al-Anbiya verse 7. For an in-depth discussion on the preservatory principles, see M.U. Chapra, "The Islamic Vision of Development in Light of Maqasid al-Shariah," Occasional Paper Series 15, *The International Institute of Islamic Thought*, pp. 1-50, 2008. Note that this article focuses only on Sunni law and its legal schools of thought.

[57] A. A. Al-Khader, "The Iranian Transplant Programme: Comment from an Islamic Perspective," *Nephrology Dialysis Transplantation,* vol. 17, no. 2, pp. 213-215, 2002.

[58] *Journal of Islamic Medical Association of North America* (IMANA), "Islamic Medical Ethics: The IMANA Perspective', paragraph 4(D), 12(4), pp. 231-237, 2005.

[59] W. Hallaq, *A History of Islamic Legal Theories: An Introduction to Sunni Usul al-Fiqh.* Cambridge: Cambridge University Press, 1997; M.H. Kamali, *Principles of Islamic Jurisprudence.* Cambridge: The Islamic Texts Society, 2003.

[60] F. A. Khan, "Religious Teachings and Reflections on Advance Directive: Religious Values and Legal Dilemmas in Bioethics: An Islamic Perspective," *Fordham Urban Law Journal*, vol. 30, no. 1, pp. 267-275, 2002; K.M. Hedayat and R. Pirzadeh, "Issues in Islamic Biomedical Ethics: A Primer for the Pediatrician," *Pediatrics*, vol. 108, pp. 965-71, 2001.

[61] M.Z. Al-Shahri, A. Al-Khenaizan, "Palliative Care for Muslim Patients", *Journal of Supportive Oncology*, vol. 3, pp. 432-436, 2005; H. Al-Jahdali, et. al., "Advance Medical Directives: A Proposed New Approach and Terminology from an Islamic Perspective," *Medicine, Health Care and Philosophy*, vol. 16, no. 2, pp. 163-169, 2013; A. R. Gatrad, A. Sheikh, "Palliative Care for Muslims and Issues Before Death," *International Journal of Palliative Nursing*, vol. 8, no. 11, p. 526-531, 2002.

[62] *Al-Mawsu'a al-fiqh al-Kuwaitiyyah* [The Kuwaiti Encyclopedia of Islamic Jurisprudence], Wizaratul al-awqaaf wa al-sh'uwn al-Islamiyyah [The Ministry of Endowments and Islamic Affairs], vol. 1-45, 1404/1984 – 1427/2006.

[63] V. Rispler-Chaim, "Islamic Medical Ethics in the 20th Century," *Journal of Medical Ethics*, vol. 1, pp. 203-208, 1989.

[64] Y. Qardawi, *Fiqh al-Muwazanah* [Comparative Islamic Jurisprudence] available at http://www.qaradawi.net/library/66/3269.html. (Accessed: December 1, 2018).

[65] S. al-Harbi, *Fiqh al-Muwazanah wa Hajat al-Imam wa al-Khatib Lahu* [Comparative Jurisprudence and the Proof of the Imam and the Neglectful Lecturer] at http://uqu.edu.sa/files2/tiny_mce/plugins/filemanager/files/4230042/feqh_almowaznah.pdf. (Accessed: December 1, 2018).

[66] S. Alsolamy, "Islamic Views on Artificial Nutrition and Hydration in Terminally Ill Patients," *Bioethics*, vol. 28, no. 2, pp. 96-99, 2014.

[67] M. Schultz et. al., "Reflections on Palliative Care from the Jewish and Islamic Tradition', *Evidence-Based Complementary and Alternative Medicine*, vol., no. 1, 2012, pp. 1-8. Available at doi:10.1155/2012/693092.

[68] Abu Dawud, al-Sijistani, *Sunan Abi Dawud* [The Canonical Collection of Prophetic Traditions by Abu Dawud al-Sijistani], Cairo, Dar al-Hadith, vol. 3, Hadith 3092, 1408/1988.

[69] Holy Quran: Sura al-Baqarah, verse 261.

[70] Holy Quran: Sura al-Nuh, verse 4.

[71] S.M. Ibn Abi Saeed, *Nur al Anwar* [The Light of Lights], Pakistan: Maktabah Rashidiyyah, 1967.

[72] According to the Office for National Statistics, 59.3% of the UK population identify themselves as Christians, 4.8% as Muslims and 0.5% as Jewish, available at https://www.ons.gov.uk/peoplepopulationandcommunity/culturalidentity

/religion/articles/religioninenglandandwales2011/2012-12-11. (Accessed: May 2, 2018).
[73] National Health Service, "Equality Objectives 2016-2020", available at https://www.england.nhs.uk/about/equality/objectives-16-20/. (Accessed: June 8, 2018).

Acknowledgements

We are grateful to Rabbi Ariel Abel, Dr Sohail Bhatti, Mufti Mohammed Zubair Butt, Reverend Canon Andrew Clitherow, Imam Yunus Dudhwala, Rabbi Dan Lieberman, and Reverend Dr Brendan McCarthy for their contribution to the multi-faith symposium which was generously funded by The Wellcome Trust [105463/Z/14/Z]. We are also grateful to the anonymous referees for their helpful comments on an earlier draft of this paper.

Reproductive Autonomy and Genetic Technologies: Ethical and Legal Implications Related to Sex Selection

Narine Harutyunyan
University of Bologna (CIRSFID), Italy

Abstract
Autonomy has always been at the center of discussions on reproduction. Development of various genetic technologies, aimed at the expansion of parental reproductive choices, resulted in heated debates on the limitations of the exercise of reproductive autonomy. The aim of this research is to analyse the arguments on the content and limitations of the concept of "autonomy" and "harm" with relevance to non-medical sex selection. In particular, I focus on interpreting "the harm principle" within the context of preconception and preimplantation sex selection.

Keywords
Autonomy; reproduction; harm; sex selection; genetic technologies.

I. Introduction

The latest advances in medicine and the increased access to modern biotechnologies have furthered discussions on the concepts of individual and reproductive autonomy in bioethics. In particular, genetic technologies like sperm sorting, pre-implantation genetic diagnosis (PGD), in-vitro fertilization (IVF), and non-invasive prenatal tests (NIPT) now allow potential parents to choose between a wide range of reproductive alternatives. One of the greatest benefits of these technologies is the ability to provide information on genetics in the context of human reproductive autonomy [1]. Such analysis can be used to inform individuals and thus reduce births of children with severely disabling genetic diseases. Some of these technologies can also be used to alleviate or cure painful or progressive genetic diseases. However, reproductive alternatives offered by new biotechnologies go beyond the medical necessity for patients seeking reproduction, as they can be used for genetic enhancement or selection of particular features of the child, such as sex, eye, hair or skin colour. The choice of these technologies is explained by the perceived notion that they are less harmful and invasive in their nature and are more reliable.

Non-medical sex selection,[1] that once was a common practice in some Asian countries (India, China, South Korea), is recently being traced in several Eastern European societies (Albania, Montenegro) and South Caucasus (Armenia, Georgia, Azerbaijan). In China and India, the harm caused by sex selection amounts to millions of missing girls [2, 20]. During the past few decades, most of these countries have adopted policies aimed at banning non-medical sex selections.

In Western Europe non-medical sex selection can mainly be traced in ethnic minority groups, but these practices remain uncommon. However, prohibitive regulations on sex selection also exist in most Western European countries (the UK, Germany, France, Switzerland) [4]. The basis for the sex selective policies in Europe is the Article 14 of the European Convention on Human Rights and Biomedicine (Oviedo Convention), which prohibits the use of techniques of medically-assisted procreation for choosing a future child's sex, except where serious hereditary sex-related disease can be avoided. Despite that some provisions of the Convention received a lot of critical and controversial feedback from the member states, there seems to be unanimity regarding Article 14. Interestingly, the prohibition of sex selection in Europe is not directly associated to "social harm" (extensive gender imbalance, increased discrimination against one sex, slippery slope to eugenic practices), and is rather a result of various ethical and legal considerations, which will be further addressed. In other countries, like the USA, Israel and the Australian state of Victoria, sex selections are not prohibited or can be practiced based on distinct regulations [5] [6].

[1] Sex selection is considered non-medical when carried out in order to satisfy parental procreative desire. It is also sometimes referred to as "social sex selection."

The aim of my research is to analyse the arguments on the content and limitations of the concept of "autonomy" related to sex selection. In particular, I focus on interpreting "the harm principle" within the context of preconception and preimplantation sex selection. For this reason, this essay will present the main arguments in favour of sex selection from scholars of the Millean line of thought and will show why those failed to be taken into consideration in countries with prohibitive regulations on sex selection. In particular, I will argue that the principle of harm is acting as a general guideline for the prohibition of non-medical sex selection, but its practical application can be problematic when dealing with enforcement of the ban on selection.

II. Theoretical overview: autonomy and harm from liberal and traditional theory perspectives

Autonomy, in general, has been equated with liberty (both positive and negative), dignity, integrity, individuality, independence, responsibility, and self-knowledge, self-assertion, critical reflection, freedom from obligation, etc. In bioethics, autonomy has most often been understood as a feature of individual persons, generally seen as a matter of independence, or at least the capacity of independent decisions and actions [1]. Meanwhile, reproductive autonomy (otherwise known as "procreative autonomy") has been extended to reflect the expansion of possibilities for individual self-determination and independence offered by new reproductive technologies [1]. While there are numerous definitions of "autonomy," the debates on limitations of the exercise of autonomy have a more practical significance.

A. Liberals on autonomy and harm

The commonly accepted basis for liberal bioethicists to define the concept of "reproductive autonomy" or "procreative liberty" is inspired by John Stuart Mill.[2] He notes that "the only freedom which deserves the name, is that of pursuing our own good in our own way [15]." Mill argues that the individual maintains sovereignty over their own body and mind – "[that] the only purpose for which power can be rightfully exercised over any member of a civilized community, against his will, is

[2] These terms are used interchangeably in bioethics.

to prevent harm to others" [7]. The principle of harm, as developed by Mill, became the foundation of arguments in favour of the unrestricted use of reproductive technologies (liberal theory of autonomy). These arguments have been further developed by several liberal scholars such as John Harris, John Robertson, and Ronald Dworkin. Following the ideology of John Stuart Mill, they argue that the use of reproductive (genetic) technologies should not be restricted, unless there is a harm caused to other people or society. In particular, John Robertson relies on the intuition that reproduction is a core human activity and that decisions on reproduction have a deep significance for personal identity and the meaning of one's life. Thus, he argues that "procreative liberty should enjoy presumptive primacy when conflicts about its exercise arise because control over whether one reproduces or not is central to personal identity, to dignity, and to the meaning of one's life…" [8]

Addressing the presumptive reproductive liberty, John Harris notes that presumption must be in favour of the liberty to access reproductive technologies and other means of founding families unless *good* and *sufficient* reasons can be shown against doing so. Thus, those who would exercise reproductive liberty would not have to show what good it would do; rather, those who would curtail freedom should not only demonstrate that it is unpopular or undesirable, but also that it is seriously harmful to others or to society; furthermore, it should be demonstrated that these harms are real and present, not probable in the future and speculative. Accordingly, the burden of proof is placed on the side which claims the existence of a very specific harm. Similar to John Robertson, Harris limits the understanding of liberty by stating that it should only be understood as a negative right. He claims that the right to reproductive freedom should be protected from the unnecessary intervention of the state (professional organizations, regulatory bodies) [9].

Dworkin rejects the abstract nature of the right to liberty; however, he points to the importance of rights to specific liberties. In particular, he acknowledges the rights of freedom of religion and freedom of speech or expression as rights protecting values of particular importance or moral and political significance [10]. For him, procreative autonomy is guaranteed by the First Amendment of the Constitution of the United States of America, which protects the freedom of religion and expression. Being able to express oneself, inter alia through the exercise of procreative liberty rights, stems from each individual's moral values. Referring to the principle of harm, Dworkin claims that if constraints on basic liberties harm us, it is not because of the constraint on liberty itself, but because of its impacts on something beyond liberty. Specifically, this implies that "what we have a right to is not liberty at all, but to the values or interests or standing that this particular constraint defeats [16] (p. 271)."

B. Traditional (feminist) theories on autonomy and harm

In contrast, more traditional ideas have been expressed by Onora O'Neill, Catherine Mills and others, who argue against the use of embryo screening and genetic manipulation for non-medical purposes.

In her book titled *Futures of Reproduction. Bioethics and Biopolitics*, Catherine Mills agrees with Harris and Dworkin that reproductive liberty is solely based on the negative right of non-intervention [11]. She notes that "any effective capacity to exercise the right to access reproductive technologies that are supposed to be protected by that reproductive liberty will require the cooperation of medical experts and others to ensure the success of that reproductive project [11] (p.39)." Referring to Onora O'Neill [1], Mills argues that contrary the arguments of Harris and Dworkin, reproductive freedom should not be a matter of self-expression; in other words, reproductive freedom should not be seen as an analogy to constitutionally-protected freedom of expression. According to Mills, the error in the analogy between the reproductive liberty and freedom of expression lies in the fact that the former aims to produce another rights-bearing individual, while the latter does not. Ignoring the difference between these two constitutes a form of "moral narcissism," as it fails to recognize the other as other and their consequent incorporation into one's own self-regard [11] (pp. 40-42).

Returning to the argument on the sole negative feature of procreative liberty, Mills argues that both Robertson and Harris fail to acknowledge the implications of self-making entailed by autonomy and reproduction; rather, their approaches stress the moment of choice. Mills adds that

> it is not simply genetic inheritance that establishes the importance of reproduction in people's lives; rather it is the

> bonds of familial attachments, and the vulnerability and responsibility that they entail, in the variety of forms they take, that ensure the existential and ethical significance of reproduction [11] (p. 45).

If this argument is true, the delineation of procreative liberty as solely negative is unsustainable. According to Mills, if freedom is understood as enacted in self-formation, it is important to note that the entanglement of reproduction, technology, and moral choice leads not only to an extension of freedom, but to its problematization. The problematization of freedom and the contestation of its limits implies that any *a priori* formulation of the limits of individual freedom will necessarily also be part of that contestation. Deriving from Foucault's construal of ethics, Mills highlights that reproductive freedom only gains meaning and reality through its enactment in everyday practices of ethical self-formation.

C. Practical implications of the application of the harm principle

The interference of the state in reproductive autonomy can be justified or not. In most cases, as already outlined, the limitation of these rights is supported by the Millean harm principle. Referring to Mill's principle of harm, Catherine Mills argues that despite the apparent simplicity, what can be considered harm is not immediately clear from the principle itself. I concur with Mills arguing that the everyday idea of harm is too vague and that harms may not always constitute the kind of moral wrong that is to be prevented by the principle. In addition, according to her the theory, harm is stretched to its limit in regard to cases involving future people [11].

Interpreting "harm," Mills refers to Joel Feinberg's distinction between harm and wronging [12]. Feinberg assumes that the principle of harm combines two understandings of harm: a non-normative notion referring to setbacks to interest, and a normative notion of harm as wrong. According to him, something constitutes harm if the interest concerned is in a "worse condition" than it would have been had the "invasion not occurred at all." The second sense of the normative notion of harm offered by Feinberg entails that "one-person wrongs another when his indefensible (unjustifiable and inexcusable) conduct violates the others rights [17] (p. 34)." By counter-arguing this interpretation, John Harris explained what constitutes harm in his view. According to him, Feinberg's focus on criminal law led to narrowing the principle of harm. Therefore, it is unclear whether such a conception of harm is appropriate to the application of the harm principle in ethics. Harris argues that it is both more economical and more plausible to say "to be harmed is to be put in a condition that is harmful," where a harmed condition "obtains wherever someone is in a disabling or hurtful condition, even though that condition is only marginally disabling and even though it is not possible for that particular individual to avoid the condition in question," or more simply, "'a harmed condition is one in which an individual is harmed or suffering in some way [18] (p. 109-110)."

This discussion shows that the idea of harm has become subjective and can hardly be established empirically. Nevertheless, in order to somewhat facilitate the discussion of the harm in the context of non-medical sex selection, I suggest to distinguish between social harm, individual harm and no harm (non-identifiable harm). This distinction is based on whether selection occurred in a country with socially widespread selective practices or it is an uncommon case within a particular society.

Most critical views on the extension of reproductive choice relate to the harmful large-scope consequences of the choice, that include "large-scale gender imbalance, increased discrimination against one sex, start of the slippery slope to more widespread selection, the child being treated as a consumer good or an inappropriate use of limited healthcare resources" [13] (p. 27). In addition, millions of the missing girls in China and India have been associated with other negative outcomes such as high crime rates, etc [2]. The interference with personal freedom in order to protect the society appears to be fully justified here.

As opposed to social harm, the issue of individual harm is more hypothetical in its nature, although not less important. Individual harm can be abortion, infanticide, or any other unwanted negative outcome for the mother or the foetus (such as post abortion health complications, psychological disorders, etc).

Catherine Mills argues that in formulating the principle of harm as the key limitation of an individual's actions (insofar as they impact on others), John Stuart Mill presupposed that the others in question were other already-existing individuals. Thus, in order to discuss the speculative harm faced by yet unborn people the concept of "non-identity problem" was developed. The non-identity problem stands in the existential theory that if individuals were not conceived in the month in which they were conceived, they would not exist – yet another, different, person would exist instead. Mills gives an example of a woman with a medical condition which implies that any child she conceives is born with a disability. However, if the woman undertakes a short treatment and then conceives, her child will be born with no disability. It appears that conceiving now would harm the child, while conceiving at a later time would result in the existence of another child. It is a choice between being born with a disability or not being born at all. The only instance in which one can say that the child in question was harmed is if the disability is so severe that the life is not worth living (thus it would have been better that the child was never born). While this dispute discusses the concept of "wrongful life," I believe it is analogous to cases when couples choose preconception or preimplantation sex selection methods: when a different embryo is conceived instead of the already-existing embryo which is of unwanted sex, the issue becomes hypothetical and leads to a regulatory dilemma [11].

III. The case of JS and LS vs Patient Review Panel

This section will discuss an important and unique decision regarding the permissibility of non-medical sex selection in the Australian state of Victoria.

The Assisted Reproductive Treatment Act of Victoria (ART) prohibits sex selection. However, it includes two exceptions: (1) sex selection can be undertaken to avoid the risk of transmission of a particular genetic abnormality or disease to a child of a particular sex, or (2) the Panel has otherwise approved the use of preimplantation genetic diagnosis (PGD) for sex selection [14]. Failing to comply with these regulations can lead to criminal prosecution. It follows, that the Patient Review Panel (or the Panel) has discretionary powers to approve non-medical sex selection on a case by case basis.

In the 2011 case of *JS and LS vs Patient Review Panel*,[3] a couple applied to the Victorian Civil and Administrative Tribunal to review the decision of the Panel, which refused to allow them PGD for non-medical sex selection [14]. The couple, who already had children of one sex, wished to conceive a child of another sex, following the death of the child of that same sex. As they applied to the Tribunal, they struggled to convince that having a child of the desired sex would help them recover from a psychological stress associated with the loss of their child. However, the Tribunal refused to grant them access to ART services based on the welfare-of-the-child principle. According to the decision of the Tribunal, sex selection would not be in the best interests of the child to be born.

Interpreting the child welfare principle cited by the Tribunal in theory is very difficult; it is even more problematic in case of its practical application. The Victorian Law Reform Commission has acknowledged its complex nature, by stating; that the health and wellbeing of a child must be given a priority. The commission found it difficult to identify ways in which the best interests of the child were served by permitting sex selection for non-medical reason. Therefore, in case of absence of available evidence about the likely effects on a child of having been selected for their sex, the commission adopted a more cautionary position.[4]

Moreover, this principle becomes more contentious when discussing the interests of unborn children, as the debates shift to mainly hypothetical domain. In our case, the Tribunal failed to explicitly mention those identifiable and established factors of risk which could in reality harm the child [14], apart from the embryo biopsy procedure as possible harm. In particular, it stated that providing a tissue for a biopsy when aiming to create an embryo (that is, a saviour sibling[5]) is more ethically (morally) justified than a biopsy with the sole purpose of establishing tissue type. Thus, according to the Tribunal's reasoning, non-medical sex selection is distinct from other types of selections, such as selection regarding serious genetic disorders

[3] *JS and LS v Patient* Review Panel[3] (Health and Privacy) [2011] VCAT 856.

[4] Final Report on Assisted Reproductive Technology and Adoption, 2007, Victorian Law Reform Commission, Victoria, 2007

[5] The saviour sibling is the child who is conceived for the purpose of providing a tissue that is compatible with that of his/her seriously ill sibling.

or saviour sibling selection, which at least have therapeutic goals. This argument seems controversial considering that the therapeutic goal of the selection is not related to the well-being of the child to be conceived, but to another, already born child. In this respect, the welfare considerations of children born as a result of selection can be identical in both cases.

It should be noted that in Australia, the protection of the parents' interests have been prioritized in many cases when the welfare of the child principle was not taken into consideration. One such example is the request to utilize the sperm of a deceased partner for the purpose of having his child without his explicit consent [14]. Such requests have been approved in the past as they would benefit the woman conceiving the baby. The welfare of the child who had to be born without one parent was not considered to be harmed. The applicants' argument failed to prioritize the welfare and interests of the unborn child by instead focusing on their own desires. However, this alone may not be a reason to deny the request, since it is not drastically different from when parents prioritize the welfare of the sick child who could be cured by using tissue of an unborn child [14]. In both cases, priority is not given to the interests of the unborn child. In case of sex selection, however, the welfare of the child to be born can be less disputable, as parents would have their desires or expectations finally met. Moreover, there is a high probability that this child will be more privileged compared to other children in the family.

Another argument of the Tribunal was that selecting the sex of the child was only a means to satisfy the parents' specific desires and the method was therefore focused on their psychological well-being, instead of that of the child. Later, the Tribunal stated that it was not proven by the applicants that they would recover from a post-traumatic stress disorder or that they would otherwise benefit psychologically from having a child of the desired sex. It stated that even if having a child of the preferred sex would have a positive effect on the couple, "their situation would fall far short of the gravity of a condition of a third party which would justify giving permission for an ART procedure to create a saviour child" [14].

Concluding, unlike in the famous case of the Masterton family[6] [19] (1), the regulatory framework of Victoria allowed the couple to apply to the Tribunal based on the possibility to circumvent the restriction of choosing the sex of an unborn child. In both cases, the couples were eventually refused to choose the sex of the child, but in the case of JS and LS, the Tribunal provided reasoning for the refusal. However, the decision of the Tribunal failed to discuss more relevant ethical and moral issues associated with non-medical sex selection. In my opinion, the comparison drawn between non-medical sex selection and selecting a saviour sibling did not provide legal and ethical clarification on why the applicants should be refused to choose the sex of their child. Moreover, the Tribunal's attempt to justify its decision, based on the necessity of pursuing a therapeutic goal, is not consistent with the ART regulation itself, since the latter provides the Panel with discretionary power to allow *non-medical* sex selection on a case by case basis. It follows that the Tribunal's reasoning on the permissibility of selection should not be associated with the therapeutic nature of the procedure.

IV. Conclusion

The prohibition of non-medical sex selection has many underlying justifications. First, it is a balancing (preventing) mechanism against a larger social harm, namely the skewed sex-at-birth ratio of a population, which has occurred in several countries with widespread selective practices [2, 20]. Second, the prohibition of non-medical sex selection is an incorporation of common moral values and principles of a certain society which considers sex selection as morally problematic.

[6] The Masterton family case is a high-profile case in the UK which involved controversial debates around the permissibility of non-medical sex selection. The family, who had lost their daughter in an accident, applied to the Human Fertilisation and Embryology Authority (HFEA) to allow them to undergo an in vitro fertilization treatment using pre-implantation genetic diagnosis, which could ensure that they conceive a daughter. The family argued that they needed a daughter and this was causing them severe psychological suffering. However, in order to apply to the HFEA, the couple was required to find a clinic which could apply for a license to undertake the treatment. Being unable to find such a clinic, the family later sought the treatment in Italy and conceived a male embryo that was donated.

As outlined, the current regulations of non-medical sex selection are generally restrictive in their nature. This is especially the case in Western Europe, where Article 14 of the Oviedo Convention plays a significant role on both legal and ethical grounds. The Article does not result in ethical dilemmas mostly because the general population does not favour sex selection. Many studies have confirmed the attitude of most Western Europeans against the moral choice of selection [22, 23]. However, there is a lack of up-to-date studies, which can confirm that this attitude has not changed in the past years.

Unlike the situation in Western European countries, non-medical sex selection in the United States is not prohibited by law [3, 21]. Some other countries (Israel, Victoria) allow sex selection if based on distinct regulations. Policies which allow undertaking sex selection under certain conditions should be well analyzed, because they are problematic with regard to their practical applicability.

Funding

This research was part of the International Doctoral Degree programme in Law, Science and Technology of the University of Bologna, funded by the European Commission.

References

[1] O. Neil, *Autonomy and Trust in Bioethics.* New York: Cambridge University Press, 2002.

[2] C. Guilmoto, *Sex Imbalances at Birth: Current trends, consequences and policy implications.* UNFPA Asia and Pacific Regional Office, 2012.

[3] M. Bayefsky, "The Regulatory Gap for Preimplantation Genetic Diagnosis." *Hastings Center Report,* vol. 45, no. 1, pp. 7-8, 2015.

[4] R. Landau, "Sex selection for social purposes in Israel: quest for the 'perfect child' of a particular gender or centuries old prejudice against women?" *Journal of Medical Ethics*, vol. 34, no. 4, 2008.

[5] G. Leiter, "What Israeli policy can teach us about elective sex selection," *Israel Journal of Health Policy Research*, vol. 3, no. 42, 2014.

[6] R. Zafran, "Non-Medical Sex Selection by Preimplantation Genetic Diagnosis: Reflections on Israeli Law and Practice," *North Carolina Journal of Law & Technology,* vol. 9, no.2/2, pp. 187-218, 2008.

[7] J. S. Mill, *On liberty and other writings.* Cambridge: Cambridge University Press, 1989.

[8] J. A. Robertson, *Children of choice: Freedom and the new reproductive technologies.* Princeton, NJ: Princeton Univ. Press, 1996.

[9] J. Harris, *Enhancing evolution: The ethical case for making better people.*Princeton, N.J: Princeton University Press, 2010.

[10] G. Dworkin, *The Theory and Practice of Autonomy.* New York: Cambridge University Press, 1988.

[11] C. Mills, *Futures of reproduction: Bioethics and biopolitics.* Dordrecht: Springer, 2011.

[12] J. Feinberg, 'The Child's Right to an Open Future'. *Whose Child? Children's Rights, Parental Authority, and State Power.* W. Aiken and H. LaFollette, eds., Totowa. NJ: Rowman & Littlefield, pp. 124-53, 1980.

[13] J. L. Scully, "Chance, choice and control: Lay debate on prenatal social sex selection," *Social Science & Medicine,* vol. 63, no.1, pp. 21-31, 2006.

[14] M. K. Smith, M. Taylor-Sands, "Comparing Non-Medical Sex Selection and Saviour Sibling Selection in the Case of JS and LS v Patient Review Panel: Beyond the Welfare of the Child?" *Bioethical Inquiry*, vol. 15, no.1 pp. 139–153, 2018.

[15] J.S. Mill, *On Liberty, London: John W. Parker and Son, West Strand*, 1859.

[16] R. Dworkin, *Taking Rights Seriously,* London: Duckworth, 1977.

[17] J. Feinberg, *Harm to Others.* Oxford, Oxford University Press, 1987.

[18] J. Harris, *Clones, Genes and Immortality: Ethics and Genetic Revolution.* Oxford: Oxford University Press, 1998.

[19] S. Wilkinson, *Choosing Tomorrow's Children: The Ethics of Selective Reproduction.* Oxford: Oxford University Press, 2010.

[20] C. Guilmoto, "The Sex Ratio Transition in Asia," *Popul Dev Rev*, vol. 35, pp. 519-549, 2009.

[21] M. Bayefsky, B.Jennings, *Regulating Preimplantation Genetic Diagnosis in the United States: The Limits of Unlimited Selection.* New York: Palgrave Macmillan, 2015.

[22] E. Dahl, K.D. Hinsch, B. Brosig, M. Beutel, "Attitudes towards preconception sex selection: a representative survey from Germany," *Reproductive Biomedicine Online*, vol. 9, no 6, pp. 600-603, 2004.

[23] F.Balen, "Attitudes towards sex selection in the Western world." *Prenatal Diagnosis,* vol. 26, pp. 614–618, 2006.

The Medicalization of Childbirth: Ethical and Legal Issues of Negative Childbirth Experience

Gergely Tari, Csaba Hamvai
Department of Behavioural Sciences, Faculty of Medicine, University of Szeged, Hungary

Abstract
As a potentially harmful process, medicalization should be analyzed in contrast to reproductive autonomy. We argue that it is crucial to advance women's control over reproduction, especially over childbirth, since disregarding autonomy could cause negative childbirth experience even without maternal or foetal morbidity or mortality. We present the results of a questionnaire-based research of our own regarding negative childbirth experiences among Hungarian women (n=506). We found that comprehensive medical information, active participation in medical decision making, freedom of movement during labour and delivery had positive effects, while the loss of control was correlated with negative childbirth experiences.

Keywords
Medicalization; reproductive autonomy; patient rights; childbirth; biopolitics; negative childbirth experience.

I. The medicalization of women' bodies

The term 'medicalization' is often used to describe an intrinsically negative phenomenon that is closely related to the philosophical problem of *power*. According to Peter Conrad, by medicalizing a phenomenon, we are describing

> a problem in medical terms, using medical language to describe a problem, adopting a medical framework to understand a problem, or using medical intervention to

> 'treat' it. This is a socio-cultural process that may or may not involve the medical profession, lead to medical and social control or medical treatment, or be the result of intentional expansion by the medical profession [1] (p. 211).

The exponentially growing medical knowledge and medical technology can be interpreted as a form of power as well, which led to a seemingly irresolvable asymmetry between doctors and lay people. In his work, *History of Sexuality I* and *History of Madness*, Michele Foucault argues that *power* at first is a 'multiplicity of force relations' that is not fixed, changes constantly, and influences our self-understanding and interpersonal relationships to a significant extent. "One needs to be nominalistic, no doubt: power is not an institution, and not a structure; neither is it a certain strength we are endowed with; it is the name that one attributes to a complex strategical situation in a particular society." [2] (p. 93). By accepting and supporting the nature of the Foucauldian definition of power, we first need to review the cultural and philosophical heritage that determines Western societies. First, regarding the issues of gender equality; and second, as a negative result of the first, the features of the paternal doctor-patient relationship.

Regarding the medicalization of women, it is essential to explore to what extent technical development affected the history of science, especially that of psychology and psychiatry. The development of medical technology and the scientific method turned the medical attention predominantly towards the *empirical* aspects of life. Somatization, as the primary concept of the Western medical approach, provided precisely the necessary epistemological background since it turned doctors' attention to the physiological side of the human being. This shift was also crucial for the survival of psychiatry, as it had to prove its objective, scientific roots.

The medicalization of women is based on several presumed ontological differences between males and females. These differences are significant for our explanation of how the Western civilization and its philosophical background created an *unequal and hierarchical* relationship between women and men – also in the delivery room.

As Foucault wrote in the *History of Sexuality*, there is a large difference between the 19th century's medical discourse of 'reproductive biology' and the 'medicine of sex.' The connection between the two is rather problematic:

> The role of the first with respect to the second was scarcely more than as a distant and quite fictitious guarantee: a blanket guarantee under cover of which moral obstacles, economic or political options, and traditional fears could be recast in a scientific sounding vocabulary [2] (p. 54).

In fact, the blanket guarantee that Foucault mentions prolonged the Aristotelian tradition into the 18th and 19th centuries, which sustained the idea of some essential differences between women and men. By the end of the 19th century, medicine served as a tool, whose truth and reliability was hardly disputable.

In Aristotle's Ancient Greek *polis,* women were not part of political decision making, and thus they could not be *zoon politicon*s. Although the feminist interpretations of Aristotle are quite diverse, it is likely that on the ontological level, he presumed a substantial difference between women and men. Such differences are also reflected in his *hylomorphistic views*. The well-known Aristotelian description of women *as 'deformed men'* suggests that the sexes do not possess the same amount of form and matter. There is strong evidence to suggest that being male and female are expressible on the level of matter and form as well, hence the two are gendered notions [3]. The explicit misogyny of the Aristotelian texts stems from the Aristotelian presumption that, biologically and ontologically, the two sexes are not equal. Feminist thinker Cynthia Freeman analyzed Aristotle's texts that characterized the relevant features of women:

> Aristotle says that the courage of man lies in commanding, a woman's lies in obeying; that 'matter yearns for form, as the female for the male and the ugly for the beautiful'; that women have fewer teeth than men; that a female is an incomplete male or 'as it were, a deformity': which contributes only matter and not form to the generation of offspring; that in general 'a woman is perhaps an inferior being [4] (p. 145-156).

II. The medicalization of childbirth and reproductive autonomy

Despite the interiorized social attitudes and the religious and philosophical tradition, civil rights movements started to demand *reproductive autonomy* in particular for males and females in the middle of the 20th century by advocating self-determination and human dignity for people in general. The term *reproductive autonomy* conceptualizes a universal and gender-neutral principle – this is especially important for women because of a simple biological obviousness: no matter how advanced medical technology is, the control over reproduction still belongs to women. Regarding medicalization, Laura Purdy argues that respecting the reproductive autonomy of women is the tool which could advance most women's welfare. So it is, since the term relates to all existentially crucial questions of reproduction, over gaining or losing control, thus highly influencing human freedom [5]. First, autonomy in this regard concerns the freedom to choose between contraceptive methods, as well as the timing of pregnancy, if the person wants to bear a child at all. Second, during their pregnancies, women must be allowed to decide on the location of delivery, the people whom they wish to be present, the diagnostic tests during each trimester, etc. Third, during labour, the woman must have the chance to maintain control over the process; that is, to receive comprehensive information from the Obstetrician, to be able to decide on the most comfortable birthing position, and so on.

In the Aristotelian perception, women were highly associated with their procreative function, which resulted in an obstacle to assertiveness. According to feminist bioethical literature, this tendency still existed in the 20th century. From this point of view, Nicolette Priaulx criticizes Abraham Maslow's theory regarding the basic and higher needs of women and men. According to her argument, by identifying basic (psychological, safety) and higher (belongingness, love, esteem) needs, Maslow prolonged the ancient Greek philosophical tradition. A deeper reading of the Maslowian texts shows, says Priaulx, that striving to boost self-esteem as a higher need for women still meant the primacy of childbearing [6].

The first document to include these rights for women was prepared at the 1994 International Conference on Population and Development under the coordination of the United Nations [5].

Following this milestone, the theory of respecting reproductive autonomy not only appears in philosophical and legal texts, but the number of obstetrics and gynaecology guidelines which integrate reproductive autonomy as a tool to reduce the vulnerability of birthing women is constantly increasing.

The International Federation of Gynecology and Obstetrics (FIGO) also acknowledges the importance to respect self-determination regarding reproductive issues:

> The principle of autonomy emphasizes the important role women should play in decision-making in respect to their health care. Physicians should try to redress women's vulnerability by expressly seeking women's choices and respecting their views [7] (p. 21).

However, the medicalization of childbirth is fed by many lay beliefs that suggest that it is better for pregnant women to renounce their agency. In a Hungarian study, Emma C. Molnár questioned over one hundred pregnant women on pregnancy-related issues [8]. The participants were asked whom they thought they should pay attention to when giving birth, and who would help them. The participants ranked the doctor as the first one, followed by the midwife and the husband, themselves ending up last [9] (p. 293). Patient control and women's involvement in decision making regarding childbirth have several psychological benefits: it reduces depressive and posttraumatic stress symptoms, and increases the overall patient satisfaction and self-esteem. [10] Considering the potentially-harmful consequences of women's passive role during labour and delivery, it is essential to abolish the paternalistic doctor-patient relationship from maternity wards. Informed consent and shared decision making are crucial for securing women's reproductive autonomy in this context.

The recommendations of the World Health Organization (WHO) emphasize women's self-determination during the different stages of delivery. In the latest edition of WHO recommendations regarding intrapartum care, over 50 recommendations were included to foster a positive childbirth experience [11] (p. 19-168). The list captures the potentially-harmful or unnecessary medical interventions that are still routinely applied in the obstetrical practice in many countries.

Despite the fall of maternal mortality rates, the WHO estimates that immediate medical intervention to prevent lifelong disabilities is still necessary in 10 to 15% of childbirths [12]. When criticizing the medicalization of childbirth, we believe that it is important to stress that medical technology and medical interventions on their own are completely value-free. The often criticized obstetrical practices such as episiotomy, Caesarean-sections or the administration of oxytocin have no inherent meaning from a moral point of view. However, applying them routinely could lead to negative outcomes by undermining the sense of human dignity. Therefore, the personalized intrapartum care should be a moral standard, since the misuse of medical interventions could subjugate women to doctors and the medical authority [13-14].

III. The Hungarian patient rights and reproductive autonomy

The Hungarian legal regulation also includes several patient rights which foster reproductive autonomy in practice. According to 'The Right to Health Care' women have the right "to choose attending a physician, with the agreement of the healthcare provider" (Title 2, Section 7(1)). Furthermore, 'The Right to Have Contact,' explicitly states that

> A woman in childbirth shall have the right to designate a person of age to stay with her continuously during labor and delivery, and after delivery, to have her new-born baby placed in the same room with her, provided it is not excluded by the mother's or the new-born baby's health condition (Section 11, 5).

According to 'The Right to Information' it is essential to provide complete information in an 'individualized form' even in the intrapartum period [15]. Hungarian Law accepts the limitation of this particular right and informed consent, if the patient does not have the capacity to decide. Nevertheless, there is no consensus in the literature whether women who experience high levels of pain while in labour have the capacity to decide, which raises medical, legal, and ethical debates [16]. The same issue applies to 'The Right to Self-determination' and 'The Right to Refuse Healthcare,' since both may only be exercised by those who have full capacity of decision.

Therefore, allowing a woman to refuse a seemingly effective medical treatment, such as an episiotomy, without being convinced of the capacity of the woman to decide, could lead to medical moral dilemmas. The situation could become even more problematic, if the couple provides the doctor with a directive (birth plan) in advance, in which the woman refuses an intervention that would be in her best medical interest in actual given situation.

Altogether, respecting patient rights rooted in the principle of reproductive autonomy is essential in order to achieve a positive childbirth experience and to avoid or minimize the occurrence of certain medical interventions.

IV. Non-medical reasons of negative childbirth experience

A. Objective and hypothesis

As argued so far, respecting women's reproductive autonomy is essential in obstetrics and gynaecology. However, the literature suggests that women are still rather vulnerable in the prenatal and perinatal stages in terms of self-determination [16-17]. Our starting hypothesis was that limiting reproductive autonomy and patient rights could lead to negative childbirth experiences, even if no maternal or foetal morbidity/mortality occurs. We also assumed that the medicalization of childbirth could cause the loss of control because of the medical practitioner's role – the guardian of safety –, which may easily overrule reproductive autonomy. We hypothesized that the following non-medical factors would have a significant connection with negative childbirth experiences: lack of information; restriction of women's movement during labour and delivery; loss of the sense of control; limitation of self-determination. We wanted to explore what women consider a negative childbirth experience and how medicalization and reproductive autonomy appear (or fail to appear) in the hospital setting.

B. Methods

We created an online questionnaire and sent it to the administrators of major Hungarian Facebook groups for women who have experienced childbirth. We collected 506 childbirth instances from women who gave birth in Hungary since 1997 and had negative experiences. We then analyzed the participants' answers to open-ended questions which asked them to describe their negative childbirth experiences. Following this stage, we processed the data gathered from closed-ended questions regarding childbirth preferences.

The questionnaire started by asking participants to score their childbirth experience in a 1 to 10-point scale, in which the lower scores represented negative experience, whereas the higher ones were linked to positive experiences. Before analyzing each childbirth instance, we filtered out the participants whose negative experiences were the result of clear medical problems. The cases of foetal and maternal morbidity or mortality were also excluded from the analysis. Furthermore, we also excluded the participants who scored their overall experience by more than 6 points. We thus reached a number of 435 childbirth experiences which we then reviewed.

C. Qualitative and quantitative analysis

Before the qualitative analysis, the main attributes of medicalization in regard of childbirth should be listed: (1) unnecessary, unwanted, or unexplained medical interventions; (2) the violation of the principle of reproductive autonomy and paternalistic doctor-patient interactions; (3) the violation of particularly important rights for women in labour, namely the right to information, the right to human dignity, the right to have contact, the right to self-determination, the right to refuse health care.

The Spearman correlation was applied to assess the relationship between negative childbirth-experience and non-medical traumatizing factors (the amount of sufficient information; the freedom of women's movement during labour and delivery; the sense of control). Participants were asked to score various statements regarding the previously-described features of reproductive autonomy (the amount of sufficient information; the freedom of women's movement during labour and delivery; the sense of control) on a 1 to 5 point scale (1 = completely agree; 2 = partially agree; 3 = not really agree; 4 = completely disagree; 5 = do not know). The respondents who could not decide were excluded from the statistical analysis. Multiple linear regression analysis was used to identify significant predictors of negative childbirth experience among the non-medical traumatizing factors. Frequency distributions were determined through Google Docs. Correlation and multiple linear regression analyses were performed in IBM SPSS Statistics 23.

D. Results

Closed-end questions suggested that comprehensive medical information ($r=0,25$; $p=<0,01$), active participation in medical decision making ($r=0,22$; $p<0,01$), freedom of movement during labor and delivery ($r=0,21$; $p<0,01$) had positive effects, while the loss of control was correlated with negative childbirth experiences.

The multiple regression model ($F=9,26$; $p<0,01$; $R^2=0,1$) explained 10% of the variance of negative childbirth experiences. The lack of sufficient information ($\beta=0,13$; $p<0,05$), the restriction of women's movement during labour and delivery ($\beta=0.11$; $p=<0,05$) were marked positive, while the loss of the sense of control was a significant negative predictor ($\beta=0,13$; $p=<0,01$). Collinearity and autocorrelation were not observed.

Hierarchical differences between practitioners and women are not favoured by women. This was further reflected in the responses given to the question on the preferred quality of doctor-patient relationship: 83% (n=348) of the respondents indicated a preference for a partner-like relationship over a paternalistic one.

While reviewing the childbirth instances (n=435), various forms of violation were identified. One of the most important one referred to the patient rights based on reproductive autonomy such as the Right to Self-determination (41%; n=181), the Right to Information (37%; n=161), the Right to Dignity (19%; n=82), the Right to Have Contact (10%; n=42), the Right to Refuse Treatment (5%; n=21).

V. Discussion

A. The Right to Self-determination

The Right to Self-determination is one of the essential patient rights, based on the well-known bioethical principle of reproductive autonomy. We collected the main freedoms and options in which this right is embodied during the intrapartum period: the possibility to choose between therapeutic options and medical interventions (artificial rupture of membranes; pelvic exam; enema; pubic shaving; episiotomy; IV-fluids; artificial oxytocin administration; vaginal birth/Caesarean-section, etc.), and the possibility to determine the position and place of labour and birth). 181 women experienced an unnecessary, unwanted, or disproportionate limitation of this right, despite that the right to self-determination and autonomy is the essential requirement of the partner doctor-patient relationship. One participant wrote: *"The doctor examined me again and without telling me or asking me anything, he ruptured the membrane!!!!!! He smiled and told me that now we would give birth. I know, but why couldn't he discuss with a frightened mother what and why he plans to perform on my body?!?!?"* Not only doctors were criticized, but also midwives. As another participant tells: *"The midwife had a completely medicalized approach, she only came to give me something, to examine me, but treated me as if I was just an object, she did not speak one word to me."* Many respondents expressed how much the burden of the loss of self-determination affected them. Indeed, the loss of self-determination is an alienating factor, hence violates the principle of autonomy and results in a paternalistic relationship. For the majority of the participants, the violation of self-determination represented the single negative aspect of childbirth, which shows how important safeguarding this right is during the intrapartum period. As another woman wrote, *"It was not the natural pain of birth which made my experience negative. But rather the medical interventions, of which I was not informed, such as membrane stretch and sweep, the Kristeller manoeuvres and other painful manoeuvres. Saying that these manoeuvres are really bad is just an understatement."*

B. The Right to Information

Personalized, objective, and understandable medical information is needed in order for women to gain control and practice patient autonomy in the delivery room. FIGO and WHO guidelines emphasize informed consent, considering it a highly effective tool in reducing the vulnerability of women during pregnancy. Providing continuous information in the intrapartum period is not only important on its own, but it also facilitates the agency of women when giving birth. Respecting the reproductive autonomy of women during childbirth can only be fulfilled if the hospital team allows them to articulate their wishes according to their preferences. Generally, women are competent during labour, therefore *any form of unnecessary or disproportionate limitation of informed consent should be recognized as a harmful by-product of paternalism as a form of medicalization.*

We reviewed the childbirth instances focusing on complaints based on the lack of communication and insufficient medical information. The results showed that the Right to Information was also frequently violated. 167 stories explicitly referred to a perceived violation of proper medical information. One of the respondents wrote: "No one informed me, or asked for my permission to perform any examination. I was given medications without being informed or giving consent. They did not treat me as a competent individual while giving birth." Another woman also reported loss of control because of lack of information: "Having insufficient information was the most difficult part of the experience, I felt like a lost child." These complaints offer a dramatic insight in the situation of women who are forced to give birth without being informed. Limitation of information and autonomy lead to a paternalistic relationship -- which is the opposite of what women need during childbirth. Throughout the instances we reviewed, the following seems to be a typical scenario: "I had a clear concept of how I would like to give birth. It just did not happen, they brought me the consent forms when I was having contractions, so I could not even assess what I was signing while I was urged that I need to sign here and there." Many instances showed that women were not informed, despite the fact that they wanted to be. The following example shows how vulnerable one can become if the minimum information is refused to them: "If I asked something, the midwife only replied in two-word sentences. Then she yelled at me derisively: 'How curious can one be…you'd better do what is your business, just like I do.' But I knew neither what to do [i.e. what was my business], now what was happening to me and why."

The 'Right to Self-determination' and the 'Right to Information' are proper tools to enhance reproductive autonomy for women who give birth. However, medicalization often results in unequal relationship in delivery rooms. The information asymmetry between the medical team (doctors, midwifes, nurses, etc.) and the women, sometimes brings unnecessary and causeless deprivation of control and reproductive autonomy. We believe that a partner-like doctor-patient relationship, shared decision making, and informed consent are the only morally acceptable tools to fill the information gap, hence paternalism as a form of medicalization must disappear from obstetrics and gynaecology. Withholding information and limiting self-determination is potentially harmful. Their violation can lead to terrible childbirth experiences, as one of the respondents emphasized: "I feel like I was deprived of the most important event of my life; it was not the most beautiful, but the most terrible. I am sure it would not have happened that way, had I been informed, prepared, and supported."

C. The Right to Human Dignity

Protection through the 'Right to Human Dignity' is crucially important for patients who are vulnerable in hospital setting. The fragile position of women in labour must be recognized, as they may feel exposed to medical practitioners due to information asymmetry. Some examples of violations of the Right to Human Dignity, identified by the respondents, include disrespectful or rude behaviour of the healthcare practitioners, as well as unnecessary, unwanted, or unexplained restrictions.

82 respondents expressed some form of human dignity violation. Objectifying, that is, reducing patients to objects in need of 'repair' is a frequent feature of medicalization. The partner-like doctor-patient relationship should be the exact opposite, as the cooperation between the two parties should be based on a simple recognition: people who seek help from medical practitioners must be perceived as subjects, not as objects, thus deserving respect and appropriate communication.

One of the respondents explained objectification it the following manner: "No one cared about my feelings, my fears. I felt humiliated and vulnerable. I was just a 'piece of work,' while giving life to my baby." Another frequent instance of human dignity violation, as expressed by the respondents, was related to alienation. Many respondents said they felt like a 'piece of meat' or an 'object': "I was just one piece of meat among the others. Not a human being. Not a woman." Such experiences were more common for patients with Caesarean-sections: "I felt like I had nothing do with the birth. I felt wretched and incompetent. […] It was just like a slaughterhouse. Rigid, metallic. And above all, inhuman. When I cried, I was told that I was not a three-year-old anymore."

D. The Right to Have Contact

The Hungarian healthcare legal regulation provides women with the right to designate someone to assist them during labour and delivery. Nevertheless, in 42 cases the Right to Have Contact was limited through a questionable manner. There were several obvious cases of misinformed women: "We arrived to the hospital at the end of an absolutely problem-free pregnancy. After that – although our doctor had previously confirmed (multiple times), that my husband can be with me during labour – we were unreasonably separated from the beginning."

Unreasonable limitation of the Right to Contact could lead to vulnerability and loss of the sense of control, therefore decreasing reproductive autonomy. "My husband was not allowed to be with me, he was sent home, and I was left alone during labour. I was not informed about anything, I felt vulnerable." Women's mental health is positively affected not only by physical and informational support, but also by emotional support [18]. As already emphasized, the emotional support during labour reduces the number of childbirth complications and postpartum depression, therefore it is crucial to allow women to choose a close person to assist them, this being an essential part of reproductive autonomy during the intrapartum period.

E. The Right to Refuse Treatment

This right ensures that competent women at birth can refuse any medical intervention that is not life-saving. A frequent issue signalled by the respondents regarded respecting women's active role in the decision-making process. The responses included cases where women were not allowed to refuse an otherwise not life-saving medical intervention. 21 participants felt that health care practitioners deprived them of practicing this form of self-determination. "They did not inform me on the upcoming examinations, somebody just came in (no introduction, nothing, he could have been a maintenance worker), touched me, and left. The midwife and the doctor neglected my requests. I was given oxytocin without information and against my will." "I had indicated in advance, that I did not want any medication to accelerate my labour, but they eventually gave me oxytocin in fear of prolongation." Disproportionate or unnecessary limitation of the Right to Refuse Treatment is potentially harmful if the right only exists on the theoretical level. If competent women are not allowed to make informed consent during labour, it opens the ground for paternalism. As the WHO and FIGO guidelines warn, it is not recommended to routinely perform medical interventions at birth, such as episiotomy. Nevertheless, some of the respondents claimed that the limitation of the Right to Refuse Treatment was due to "hospital protocol." If this is the case, health care providers must protocols and recommendations on labour and delivery.

F. Paternalism: an unnecessary, disproportionate limitation of reproductive autonomy and patient rights

As argued above, the medicalization of childbirth manifests in various forms. However, the results are similar, since paternalism is an obvious consequence of the misuse of medical power. Unnecessary or disproportionate limitation of reproductive autonomy is a result of paternalism. Nevertheless, the intention of the healthcare practitioners could be intrinsically good in most situations if the limitation of autonomy is based on a medically-sound reason. Overriding self-determination could serve the patient's best medical interests, since immediate medical help can prevent life-long disabilities or even maternal or foetal death. Paternalism, as the suppression of patient autonomy and self-determination is not immoral in itself. However, it is crucial to emphasize that it is problematic to disregard the will of an otherwise competent individual.

Within the respondents' answers, paternalism was also a major complaint. 141 women to have felt vulnerability, self-abandonment, and infantilization because of the medical environment: "When I arrived at the hospital, I became a puppet. I seemed to not be competent at all; they arranged everything without my will. They did not inform, ask me anything, and I was laughed at if I asked something." Alienation, as another result of paternalism, was also a complaint: "The doctor's and midwife's lack of communication and empathy made it horrible. They discussed important issues without me, issues in which I should have been involved, and they asked each other questions that I should have been asked as I was the one who knew the answer. They did not explain what was happening and I felt ordered around."

Paternalism and its forms were the most common complaints of the respondents. Again, parentalism has many features, but has similar consequences: alienation from birth, vulnerability, and the medicalization of an otherwise natural physiological process: "I felt like someone who is not capable of assessing her own condition and feelings, just like an incompetent patient." Transforming healthy people into sick ones, using medical technology to expand control over the unpredictable features of birth (duration of labour and delivery, the emotions of the woman in labour, the need of birth-experience personalization) are essential parts of medicalization.

Our results are in accordance with the findings of other studies which explore the ethical and legal aspects of obstetric care. Similarly, a recent study also found that the Right to Information and Self-determination are often violated in obstetrics practice in Hungary [19]. The impact of reproductive autonomy in the sense of women playing active roles during labour and delivery in order to facilitate a positive childbirth experience is also confirmed by recent studies [20, 21, 22].

VI. Conclusion

This essay gave an insight to the medicalization of childbirth. The most well-known guidelines regarding pregnancy and childbirth (such as those of the FIGO or WHO) emphasize the importance of reproductive autonomy. However, both the FIGO and WHO warn that there are countries in which medicalization is an issue to be resolved, because of the unprecedentedly high prevalence of medical interventions such as Cesarean-sections or episiotomies.

Our survey explored the manners in which childbirth experiences are affected by medicalization. The results showed that a significant number of types of medicalization are potentially responsible for negative childbirth experiences. By reviewing the respondents' answers, several non-medical reasons were found which affect women's satisfaction at birth. Unwanted, unexplained, or unnecessary medical interventions, disproportionate limitation of self-determination are the most common ethical and legal issues of childbirth. It should also be highlighted that the majority of the respondents prefer a partner-like doctor-patient relationship and shared decision making instead of paternalism. Since hierarchical differences between healthcare practitioners and women increase vulnerability during labour and delivery, it is essential to allow reproductive autonomy and self-determination dominate the intrapartum care. Therefore, in line with informed consent, women must be involved in the decision making process after being given comprehensive medical information in an understandable manner. Practitioners must consider women's preferences regarding the nature and timing of medical interventions, position during labour and delivery, and choosing a person/family member to assist them during labour.

References

[1] P. Conrad, "Medicalization and Social Control," *Annual Review of Sociology,* vol. 18, pp. 209-232, 1992.

[2] M. Foucault, *The History of Sexuality.* New York: Random House, 1978.

[3] C. Witt, L. Shapiro, *Feminist History of Philosophy.* https://plato.stanford.edu/archives/spr2017/entries/feminism-femhist: Metaphysics Research Lab, Stanford University, 2017. (Accessed: September 10, 2018).

[4] C. Freeland, "Nourishing Speculation: A Feminist Reading of Aristotelian Science," *Engendering Origins: Critical Feminist Readings in Plato and Aristotle,* Bat-Ami Bar On, eds., Albany: State University of New York Press, 1994.

[5] L. Purdy, "Women's reproductive autonomy: medicalisation and beyond," *Journal of Medical Ethics,* vol. 32, no. 5, pp. 287-191, 2006.

[6] N. Priaulx, "Rethinking Progenitive Conflict: Why Reproductive Autonomy matters," *Medical Law Review,* vol. 16, no. 2, pp. 169-200, 2008.

[7] FIGO, *Ethical Issues in Obstetrics and Gynecology.* London: FIGO, 2015.

[8] E. C. Molnár, *Az anyaság pszichológiája* [The Psychology of motherhood]. Budapest: Akadémiai Kiadó, 1996.

[9] V. Sallay, T. Martos, E. Hegyi, "Fiatal nők szüléssel kapcsolatos attitűdjei: a Szülés-attitűdök Kérdőív kidolgozása és első eredményei," [Young women's attitudes towards childbirth: Developing and introducing the first results of the Birth-attitude Questionnaire], *Létkérdések a születés körül* [Vital issues on chilbirth], K. Barbara, ed., pp. 291-313. Budapest: L'Harmattan, 2015.

[10] H. Goldberg, "Informed Decision Making in Maternity Care," *The Journal of Perinatal Education,* vol. 18, no. 1, pp. 32-40, 2009.

[11] WHO, *WHO recommendations: intrapartum care for a positive childbirth experience.* Geneva: World Health Organization, 2018.

[12] R. Johanson, M. Newburn, A. Macfarlane, "Has the medicalisation of childbirth gone too far?" *BMJ: British Medical Journal,* vol. 324, no. 7342, pp. 892-895, 2002.

[13] R. Kukla, K. Wayne, *Pregnancy, Birth, and Medicine.* (E. N. Zalta, ed.) Stanford University: Metaphysics Research Lab, 2018.

[14] A. D. Lyerly, "Shame, Gender, Birth," *Hypatia*, vol. 21, pp. 101-118, 2006.

[15] Act CLIV 1997 on Health.

[16] J. M. Torres, R. G. De Vires, "Birthing Ethics: What Mothers, Families, Childbirth Educators, Nurses, and Physicians Should Know About the

Ethics of Childbirth," *The Journal of Perinatal Education*, vol. 18, vol. 1, pp. 12-24, 2009.
[17] M. S. Rosenthal, "Socioethical Issues in Hospital Birth: Troubling Tales from a Canadian Sample," *Sociological Perspectives*, vol. 49, no. 3, pp. 369-390, 2006.
[18] M. Iliadou, "Supporting women in labour," *Health Science Journal,* vol. 6, no. 1, pp. 385-391, 2012.
[19] I. Szebik, E. Susánszky, Zs. Szántó, A. Susánszky, N. Rubashkin, "Ethical implications of Obstetric Care in Hungary: Results from the Mother-Centred Pregnancy Care Survey," *EJMH*, vol. 18, no. 1, pp. 51-69, 2018.
[20] A. Karlström, A. Nystedt, I. Hildingsson, "The meaning of a very positive birth experience: focus groups discussions with women," *BMC pregnancy and childbirth,* vol. 15, pp. 251, 2015.
[21] O. T. Oladapo, Ö. Tunçalp,M. Bonet, T.A. Lawrie, A. Portela, S. Downe, A.M. Gülmezoglu, "WHO model of intrapartum care for a positive childbirth experience: transforming care of women and babies for improved health and wellbeing," *BJOG: an international journal of obstetrics and gynaecology*, vol. 125, no. 8, pp. 918-922, 2018.
[22] M.B.C. Ferrer, M.C. Jordana, C.B. Meseguer et al., "Comparative study analysing women's childbirth satisfaction and obstetric outcomes across two different models of maternity care," *BMJ Open*; 6:e011362, 2016.

Prophylactic Mastectomy on Demand

Csaba Hamvai,[1] Gergely Tari,[1] Melinda Csenki[2]

[1] University of Szeged, Department of Behavioural Science, Hungary
[2] University of Szeged, Department of Oncotherapy, Hungary

Abstract

This paper presents the case of a breast cancer patient who demanded prophylactic mastectomy after her contralateral breast cancer was treated. Removing the otherwise healthy organ was contraindicant according to both the surgeon and the oncologist, since the patient was neither a BRCA mutation carrier, nor did she have a strong family history with this disease. This situation resulted in the clash of 'autonomy' and 'nil nocere' bioethical principles. Besides the bioethical aspect, this piece will discuss legal and psychological issues of the case.

Keywords

Prophylactic mastectomy; Sword of Damocles syndrome; informed consent; principle of autonomy; principle of *nil nocere.*

I. Prophylactic mastectomy

Prophylactic mastectomy (PM) is the removal of one or both breasts to reduce the risk of developing breast cancer. The procedure itself has a long history. In the first century AD the Greek physician Leonides already used a special technique, the 'Escharotomy' to operate breast cancer. Hot poker and incisions were successively used in this method until the breast was fully burnt off. Of course breast cancer surgery has developed a lot since Leonides' first efforts. An important turning-point took place in the late 19th century, when the American William Halsted described the steps of the radical mastectomy and emphasized the importance of aseptic methods and anesthesia. Due to Halsted's contributions, operative mortality rate dropped dramatically, although the survival rates did not prove to increase after the radical mastectomy.

In the 1980s the Halsted dogma on radical mastectomy was questioned step by step, and the surgeons started to turn towards the possibilities for breast conversation (lumpectomy). The latest advances in the evolution of mastectomy are the nipple-preservation techniques [1].

PM is recommended only in particular conditions. The risk assessment takes into consideration several factors. For example the *BRCA1* and *BRCA2* gene mutations might indicate an increased risk for developing breast cancer thus PM is advisable in *BRCA* positive mutation tests. Strong family history is also among the main indications [2]. The medical team also have to consider the impact of the operation on disease free survival and overall survival. In this regard, the field has not had any convincing clinical data yet [3].

As we see, PM is advisable only for a particular risk population. Nevertheless, several studies show an increasing tendency for contralateral PM among women who underwent a surgical treatment for their unilateral breast cancer [4]. For example, Wong et al [5] found that during their study period, the rate of contralateral PM use more than tripled: the proportion of women who received this treatment increased from 3.9% in 2002 to 12.7% in 2012. The number of PM grew despite the fact that no significant improvement in breast cancer-specific survival and overall survival was found for women undergoing this more invasive, radical treatment when it was compared to breast conversing therapy. It seems that several women received a therapy that otherwise did not have any benefit over another less invasive surgical methods. Several reasons account for the increased rate of PM. Media conveys stories about celebrities undergoing mastectomy, strengthening the image of the preventive and life saving effect of this medical option. Heuristics, intuitive lay beliefs also intensify the willingness on more radical procedures. Women try to reduce the uncertainty of the future, and regain the sense of control by demanding PM. They might also manage the so-called future regret, the belief according to which, "I want to do everything I can now, because if I get recurrence, I feel that I did all that I could" [3] (pp. 3210).

This raises several bioethical dilemmas and questions which we would like to discuss and illustrate with a clinical case. First, the case will be presented, after which the bioethical, legal considerations, and guidelines will be outlined. Finally, the psychological background of demanding a PM will be also discussed.

II. The clinical case

The 35 years old female patient discovered the lump in her left breast herself. The mammography and core biopsy confirmed a malignant breast cancer (*neoplasma malignum mamme*). The patient first received neoadjuvant chemotherapy, then the tumour was operated and removed. Later, she also received postoperative radiotherapy. Finally, the treatment resulted in a complete remission.

Four years later the follow-up mammography and core biopsy showed the recurrence of the tumour. This time complete mastectomy was carried out and the axillary lymph nodes were also removed. Her treatment then finished with chemotherapy.

One year later, the patient demanded contralateral PM for reducing the risk of developing cancer in the other breast. Neither the oncologists nor the surgeon found this treatment advisable in her case for several reasons. First of all, the patient's *BRCA* tests proved to be negative, thus she was not a *BRCA* mutation carrier. Furthermore, she did not have a strong family history with similar diseases. The surgeon also informed the patient that the operation could cause unnecessary complications and harm with no likely benefit. He further argued against the procedure by emphasizing that it would remove a completely healthy organ which is also part of her femininity. Consequently, although the patient insisted on her request, the surgeon denied it.

Next, the patient was offered psychological and psychiatric consultation, which she accepted. It resulted that the patient was suffering from extreme distress (the psychological aspects of the case will be detailed later). A bioethical specialist was also involved in the case and in the end, considering the patient's psychological state, a different surgeon carried out the PM.

III. Ethical and legal aspects of the case

Like any other medical treatment, PM implicates moral issues. Bioethical principles provide useful guides in the medical practice to resolve such ethical dilemmas [6]. In the case of risk-reducing mastectomy there are usually two conflicting bioethical principles: the principle of beneficence and the principle of autonomy. The principle of beneficence means that the health care provider should assure the benefit of the patient. This includes both preventing harm and protecting the patient. Within the medical context, the principle of autonomy means that patients are rational agents who are involved in making informed and voluntary decisions [6]. This also implies that only those agents who can understand and choose between the offered therapeutic options can be considered autonomous. In the case of the impaired or those lacking cognitive or volitional capacities, autonomous action is reduced or impossible [7].

The two bioethical principles typically clash when more medical options are plausible for treating breast cancer, such as a radical risk-reducing mastectomy or a conserving surgery. The patient may choose the more invasive mastectomy over the conserving surgery. Nevertheless, the surgeon's aim is to achieve the same outcome with a less harmful conserving surgery which can also reduce the risk of complications.

In our case, a different kind of bioethical dilemma can be identified: the opposition of the patient's autonomy and the *nil nocere* (do not harm) principles. *Nil nocere* or non-maleficence requires the health practitioner to not cause any form of intentional harm either through acts of commission or omission [6]. The complexity of the situation stems from the fact that the patient demanded the removal of an otherwise healthy organ. The surgeon argued against the PM and most of his arguments reflected on the principle of *nil nocere*. From the medical perspective, the surgeon's standpoint was decent and correct. Nevertheless, he also had to suppress an otherwise competent patient's request, and violate her autonomy. Angelos et al. [3, p. 3210] highlighted the essence of the dilemma:

> A central controversy is the following: Are we seeing the appropriate extension of respect for patient autonomy in going along with all requests for CPM?* Or has the pendulum swung too far such that physicians no longer have input on what will benefit patients? Patient choice is critical in breast cancer surgery. **contralateral prophylactic mastectomy*

Eventually, the surgeon's decision resulted in the so-called benevolent paternalism. The roots of the paternalistic attitude can be found in the Hippocratic Oath, the ethical code which implicitly ruled worldwide medical practice for centuries. The dominant principle in the oath is the principle of the previously-discussed beneficence. The overemphasis of this principle and the subordination of the other assumptions to it, leads to medical paternalism. The paternalistic attitude implies that every medical act is morally correct if it serves the patient's well-being. It is also based on the presumption that the patients do not want to know their own personal good, and they would automatically consent the physician's option as being the best [8].

The possible legal guidelines for similar problems should be outlined at this point. The "Act CLIV of 1997 on Health" [9] of the Hungarian law includes two relevant sections: "The Right to Self-determination Section 15. §" and "The Right to Deny Care Section 131. §". The former assures the patient's autonomy in some cases:

> (1) The patient shall have a right of self-determination, which may only be restricted in the cases and in the ways defined by law.
> (2) Within the framework of exercising the right of self-determination, the patient is free to decide whether he wishes to use health care services, and which procedures to consent to or to refuse in the course of using such services...

Two remarks should be stressed here. Firstly, the section refers to the situations in which the patient's autonomy can be limited: public health emergency, life-threatening emergency situations, or the cases of incompetent patients (minors, mentally impaired, unconscious etc.) are the most typical examples. Second, the section warrants the patients' self-determination only for accepting or refusing a treatment, while not providing a similar right to demand a treatment. This is in line with the "The Right to Deny Care Section 131. §":

> (3) A physician may refuse to provide care for a patient only following an examination, if in the course of the examination he determines that a) the patient's health status does not require medical care, b) the treatment requested by the referring physician or the patient is not justified professionally.

In summary, the law seems to represent only a partial patient autonomy in cases when the treatment is already recommended. The final choice is entirely in the physicians' hands when the patient himself or herself initiates the treatment. So it is still a question of how the patient autonomy can be assured in this situation.

Although the right to self-determination is considered the cornerstone of the patient-centred care, the legal framework also emphasises the rights of the health care provider. Altogether, the Hungarian patient right system follows the global trend, thus a comprehensive medical information grants autonomy in practice. However, the Hungarian patient right system that links Hungary to all the other countries that facilitate the partner-like doctor-patient relationship with the declaration of the right to self determination has only been valid since 1997. Consequently, it still takes time more to abolish the potentially-harmful practice of paternalism and change the mindset of both doctors and lay people to routinely involve patients in the decision-making process.

IV. The shared decision making, the waiting time, and the Manchester guideline

The most important pragmatic manifestation of patient autonomy is the Shared Decision Making. Shared Decision Making is "an approach where clinicians and patients share the best available evidence when faced with the task of making decisions, and where patients are supported to consider options, to achieve informed preferences" [10] (p. 1361). In general, Shared Decision Making can be a three-step-process. During the step of the *choice discussion,* the patient's awareness of the existing alternatives is raised. During the *option discussion*, the patients receive more detailed information on treatment options. Finally, the *decision discussion* includes the consideration and debate of the preferences as well as the final choice [10]. This strategy also helps maximize the patient's autonomy in breast cancer decisions by synchronizing the patients' own values with the physicians' expertise which is based on up-to-date medical science [11]. In an optimal situation, the patient, in the end, gives an informed consent.

The use of the so-called "cooling off" period might also promote informed consent. As patients are usually overwhelmed or shocked by the different alternatives, it can be effective to postpone the final decision. Thus bad and premature decisions, born under the emotionally-overloaded situation, can be prevented.

The *Manchester Guidelines for Contralateral Risk-Reducing Mastectomy* is a 5-step recommendation developed by a multidisciplinary team especially for cases where women request PM [12]. The guideline integrates the aforementioned autonomy promoting the strategies of Shared Decision Making and waiting period. the importance of guideline should be stressed here, as the clinical team dealing with our case study managed the ethically-challenging case in a similar way. Step 1 is observing the history which includes the clarification of the major reason for the PM request. Our clinical team repeated this phase repeated several times. The surgeon and two oncologists explored the patient' priorities and discussed with her the different alternatives. During step 2, the risk of developing contralateral breast cancer is assessed. As we mentioned, the patient was not assigned to the risk population according to the risk analysis. Step 3 represents the cooling off period, although this phase can be introduced at any time of the process. It is important to note here that the guideline does not give any specific recommendation regarding the length of the cooling off period. In our case, consultations with various specialists automatically provided repeated waiting periods for the patient. It took approximately 5 months until the patient was finally operated. During step 4, a multidisciplinary team, including a breast care nurse, a pathologist, a breast surgeon, and a radiologist should assess and discuss the case. The presence of a psychologist or psychiatrist is not mandatory but additional psychology exploration can also be necessary in some cases. Our case was reported and discussed within an onco-team only comprised of physicians (oncologists, radiologist, surgeon). Later, when the communication with the patient reached a blockage, a psychologist, a psychiatrist and a bioethical specialist were also involved in the case. During the waiting period, the patient visited 6 times the psychiatrist who was specialized in oncological cases. During step 5, the specialist first summarizes the positive and negative aspects, the possible outcomes and side effects, followed by the patient who signs the informed consent form.

V. A psychological assessment protocol for women requesting prophylactic mastectomy

Since psychological consultation and exploration with our patient led to a better understanding of her values, priorities, and motivation, it would be useful to present a psychological assessment protocol which can be used in similar situations. The patient's psychological status will be also described in this section, giving an example on the motivations which result in the wish to undergo PM.

The protocol addresses the issues below during the psychological consultation [13] (pp. 712). Many of them were discussed with the patient:

- The impact of family history and risk understanding
- Motivating factors and the process of decision making
- Sexuality and body image
- Interaction with partner, close family members and their reaction to potential surgical procedure
- Past psychiatric history and current mental health
- Impact of family history and bereavements on mental health
- Expectations of surgery and discussion of potential post-operative complications

The patient's major motivation for undergoing the risk-reducing mastectomy was managing her extreme fear of developing contralateral breast cancer. She explained that she had recently discovered a lump in her breast. Although it was proved to be benign and harmless, it intensified her anxiety and fear. She wanted to do everything to avoid the shock she underwent when her first malignant tumour was diagnosed. This is known as the Sword of Damocles syndrome, the extreme fear of cancer reoccurrence or development [14]. It is in line with several findings according to which reducing future risk [15], or reducing the anxiety of developing cancer [16] were among the women's main reasons to request PM. Regarding her sexuality and body image, the patient stated that she was ready to accept further changes on her body, it was irrelevant to her, although this was one of the surgeon's main arguments. The patient had strong family ties who supported her demand for the PM. Either her mother or her husband were also present during the medical consultations. The patient did not have any psychiatric history, and her mental status (appearance, cognitive functions, thought process, level of consciousness) seemed normal, except for her mood and emotional state. Regarding the latter, the patient was very anxious and frustrated, and reported suicidal thoughts. Her frustration came from the feeling that she had been treated as a child who could not make a decision about her own body. Based on these symptoms, she was diagnosed with "adjustment disorder, unspecified" (ICD-10 Code: F 4320). There was no aspect in the patient's family history (including bereavement) which would have influenced her mental status. Finally, the patient had clear and realistic expectations of the surgery and the potential post-operative complications. Many patients' decision is driven by their misconceptions and heuristics about mastectomy [3]. Kimberly Van Zee, surgical oncologist wrote [17] (p. 1722) that "there is a huge misconception out there. Many women think that removing their other breast will cure them of the disease they already have, but there is no survival advantage."

This was not the case with our patient. She was aware of the possible complications, the long and painful recovery period and the worst scenario: the cancer can come back in another form with or without the mastectomy.

In summary, the patient considered the PM as a means of coping with her anxiety caused by an uncertain future and the possibility of developing another cancer. After realizing this, the medical team was flexible enough to revise their first decision.

After the surgery, the patient had follow-up visits at the psycho-oncology ambulance. She was satisfied with her decision, and her anxiety and distress were dramatically reduced, while her current quality of life was significantly improved. Her psychotherapy had a new focus: return to work. gradually, she changed her stressful job for a less exhausting one.

VI. Discussion and conclusion

We presented a case in which a female patient requested a PM. At the beginning, her request was denied on the basis of several medical assumptions causing a clash between the bioethical principle of *nil nocere* and autonomy. Finally, the patient's autonomy overwrote *nil nocere* and the patient was satisfied with the decision. Nevertheless, the case still raises questions and issues.

First of all, can psychological factors, such as the patient's Sword of Damokles syndrome, weigh more than medical concerns which deem the relative benefit of the treatment too low compared to its possible complications. In our case, the answer was yes. Nevertheless, future protocols of PM should give more focus on psychological assessment too.

Another conclusion of this case stands in the difficulty of actually carrying out the Shared Decision Making. Although the patient's priorities and values were obvious at the beginning, it was hard to establish their agreement with the medical concerns.

The final issue raised by the case is the patient's long-term satisfaction with her decision, i.e. could a patient psychologically profit form a relatively ineffective, and inadvisable surgical treatment? And if yes, could favourable psychological outcomes compensate for the possible medical complications? According to empirical evidence, the majority of the women who had PM were highly satisfied with their decisions [18-19]. Furthermore, they also reported reduced emotional concern about developing breast cancer [19].

References

[1] C. Laronga, J. D. Lewis, MD, P. D. Smith, "The Changing Face of Mastectomy: An Oncologic and Cosmetic Perspective" *Cancer Control*, vol. 19, no. 4, pp. 286-294, Oct. 2012.

[2] M. E. Carter, K. Schwarz, "Prophylactic Mastectomy: Indications, Options, and Reconstructive Alternatives," *Plastic and Reconstructive Surgery*, vol. 115, no. 3, pp. 891-909, Apr. 2005.

[3] P. Angelos, I. Bedrosian, D. M. Euhus, V. M. Herrmann, S. J. Katz, A. Pusic, "Prophylactic Mastectomy: Challenging Considerations for the Surgeon," *Ann Surg Oncol*, vol. 22, no. 10, pp. 3208-3212, Oct. 2012.

[4] Y. Kim, A. M. McCarthy, M. Bristol, K. Armstrong, "Disparities in contralateral prophylactic mastectomy use among women with early-stage breast cancer," *NPJ Brest Cancer*, vol. 3, no. 2, pp. 1-6, Jan. 2017.

[5] S. Wong, R. A. Freedman, Y. Sagara, F. Aydogan, W. T. Barry, M. Golshan, "Use of Contralateral Prophylactic Mastectomy Despite no Improvement in Long-term Survival for Invasive Breast Cancer," *Annals of Surgery*, vol. 265, no. 3, pp. 581-589, March 2017.

[6] T. R. McCromick, "Principles of Bioethics" https://depts.washington.edu/bioethx/tools/princpl.html (Accessed: May 05.2018).

[7] O. O'Neill, "Paternalism and partial autonomy," *J. Med. Ethics*, vol. 10, no. 4, pp. 173-178. Dec. 1984.

[8] N. H. S. S. Tan, "Deconstructing Paternalism – What Serves the Patient Best?" *Singapore Med. J.*, vol. 43, no. 3, pp. 149-151, Mar. 2002.

[9] "Act CLIV of 1997 on Health" http://www2.ohchr.org/english/bodies/cescr/docs/E.C.12.HUN.3-Annex10.pdf (Accessed: May 28. 2018).

[10] G. Elwyn, D. Frosch, R. Thomson, N. Joseph-Williams, A. Lloyd, P. Kinnersley, E. Cording, D. Tomson, C. Dodd, S. Rollnick, A. Edwards, M. Barry, "Shared Decision Making: A Model for Clinical Practice," *J. Gen. Intern. Med.*, Vol. 27, no. 10, pp. 1361-1367., Oct. 2012.

[11] K. A. Martinez, A. W. Kurian, S. T Hawley, R. Jagsi, "How can we best respect patient autonomy in breast cancer treatment decisions?" *Breast Cancer Manag.*, vol. 4, no. 1, pp. 53-64, Nov. 2015.

[12] N. N. Besu, G. L. Ross, D. G. Evans, L. Barra "The Manchester guidelines for contralateral risk-reducing mastectomy," *World J. Surg. Oncol.*, vol. 7, no. 13, pp. 237. Aug. 2015

[13] F. Lalloo, A. Baildam, A. Brain, P. Hopwood, D. G. R. Evans, A. Howell, "A protocol for preventative mastectomy in women with an increased lifetime risk of breast cancer," *Eur J Surg Oncol.*, vol. 26, no. 7. pp. 711-713, Nov. 2000.

[14] A. J.Curda, "The Damocles Syndrome: where we are today," *Canc. Educ.* vol. 26, no. 2, 397-398., Jun. 2011.

[15] D. F. Baptiste1, E. L. MacGeorge, M. K. Venetis, A. Mouton, L. B. Friley, R. Pastor, K. Hatten, J. Lagoo1, S. E. Clare, M. W. Bowling, "Motivations for contralateral prophylactic mastectomy as a function of socioeconomic status," *BMC Women's Health*, vol. 17, no. 10, Feb. 2017. doi: 10.1186/s12905-017-0366-2.

[16] M. B. Tan, E. M. Bleiker, M. B. Menke-Pluyemers, A. R. Van Gool, S. van Dooren, B. N. Van Geel, M. M. Tilanus-Linthors, K. C. Bartels, J. G. Klijn, C. T. Brekelmans, C. Seynaeve, "Standard psychological consultations and follow up for women at increased risk of hereditary breast cancer considering prophylactic mastectomy," *Hered Cancer Clin Pract.*, vol. 7, no. 1, Mar. 2009. doi: 10.1186/1897-4287-7-6

[17] C. Printz, "Risk assessment: Physicians discuss motivations for rising rates of prophylactic mastectomies," *Cancer* vol. 121, no. 11, pp. 1721-1722 Jun. 2015.

[18] C. Anderson, J. Y. Islam, M. E. Hodgson, S. A. Sabatino, J. L. Rodriguez, C. N. Lee, D. P. Sandler, H. B. Nichols, "Long-Term Satisfaction and

Body Image After Contralateral Prophylactic Mastectomy," *Ann Surg Oncol*, vol. 24, no. 6, pp. 1499–1506, Jun. 2017.
[19] M. H. Frost, D. J. Schaid, T. A. Sellers, J. M. Slezak, P. G. Arnold, J. E. Woods, P.M. Petty, J. L. Johnson, D. L. Sitta, S. K. McDonnell, T. A. Rummans, R. B. Jenkins, J. A. Sloan, L. C. Hartmann, "Long-term satisfaction and psychological and social function following bilateral prophylactic mastectomy," *JAMA*, vol. 284, no. 3, pp. 319-324, 2000.

Truth Disclosure in the Age of Technologized Medicine

Coralia Cotoraci, Alciona Sasu, Mircea Onel, Cristina Ghib-Para
Department of Bioethics and Medical Deontology,
"Vasile Goldis" Western University of Arad, Romania

Abstract
Truth disclosure refers to informing the patients and their families on the truth about the diagnosis and everything linked to the disease itself (prognosis, evolution, treatment). This study aims to reveal the medical students' opinion on truth disclosure. We selected a lot of 305 medical students, from the Faculty of Medicine, "Vasile Goldis" Western University of Arad, Romania. They were given a short questionnaire containing questions on how much they agree or not with patient truth disclosure. Most students agree with truth disclosure, but there are also a category of students, mainly from non-EU countries who do not agree with truth disclosure, as well as students who agree with truth disclosure only in special circumstances.

Keywords
Diagnosis; patient; communication; truth disclosure; doctor patient relationship.

I. Introduction

Patients are always linked to doctors and diagnoses. Diseases usually change the patient's life, either temporarily or definitively. Dramatic changes mostly occur in oncological cases. An oncological patient, mostly because of the decline of his quality of life, feels that there may be a pathological reason for his altered condition. In this situation, whether the patient is admitted or not into hospital, truth disclosure comes into question. The idea of truth disclosure refers to informing the patient and the family on the truth about the diagnosis and everything linked to the disease itself: prognosis, evolution, treatment.

There have been various approaches to this issue overtime. In the paternalistic era, that is until the early 20th century, the doctor was the decision-maker and truth disclosure was considered unnecessary if the patient was thought to be incapable of understanding the information [1]. Nevertheless, starting with the 20th century, medical ethics introduced and focused on the notion of patient autonomy: the patient decides [2]. However, this type of decision-making process is also closely related to the doctor himself, as the doctor is the one who informs the patient on all alternatives before any decision can be made [3]. This process is supported by a new and modern doctor-patient relationship in which the patient has more independence and which assumes that the patient is almost always able to choose between the available medical options. Laws and regulations support the patient right to know the truth. The European Charter of Patients' Rights [4] (Rome, 2002) underlines the right to continuous information for the patient, as well as the right to free choice. Moreover, the WMA Declaration of Lisbon on the rights of the patient, revised in 2015 [5], gives the patient the right to receive all information recorded in any of his/her medical records, and to be fully informed on his/her health state, including the medical facts about his/her condition.

However, future medicine is already envisioned as high technology medicine, in which the doctor plays either an insignificant role or no role at all [6]. In the proximal future, the patient might be diagnosed by an intelligent computer and might receive the diagnosis and treatment by mail. In these cases, truth disclosure is a technological intervention where the benefits of communication are missing [7, 8].

Misunderstandings during the informative process can become the cause of suboptimal communication between patient and doctor. They are also a common justification for partial or even false disclosure of the diagnosis and prognosis.

Until the 1990s, doctors usually withheld the truth from cancer patients in order to give them hope. Surveys conducted around the year 1990 showed that less than 50% of cancer patients were accurately informed on their diagnoses (Africa, France, Hungary, Italy, Japan, Panama, Portugal, and Spain). [9]

Analyzing the doctor-patient communication around the world, researchers demonstrated that most patients seek complete diagnostic information. However, surveys conducted in the Mediterranean countries showed levels of awareness which hardly reached 50%, while only 40% of the Norwegian patients seem to be informed on their prognosis [10]. Concerning the situation in the USA, Hlubocky et al. discussed patients enrolled in trials for cancer, showing that only 52% reported having had a prognosis discussion with their physician, while 45% denied having received any information at all [11]. A 2018 Jordanian study showed that 37 physicians usually withheld the diagnosis of "serious illness" from patients. Non-disclosure was primarily motivated by request from the patient's family [12].

In a study published in 2017 by Erer et al. on Turkish cancer patients, conducted in 2005, 86.5% of participants believed that patients have the right to know the truth and be correctly informed on the treatment; 76.9% believed that physicians must disclose the truth to patients [13].

Data from Chinese physicians and surveys among their cancer patients showed five reasons for poor truth disclosure: lack of awareness of patients' right to knowledge, cultural influences, insufficient medical resources and training, the families' financial concerns, and the need to protect doctors from violence [14].

"The Patients' rights in the European Union; Mapping eXercise: final report – Study (2017)" [15] lists countries which have laws for patient's rights. The document also includes countries which rule the patient's right to not be informed on the disease. Therefore, countries such as France, Germany, Hungary, the United Kingdom, and Romania have distinctive Patient Laws, whereas countries such as Italy, Tunis, Algeria, Morocco do not have any specific laws, only regulations concerning patients' rights. Romania has Law 46 of 2003 and Law 95 of 2006 [16] which regulate the patient's rights to consent and complete information.

Informed consent is now a legal and deontological requirement. Patients expect and receive more complete information and are progressively involved more in their medical care. Modern medicine involves telemedicine, clinical healthcare from distance, telemedical devices to support in-home care, E-health, healthcare practice supported by electronic processes, and cybermedicine, that exclude direct human to human interaction. Nevertheless, medical studies are still traditional everywhere, preparing future doctors to actually face patients and successfully communicate with them. In the followings, we present the outcomes of a study based on this particular fact of the doctor-to-be learning how to face future patients.

II. Objectives

This paper aims to present medical students' opinions on truth disclosure. Truth disclosure was defined as information given to the patient concerning his diagnosis and its implications (prognosis, evolution of disease, treatment). Modern medicine involves both doctor and patient in the decision-making process on diagnosis procedures and treatment, thus a proper communication of medical data represents the basis for an optimal patient management. This study was carried out with the premise that medical practice should begin with accurate information and principles on the patients' rights, including truth disclosure which is a mandatory requirement.

III. Material and methods

We selected a lot of 305 medical students, undergraduates in the 5th (23%) and 6th year (77%) of study, from the Faculty of Medicine, "Vasile Goldiş" Western University of Arad (Romania). All students enrolled in the 5th and 6th years of study were invited to participate in the study, but only part of them accepted to fulfil the questionnaire. The subjects were male (51%) and female (49%), aged between 21 and 45. 122 subjects were Romanian and the international students were as following: 67 from Italy, 32 from Israel, 27 from Morocco, 27 from Tunis, 15 from France, 8 from Algeria, 3 from Germany, 2 from Nigeria, 1 from the USA, and 1 from Ghana. A short questionnaire was applied to all voluntary participants before one of their weekly course. The questionnaire included some demographic information on age, gender, and country of origin. The participants was asked two questions: "As far as you know, are there laws in force or regulations on truth disclosure to patients in Romania or other EU countries?" and "Do you agree with truth disclosure to patients?" The latter question had three possible answers: "yes, always," "yes, only in certain situations," or "no." Moreover, the students were asked to write which should be the conditions to truth disclosure, if any.
The resulting data was introduced in a Microsoft Excel database and thus statistic results were obtained.

IV. Results

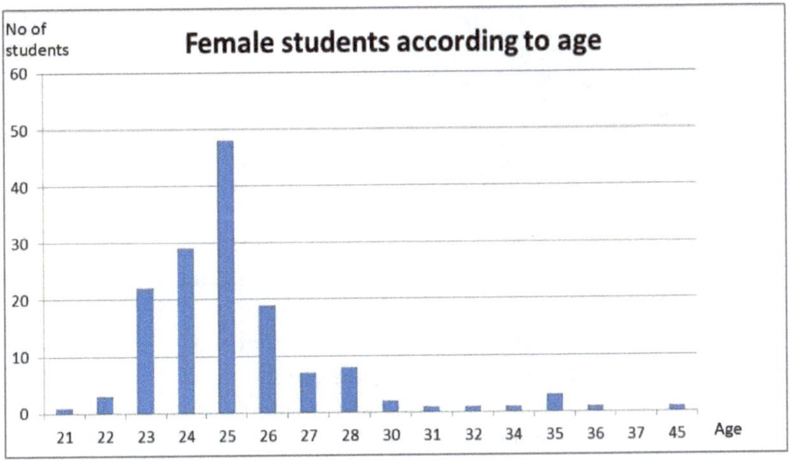

Fig. 1.a. Responders according to gender and age

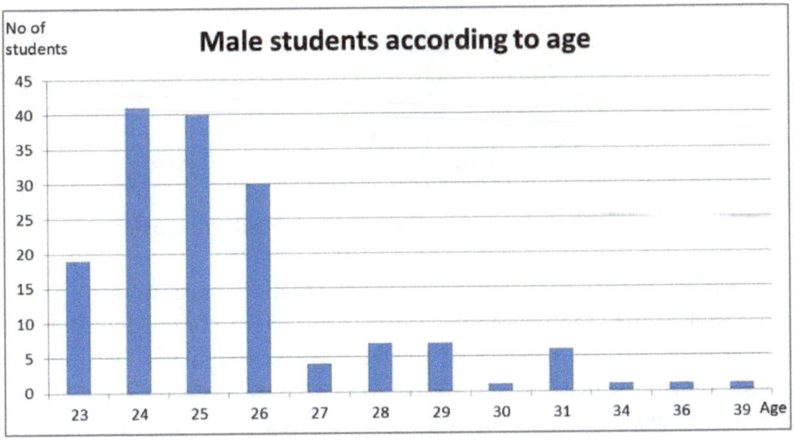

Fig. 1.b. Responders according to gender and age

Most responders were aged between 23 and 26 (Fig.1.a and 1.b.).

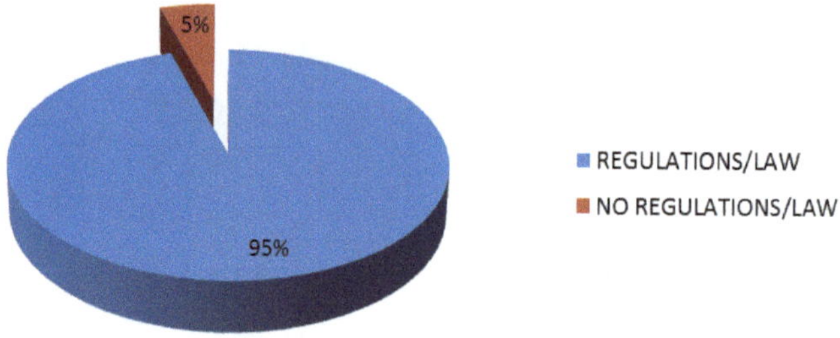

Fig. 2. The responders' knowledge on existing regulations/laws on truth disclosure

95% of students were aware of the existence of regulations/laws regarding patients' rights in Romania and other countries (especially EU countries). Still, 5% responded that there are no such regulations in Romania or other EU countries (Fig. 2).

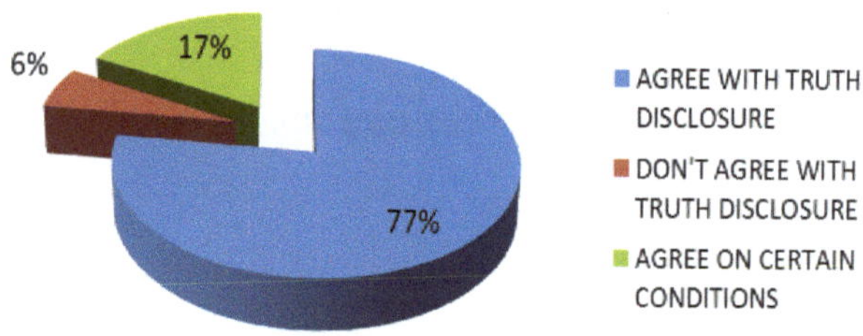

Fig. 3. The responders' opinion on truth disclosure

Most of responders agreed with truth disclosure to any patient, anytime. 17% of students emphasized that certain conditions should be fulfilled before telling the truth to the patient, which will be presented below. A smaller percentage, although not to be neglected, was that of responders who do not agree with truth disclosure to patients (Fig. 3).

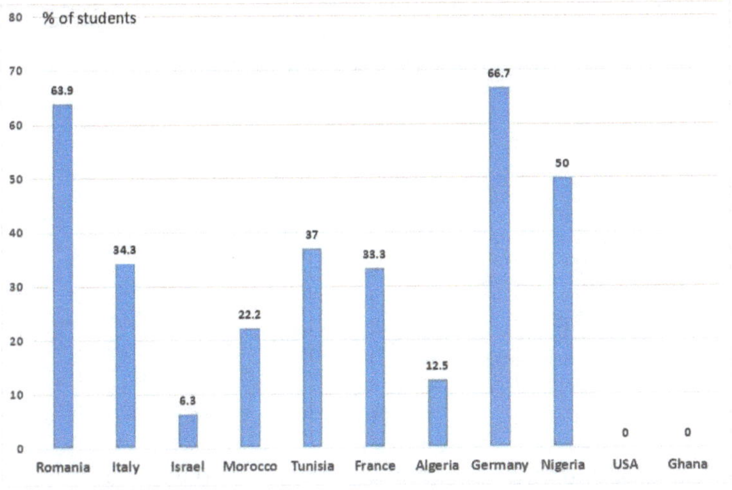

Fig. 4. Percentage of female students who agree with truth disclosure, according to country of origin

The results show that the highest percentages of female students who agree with patient truth disclosure are from Germany and Romania (Fig. 4).

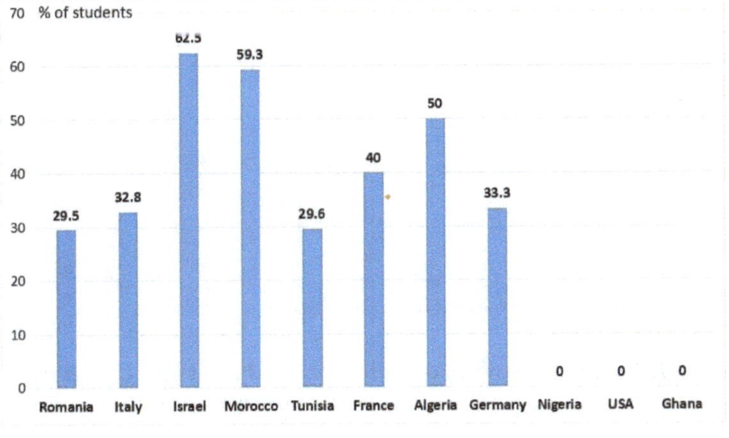

Fig. 5. Percentage of male students who agree with truth disclosure, according to country of origin

The highest percentages of male students who accept truth disclosure are from Israel and Morocco. A high percentage is also represented by students from Algeria and France (Fig. 5).

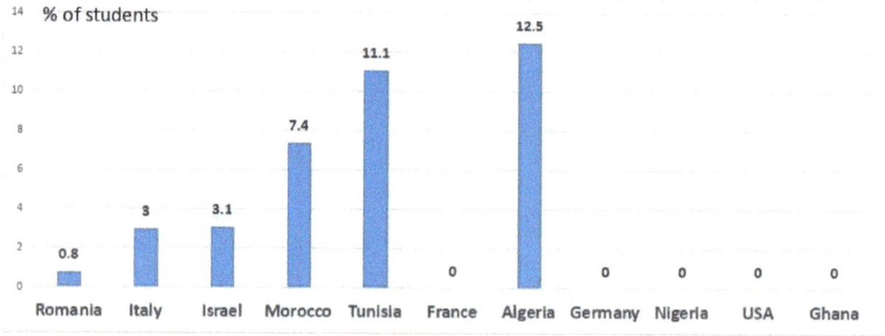

Fig. 6. Percentange of female students who do not agree with truth disclosure, according to country of origin

Female students coming from Algeria, Tunis, Morocco, and Israel represent the highest percentages in the category which does not agree with patient truth disclosure (Fig. 6).

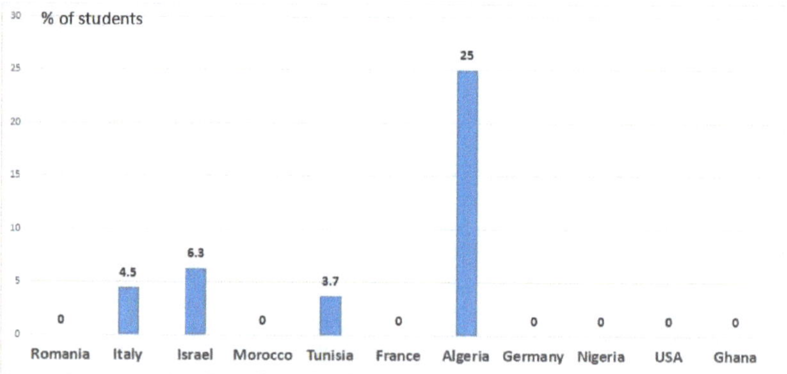

Fig. 7. Percentage of male students who do not agree with truth disclosure, according to country of origin

Male students who did not agree with truth disclosure come mostly from Algeria, Israel, and Italy (Fig. 7).

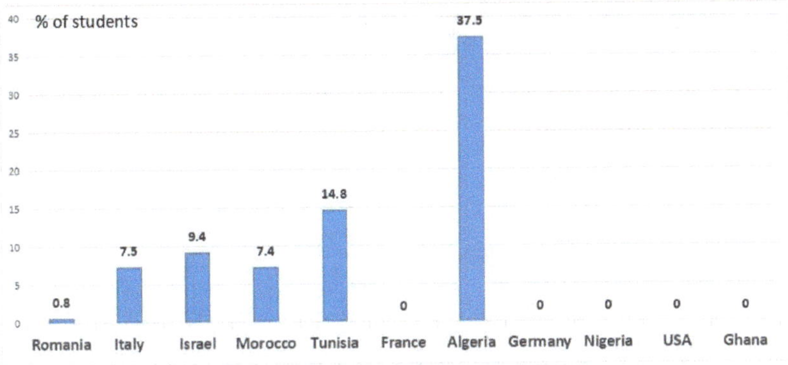

Fig. 8. Percentage of students (male and female) who do not agree with truth disclosure, according to country of origin

The statistics of the total number of students / country (male and female students) shows that the highest percentages of students which do not agree with truth disclosure belong to Algeria, Tunis, and Israel (Fig. 8).

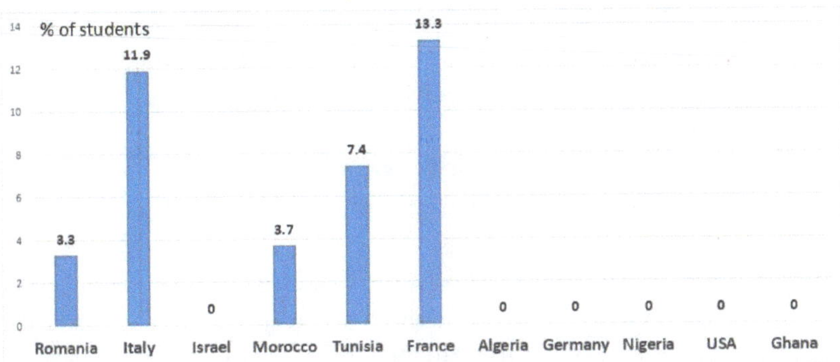

Fig. 9. Percentage of female students who agree with truth disclosure only in certain situations, according to country of origin

The highest percentage of female students who accept patient truth disclosure only in certain situations belong to students coming from France and Italy (Fig. 9). The conditions they suggested are listed below.

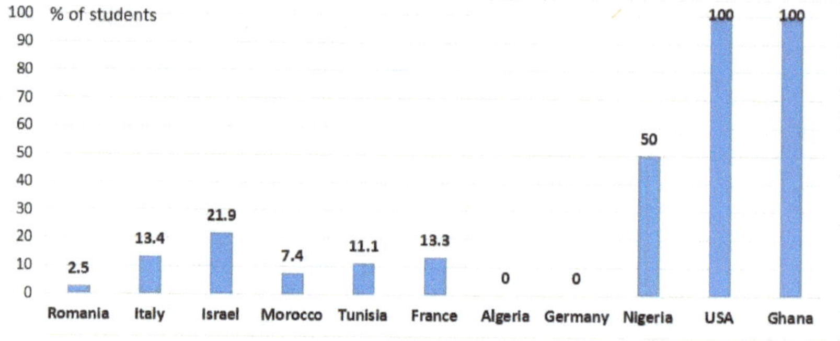

Fig. 10. Percentage of male students who agree with truth disclosure only in certain situations, according to country of origin

The highest percentages of male students agreeing with truth disclosure only in certain situations belonged to the USA, Ghana, and Nigeria (Fig. 10)

The conditions outlined by the students following the question on whether they agree or not with patient truth disclosure are as following:

- the patient must be in the best psychological conditions in order to accept the diagnosis;
- special attention must be paid to the psychologically fragile patient;
- the patient must be emotionally and mentally capable of accepting the diagnosis;
- the patient is not a psychiatric patient;
- doctors should communicate only the information that the patient wants to know;
- only if the patient wants to know the diagnosis;
- the information should not be disclosed to patients diagnosed with depression; in this case, the patient's family is informed;
- only if the patient can accept the diagnosis;
- only if the patient is not in the terminal phase; in this case, the family has to be informed;
- only if the patient does not have an increased emotional instability.

V. Discussion

The results showed a difference of opinion on truth disclosure which mostly varied according to the country of origin. The subjects included mostly students from Romania, and there were some countries represented by only one or a few students. The students from the EU mostly accepted patient truth disclosure. Nevertheless, not all responders agreed with truth disclosure, which is an interesting find, as patient autonomy and bioethical information is given and highlighted to the medical students in our University. Most of the EU countries have specific laws regarding patients' rights, thus students from countries which do not have such regulations (Alger, Tunis, Morocco) showed the highest percentages in the category of responders who did not agree with truth disclosure.

Regarding the acceptance of truth disclosure on certain conditions, the highest percentages belonged to female students from Europe (France and Italy) and to male students (including the student from the USA). In Europe and the USA, the communication with the patient is of highest importance: the laws in force regarding patients' rights are generally known by patients, thus doctors need to inform their patients properly. Most of the students knew about the existence of these regulations on truth disclosure – also because in our university patients' rights is taught from the second year, including the patients' right to information, the right of the patient to decide to not be informed, or to the right to choose a different person to be informed instead. Among Romanian doctors, there are still professionals who believe that the information on diagnosis (especially if oncologic) is a too heavy burden for the patient and thus the potential psychological consequences could dramatically alter the patients' quality of life. Communicating the truth to an oncological patient requires experience, time, and sometimes the help of a psychologist. The students listed realistic reasons which could challenge the doctor in truth disclosure. Of these, the psychological ones are the most frequently encountered and the physician must take them into consideration. All these issues must be part of the complex education given to all medical students.

VI. Conclusions

Students in the final undergraduate years, as future doctors, have partially settled their medical principles. The new generation of doctors appear to be divided into future doctors who agree with truth disclosure, and future doctors who still favour medical paternalism. Most of the students who did not agree with truth disclosure were from Alger, Tunis, Morocco, as some of these countries do not have any special laws for patients' rights. Thus medical bioethical education is important in today's universities because it clarifies the aspects related to legacy and helps the student think in ethical terms.

References

[1] A.L. Caplan, B.H. Lerner, "Is There Still a Place for 'Paternalistic' Doctors?" July 08, 2014, available at https://www.medscape.com/viewarticle/827006 (Accessed: July 20, 2018).

[2] L. Murgic, P.C. Hébert, S. Sovic, G. Pavlekovic, "Paternalism and autonomy: views of patients and providers in a transitional (post-communist) country," *BMC Med Ethics*, vol. 16, no. 65, 2015. Published online 2015 Sep 29. doi: 10.1186/s12910-015-0059-z, PMCID: PMC4589086, available at https://bmcmedethics.biomedcentral.com/track/pdf/10.1186/s12910-015-0059-z

[3] A. Aggarwal, J. Davies, R. Sullivan, "Nudge" in the clinical consultation – an acceptable form of medical paternalism?" *BMC Medical Ethics*, 201415:31, https://doi.org/10.1186/1472-6939-15-31

[4] European Charter on Patients Rights, http://ec.europa.eu/health/ph_overview/co_operation/mobility/docs/health_services_co108_en.pdf

[5] WMA Declaration of Lisbon on the Rights of the Patient; https://www.wma.net/policies-post/wma-declaration-of-lisbon-on-the-rights-of-the-patient/(Accessed: June 20, 2018).

[6] J. Stockert, "Artificial intelligence is coming to medicine — don't be afraid", August 18, 2017, available at https://www.statnews.com/2017/08/18/artificial-intelligence-medicine/(Accessed: July 2, 2018).

[7] "10 Bioethical Issues in the Future of Digital Health," http://medicalfuturist.com/bioethical-issues (Accessed: July 13, 2018).

[8] "Eight Bioethical Issues to Know in 2017," https://healthlaw.hofstra.edu/resources/articles/eight-bioethical-issues-to-know-in-2017/, Accessed: July 20, 2018.

[9] J.C. Holland, N. Geary, A. Marchini, S. Tross, "An international survey of physician attitudes and practice in regard to revealing the diagnosis of

cancer," *Cancer Invest*, vol. 5, no. 2, pp.151-4, 1987. available at https://www.ncbi.nlm.nih.gov/pubmed/3607572

[10] G. Numico, M. Anfossi, G. Bertelli, E. Russi, G. Cento, N. Silvestris, C. Granetto, G. Di Costanzo, M. Occelli, E. Fea, O. Garrone, M. Gasco, I. Colantonio, M. Merlano, "The process of truth disclosure: an assessment of the results of information during the diagnostic phase in patients with cancer," *Annals of Oncology*, vol. 20, no. 5, pp. 941–945, 2009, available at https://doi.org/10.1093/annonc/mdn709

[11] F.J. Hlubocky, N.E. Kass, D. Roter, S. Larson, K. E. Wroblewski, J. Sugarman, C.K. Daugherty, "Investigator Disclosure and Advanced Cancer Patient Understanding of Informed Consent and Prognosis in Phase I Clinical Trials,", *Journal of Oncology Practice*, vol. 14, no. 6, pp. 357-367, available at https://doi.org/10.1200/JOP.18.00028

[12] S.M. Borgan, J.Z. Amarin, A.K. Othman, H.H. Suradi Y.Z. Qwaider, "Truth Disclosure Practices of Physicians in Jordan," *Journal of Bioethical Inquiry*, vol. 15, no. 1, pp. 81–87, 2018, available at https://link.springer.com/article/10.1007%

[13] S. Erer, E. Atici, A.D. Erdemir, "The views of cancer patients on patient rights in the context of information and autonomy," *J Med Ethics*, vol. 34, no. 5, pp. 384-388, 2008, doi: 10.1136/jme.2007.020750. [PubMed: 18448722], available at https://onlinelibrary.wiley.com/doi/full/10.1002/pon.4545

[14] Y. Gan, L. Zheng, N. Xiaonan Yu, G. Zhou, M. Miao, Q. Lu, "Why do oncologists hide the truth? Disclosure of cancer diagnoses to patients in China: A multisource assessment using mixed methods," *PsychoOncology*, vol. 27, pp. 1457–1463, 2018, First published: 23 August 2017, https://doi.org/10.1002/pon.4545, available at https://onlinelibrary.wiley.com/doi/full/10.1002/pon.4545

[15] EU publications, Patients' rights in the European Union, Mapping eXercise: final report – Study, Published: 2018-01-25, Corporate author(s): Directorate-General for Health and Food Safety (European Commission), Personal author(s): KU Leuven; European Observatory on Health Systems and Policies; Maastricht University, https://publications.europa.eu/en/publication-detail/-/publication/8f187ea5-024b-11e8-b8f5-01aa75ed71a1/language-en

[16] Law no. 46/2003, Law on patients rights, Romania, available at http://www.dreptonline.ro/legislatie/legea_drepturilor_pacientului.php (Accessed: June 20, 2018).

Ethical Challenges Related to Marketing Drugs

Miroslav Radenković,[1] Ivana Lazarević,[2]
Marko Stojanović,[1] Tanja Jovanović[2]

[1] Department of Pharmacology, Clinical Pharmacology and Toxicology,
Faculty of Medicine, University of Belgrade, Serbia
[2] Institute of Microbiology and Immunology,
Faculty of Medicine, University of Belgrade, Serbia

Abstract

Drugs have unquestionably become a key factor in the progress of human health and the considerable prolongation of life extent. There are numerous signs showing that the advertising of medicines can actually affect understanding, attitudes, and expectations of patients in relation to the specific drug. It is therefore crucial that the pharmaceutical industry implements the highest ethical standards of good laboratory, clinical, and manufacturing practice. The health-care and well-being of patients should be the first priority. The promotion of drugs must be done in an ethical, accurate, and balanced way, while not being misleading. The proper assessment of the risks and benefits of the product use must be ensured. Medical doctors should always consider evidence-based medical facts and medical scientific knowledge in order to avoid prescribing specific drugs uncritically and under the influence of the pharmaceutical industry.

Keywords

Drugs; prescribing; ethics; patients; marketing.

I. Introduction

Drugs have unquestionably become a key factor to the progress of human health and the considerable prolongation of life extent. Nowadays, the pharmaceutical industry uses the highest standards of good laboratory; clinical, and manufacturing practice, thus providing users with adequate data on their products. This way, significant steps have been taken to ensure a compliant and ethical interaction between physicians and patients on one side and the representatives of the pharmaceutical industry, on the other [1].

The most profitable pharmaceutical products require notable annual financial expenditure solely for marketing. Moreover, at the end of the process, marketing expenses can notably increase the initial development costs of the drug. This implies that the manufacturer must be continuously involved in scientific research, medical education, health policy, the media, professional associations, patients' associations, etc.

It is now clear that advertising medicines can greatly influence the understanding, attitudes, and expectations of drug users in relation to the specific drug. Therefore, in some circumstances, drug advertising can be considered inappropriate and unnecessary, considering that the focus of this type of promotion is usually based on the emotional experience of the consumer in relation to an existing illness. Thus it is necessary to prevent the inappropriate increase of patient expectations associated with the potential therapeutic efficacy of the prescribed drugs.

Providing scientifically valid, accurate, and balanced information on medical products should be paramount for any pharmaceutical company. Promotion must therefore be ethical, balanced, accurate, and not deceptive. The information in promotional materials must support an adequate estimation of the risks and benefits of the drug use and must also inform on its proper and timely use [1].

An intense promotion of the drug immediately after its registration (including promotional, but not always scientific information) was associated with inappropriate drug–prescribing activities amongst physicians. Therefore, doctors should always take into account accepted medical facts and related scientific knowledge in order to avoid prescribing specific drugs uncritically, and under the pressure of the pharmaceutical industry. Thus, it is essential to have an adequate control (done by relevant professional associations) of possible fees, substantially-expensive gifts, or other privileges provided by pharmaceutical companies to doctors.

There is no doubt that pharmaceutical companies must always maintain the highest ethical standards and, when promoting drugs, professional advertisers should adhere to the highest ethical standards and maintain objectivity based on medical facts [2]. It is essential that the drug information is up-to-date, which means that any information on therapeutic efficacy, adverse drug reactions and contraindications must mirror up-to-date scientific views [2]. The selection and further interpretation of scientific evidence regarding an individual medical product should receive the highest level of expertise. It is therefore necessary to ensure that any material used in the promotion does not conflict with the existing advertising regulations of a particular country. In addition, it is necessary to have a balance between motivating and appealing drug promotion in relation to the medically-supported accuracy of the information and messages provided. Finally, it should be underlined that non-critical drug prescription always carries the risk of denying drugs intended for patients in need of urgent health care.

II. Patients and drug marketing

There are increasing methods that the drug advertising industry uses to significantly influence the attitudes and expectations of patients. In most cases, the public is responsive to rapidly increasing health information resources (conventional news media, internet advertising, or social networking); however, it often has no reliable way of assessing the accuracy of health information presented in the media or in direct-to-consumer advertising (DTCA) [3]. The problem is not the availability of information, but rather in reliable information. Consequently, although the public has more access to health information than ever, patients are increasingly vulnerable to potentially-deceptive information [3].

The term "personalized medicine" implies that the right drug should be prescribed to the right patient at the right time [4]. Therefore, assertive drug marketing (which usually leads to uncritical prescription of drugs) should be restricted, while inappropriate stimulation of patient expectations should be prevented. Still, autonomy is a fundamental ethical foundation of current medicine and highlights the necessity of each individual to be in charge of their own well-being [5]. As a result, drug consumers have more privileges and opportunities considering preferential treatment choices. Accordingly, many patients have certain expectations regarding the care they want to receive [3]. This choice depends on personal or family wishes, beliefs, past experiences of the self and others although the patient often makes no genuine effort to understand evidence-based medical facts. Moreover, in today's electronic environment, an increasing number of patients use the Internet in anticipation of more comprehensive and accurate drug-related information [6].

Direct-to-consumer advertising (DTCA) mirrors the patient's growing role in making decisions about the prescribed drug, the need for more detailed information, and ultimately the distrust in providers. It is therefore not surprising that the audience of DTCA sees it as a fitting extension of patient empowerment, whereas opponents discard it as unsuitable and misleading [7]. Specifically, supporters claim that DTCA: (a) provides educational information, (b) improves doctor-patient relationships, (c) enhances adherence to prescription drug protocols, (d) stimulates competition, and (e) leads to lower drug prices. Opponents argue that: (a) consumers lack the expertise to evaluate the quality of ad content, (b) the promotional feature of advertising will not efficiently transfer all risk information, (c) it will impair the doctor-patient relationship, (d) it will lead to increased drug selling prices, (e) it will lead to increased liability and litigation; and (f) it will promote increased use of drugs [7]. It seems that DTCA risks creating unrealistic expectations for patients and conflicts of interest for physicians [8].

Three DTCA categories are usually recognized and overall accepted: (a) advertisements where a specific product must be presented through a brief summary with a fair balance of benefits and risks; (b) help-seeking advertisements where information on diseases or conditions is provided to encourage consumers to consult their doctors on treatment options; however specific prescription drugs or treatments cannot be mentioned, nor can the advertisement contain links to materials identified with a specific product; and (c) reminder advertisements where materials provided cannot contain or suggest the clinical role of the product [7].

DTCA-related ads are often persuasive, as ads connect to their target audiences by means of emotions [9, 10] and by emphasizing pharmacological efficacy, simple use, adequate symptom control, and innovative features – although not the costs [11]. Because of the strong subjective influence of drug marketing in choosing a specific pharmacological treatment, problems in the doctor-patient relation in the drug selection process seem inevitable; however, physicians who listen carefully, communicate effectively, and show compassion and concern are nevertheless important for the decision of each and every patient [12].

III. Medical doctors and the drug marketing

As treatment options are constantly increasing, medical doctors need to be up to date with scientific innovations, including new pharmaceuticals. Thus the interaction of physicians with pharmaceutical companies can commonly take several forms, such as detailing (providing information on new drugs through direct visits by pharmaceutical representatives to physicians), industry gift-giving, dispersing free drug samples, or attending industry-sponsored meals and events. In addition, medical doctors are involved in industry-funded research, they obtain royalties to recommend or use medical devices, give consultancy services, play an active role in advisory boards, prepare industry-sponsored presentations, or attend industry-supported medical educations [13]. Communication between pharmaceutical companies (key players in the development and manufacture of medicines) and the doctors who prescribe their drugs is essential for the appropriate, effective, and rational use of prescription pharmaceuticals [1].

It is essential that the information provided by pharmaceutical companies is scientifically accurate and fair; moreover, all interactions between pharmaceutical companies and doctors should be transparent at all times and be in the benefit of the patient care [1]. Unfortunately, inappropriate drug prescribing is largely the result of the aggressive drug promotion immediately after registration, such as organizing various forms of educational meetings with a promotional, but not always a scientific character. However, aggressive marketing could be diminished by stimulating the off-label use of drugs [14]. This would ensure that a doctor's practice is not influenced in any way, such as by financial support from pharmaceutical or other commercial companies [15].

It is therefore obvious that the relationship between the pharmaceutical industry and the medical profession enhances the potential of doctors to be involved in conflicts of interest [14].Gifts from the pharmaceutical industry most probably generate potential conflicts of interest that can affect the doctors' decision making, jeopardizing professional objectivity and potentially compromising patient welfare [16]. Doctors who do not primarily take into consideration the acknowledged medical facts and scientific achievements of a particular drug, but prescribe it without criticism under the influence of the pharmaceutical industry, are a typical example of medical conflict of interest.

The conflict of interest is the dichotomy between the professional responsibility and the personal interest of a person in a position of trust. Conflict of interest is typically characterized by a set of circumstances where conflicts arise between two duties (two value systems). These would include: (a) primary, professional duty (in medicine for patients and general public welfare), and (b) secondary, personal interest (as a rule economic, but also professional, including expert recognition, scientific project support, or academic advancement funding). In such circumstances, a physician may decide in accordance with their personal rather than the general interest, which usually depends on the assessment of the likelihood of the secondary interest affecting and damaging the professional duty [17].

The management of conflicts of interest includes several points. First, transparency should be paramount through a straightforward declaration of conflict of interest. So, if the pharmaceutical company pays the physician directly or indirectly for medical research, this should be disclosed in any publication containing relevant research findings [15]. Any potential conflict of interest should be disclosed in case of financial support connected to professional activities, including lectures, presentations, publications, or development of clinical services. Secondly, sponsoring research and educational programs should involve the existence of several sponsors in order to avoid direct payments to individuals. Thirdly, accommodation, trip expenses, and other comparable activities related to educational/scientific programmes should be modest; however, small value gifts can be allowed.

Because pharmaceutical companies invest significant financial resources to promote new products to medical doctors (by means of gifts, free samples, educational seminars, entertainment, consulting arrangements, etc.), an adequate control by relevant professional associations is required [18]. Professional associations should provide their members with assistance and guidance at any times, on matters such as the suitable level of cash benefits, gifts, etc. It would also be beneficial to establish a relevant public registry. After all, the public must be better informed about the influence of the pharmaceutical industry on the medical profession and advocate the necessary adjustments [14].

IV. Marketing drugs and ethical standards – the promotion

In reaction to the changing business environment, the pharmaceutical industry has made notable efforts to ensure that their marketing and promotion routines are compliant with ethical standards [19]. Marketing can be any information, activity, practice, or communication intended to increase the prescription rate, supply, sale, or consumption of a medicinal product [20]. The promotion *per se* presents the composition, effect, and administration of a medicinal product, and it is only performed for and by healthcare professionals who are qualified to prescribe and sell medicinal products [20]. The most frequent promotions come in the form of written or verbal information on drugs. Written drug information refers to information communicated in written words, pictures or sound, via all media, whereas verbal drug information covers information transmitted in person by representatives of pharmaceutical companies [21]. This means that such information can be provided in combination with personal visits, visits to clinics, symposia, training seminars, conferences, and different other meetings.

The most common way of passing relevant information on a specific drug is detailing. The traditional function of detailing is to provide a physician with medical information, which ranges from simple awareness-building to detailed technical information [22]. Professionals involved in the promotion process should comply with the highest ethical standards and maintain objectivity based on medical facts. The selection and elucidation of scientific facts relating to a particular pharmaceutical or other medicinal product should be at the highest level. For example, this aspect is clearly reflected in the assessment of the benefit versus risk ratio in the specific drug prescribing process, especially if a drug with a significant number of potential adverse reactions is prescribed. This is why any promotion must support the rational use of drugs by objectively presenting them without amplifying their properties [23]. In other words, without a justified reason, the marketing must not use the word *safe* or imply that the product has no adverse effects, or that its use is not linked to the risk of addiction [24].

In order to enable physicians to build their own opinion on the therapeutic worth of the medicinal product of interest, the information and documentation supplied with the medicinal products during marketing practice must be exact, balanced, fair, objective, and suitably complete. The information must be based on the state-of-the-art assessment of scientific evidence [20]. It must be ensured that any material used in the promotion does not conflict with the existing national advertising regulations of a particular country. This means that marketing must enclose neither violence, sexual behaviour, criminal actions (or any reference to them), nor offensive or tasteless expressions [24].

The distribution of free drug samples is a well-known method of promotion. Free samples of pharmaceuticals can only be disseminated to individuals eligible to prescribe or dispense them, who actually prescribe the drug of interest, consequently benefiting from the chance of becoming familiarized with the pharmacological product [24]. Several conditions must be fulfilled during this process [15, 23]. Namely, free samples are only provided on an exceptional basis and for the purpose of gaining experience with such a product; each sample cannot be larger than the smallest product presentation on the market; each sample must be marked with *free medical sample - not for sale*; and, finally, each sample must be accompanied by a copy of the summary of product characteristics (SmPC).

When promoting a new pharmaceutical product, the balance between interesting and attractive promotion must be found in relation to the medically-based accuracy of the information and messages delivered. It is necessary to ensure that the public, doctors, pharmacists, and patients always have accurate information on the medicine since non-critical drug prescribing always carries the risk of depriving drugs from patients who actually need health care. Accordingly, the extension of previously-established pharmacological indications to mild or transient disorders or the identification and classification of new clinical entities without scientific evidence should be avoided. The distribution of medicines over the Internet, especially those that avoid the legal frameworks adopted, should be under continuous control. Moreover, mailing lists must be always updated, and any request from physicians to be removed from promotional mailing lists must be promptly answered [23]. When conducting market research, particular attention must be paid to protecting the privacy of all the participants who participated in the research, and the questions asked must respect the principle of objectiveness. This also implies that market research should not affect individual patients' treatment [24].

Individuals and institutions responsible for protecting the general population's health should not become an extended hand of drug manufacturers. It is therefore necessary that the legislation be continuously a key element in the management of all irregularities related to: (a) the process of drug registration (for example, lack of necessary information); (b) intentionally slowing down the registration process in order to obtain additional funding from suppliers; (c) bias in the selection of members of drug registration commissions and other regulatory bodies; and (d) theft and fraud in the drug distribution chain, including health facilities. In order to prevent the above irregularities, it is necessary to stimulate legislative reforms aimed at the formation and active implementation of anti-corruption laws, as corruption can radically affect the well-being and everyday lives of millions of people worldwide [17].

V. Advertising drugs and ethical issues - the general public

Only non-prescription drugs can be marketed to consumers, with the exception of distributing public information on vaccination campaigns [24]. Generally speaking, marketing communication is any activity and provision of information that is directly relevant to the business of a company, irrespective of the form and means in which it is delivered and performed in order to influence the attitude and conduct of the communication addressee [20]. For example, recent Canadian legislation supports two types of advertisements to the general public: (a) reminder ads, which include only the brand name without mentioning health claims or product use; and (b) disease-oriented ads, which discuss a specific condition, but unclear treatment [25].

An advertisement for a pharmaceutical must contain the following information [24]: (a) the name of the medicine along with the name of the pharmacologically active substance (if the drug of interest contains only one active ingredient), all in accordance with the national legislation on medicines and medical devices; (b) indications for use; (c) the information necessary for the correct and safe use of the medicine, as well as any special precautions, interactions, and adverse effects; (d) a clear, readable, and comprehensively written, drawn or spoken warning to the patient or user to carefully read the instructions , and to consult a physician or pharmacist about the potential risks, as well as adverse reactions; and (e) the name of the marketing authorization holder, importer or marketer.

Marketing practice involving the general public must always implement the highest ethical standards [20]. This means that marketing practice will: (a) never discredit the pharmaceutical industry; (b) be executed with a focus on the particular nature of a medicinal product and the professional standing of the recipient; (c) not be abusive, offending, decisive, or aggressive; (d) be concise and consistent with the grammatical rules and orthography, as well as proper style of the language used; (e) be easily understood by consumers for whom the advertising is intended; and (f) be subject to reasonable self-control and moderation.

When advertising a drug or medical device, the general public should not be led to the impression that: (a) the drug does not have adverse reactions; (b) it is not necessary to consult a physician before taking the medicine; (c) medical examination, advice, or surgical intervention may be avoided by using the drug; (d) taking the medicine guarantees success in treating the disease; (e) particular medicine is better than another drug; (f) medicine should be taken even when there are no signs of illness, implying that it improves health; (g) the health of individuals not taking the medicine will be impaired, (h) because of its natural origin, the medicine is harmless and effective; (i) the medicinal product represents food, cosmetics, or other objects of general use; and (j) the recommended drug may be replaced by another drug [20, 24].

Currently, the United States and New Zealand allow advertising of prescription drugs [26]. On the other hand, the European Union restricts all advertising of prescription drugs, including television and print advertisements [27]. In particular, it should be forbidden to advertise the following drugs to the general public: (a) medicines that are issued with prescription; (b) the medicines issued at the expense of the health insurance fund; (c) drugs containing opioid or psychotropic substances; (d) drugs for tuberculosis; (e) drugs for sexually-transmitted diseases; (f) drugs for infectious diseases; (g) medicines for chronic insomnia; and (h) medicines for diabetes and other metabolic diseases.

Finally, in advertising medicinal products to the general public is not permitted if it displays children taking medicine without the presence of adults. Moreover, advertising of the medicinal product to the general public should not be exclusively or mainly directed at children. Furthermore, it is not permitted to collect and disseminate personal data on the disease of a particular person, diagnoses, therapeutic procedures, especially medicines that have been used.

VI. Advertising drugs and ethical issues - the professional public

A worldwide and important ban related to advertising pharmaceuticals or introducing a new use of an existing medicine, is that the drug cannot be promoted before the regulatory marketing authorization was issued [1]. It is broadly accepted that advertising materials for the professional public must be labelled with the wording "solely for the professional public." Moreover, marketing must be based on the most recent summary of product characteristics (SmPC), which encompasses the factual basis for information on the drug [2, 24]. Nonetheless, doctors should not rely merely on promotional literature from pharmaceutical representatives for information on specific drugs, but should seek independent and evidence-based sources of information on the benefits and risks before actually prescribing them [15]. It is certain that patients can only benefit from medical doctors who are aware of the recently-approved and clinically-effective treatments that can, for example, induce fewer adverse reactions or improve adherence [28].

Professional associates (advertisers with usually a degree in medicine, dentistry, pharmacy, or veterinary medicine) can promote the medicinal product to the professional public. Advertisers must receive special training in basic and clinical pharmacological knowledge on the medicines they promote. The pharmaceutical company is obliged to provide continuous education for the professional associates who perform the promotion. And *vice versa*, the advertisers promoting the medicinal product are obliged to provide all information concerning any reported adverse effects on the medicinal product to the pharmaceutical company. Importantly, sales representatives should not use any incentive or tactic to get an interview with a doctor; furthermore, under no circumstances, should they pay for access to a healthcare professional [15]. While promoting a product to the professional public, it is neither allowed to encourage prescribing, issuing, or purchasing specific medicines, nor to offer money in cash, high-value gifts, or to promise or provide privileges or rewards. In case the relationship also involves medical research, the doctor must make sure that the relationship with the pharmaceutical company neither affects the study itself, nor its design or the interpretation of any obtained findings. This relationship should be reported to the relevant ethics committee, and explicitly quoted in future publications, and at the relevant scientific meetings [15].

As for scientific meetings, pharmaceutical companies have substantial financial resources for payments to medical doctors in the form of meals, gifts, and educational materials, as well as speaker and consulting fees. These payments usually compensate physicians for their clinical and scientific input and support them in educational meetings, but they are also used for promotional purposes [28]. Many national codes cover issues related to the support of healthcare professional attendance at medical conferences. In particular, the main purpose of the meeting must be scientific and professional in nature, and any conference packs/meals/conference gifts provided must be secondary to that purpose. The venue itself should encourage the scientific or educational purpose of the meeting [1, 5]. In other words, the main focus of the event must be on research and pharmaceutical information, or other relevant medical training, meaning that participants should spend most of the time in the scientific programmes/sessions/trainings offered. The event must not be organized in venues celebrated for entertainment or luxury [24]. Thus, the choice of the location must be rational with regard to the purpose of the meeting, and leisure resorts, as well as places known for their exclusivity, must be avoided [2]. Finally, any hospitality associated with an event shall be reasonable and restricted to the economical travel arrangements, meals, simple accommodation, and genuine registration fees [23, 29].

There is a considerable ethical concern that gifts from the pharmaceutical industry to individual healthcare providers could compromise their objectivity and integrity associated with their primary ethical duty of putting the interests of patients first [16]. Although physicians may generally express negative to neutral attitudes towards pharmaceutical representatives, in addition to creating the so-called reminder effect, a constant interaction develops a routine of benevolence between a sale representative and the medical doctor, which may be later translated to a positive physician prescription performance [22]. This is not surprising, as the recipient of a gift often experiences three types of responsibilities towards the giver: (a) grateful conduct (i.e., acceptance of the gift and expression of gratitude); (b) grateful use (i.e., in accord with the giver's intention), and (c) reciprocation [16]. Moreover, one should bear in mind that gifts place people in binding personal relationships that generate vague and undefined moral obligations [16]. For all preceding reasons, doctors should not accept gifts (including hospitality) from pharmaceutical or other commercial companies, excluding reasonable fees for any work they do as part of a contractual arrangement with a company [15]. Moreover, providing low-value promotional aids (pens, pads, etc.) has long been a routine of pharmaceutical advertising, and international rules still permit inexpensive promotional aids, provided they are applicable to the practice of the healthcare provider [1].

VII. Conclusion

Instead of the classical concluding remarks, we would like to list the eight principles of the International Federation of Pharmaceutical Manufacturers and Associations (IFPMA) which, in our opinion, reflect the minimum standards of many worldwide codes regardig the pharmaceutical industry's ethical interaction during the drug marketing process [30]. The eight principles are as following: (1) The health-care and well-being of patients are the first priority of pharmaceutical companies; (2) Pharmaceutical companies shall comply with high standards of quality, safety, and efficacy as determined by regulatory authorities; (3) Pharmaceutical companies' interactions with stakeholders must always be ethical, appropriate, and professional; nothing should be offered or provided by a company in a manner or on conditions that would have an inappropriate influence; (4) Pharmaceutical companies are responsible for providing accurate, balanced, and scientifically-valid data on products; (5) Promotion must be ethical, accurate, balanced, and must not be misleading; information in promotional materials must ensure proper assessment of the risks and benefits of the product and its appropriate use; (6) Pharmaceutical companies shall respect the privacy and personal information of patients; (7) All clinical trials and scientific research sponsored or supported by companies will be conducted with the intent to develop knowledge that will benefit patients and advance science and medicine; pharmaceutical companies are committed to the transparency of industry-sponsored clinical trials in patients; and (8) Pharmaceutical companies should adhere to the applicable industry codes in both spirit and letter; to achieve this, pharmaceutical companies will ensure that all relevant personnel are appropriately trained.

Acknowledgements

The Ministry of Education and Science of the Republic of Serbia and the Fogarty International Centre at the U.S. National Institutes of Health (NIH) supported this article through grants 175073, and 2 R25 TW008171-06A1, respectively.

References

[1] J. Francer, J.Z. Izquierdo, T. Music, K. Narsai, C. Nikidis, H. Simmonds, P. Woods, "Ethical pharmaceutical promotion and communications worldwide: codes and regulations," *Philos. Ethics Humanit. Med.,* vol. 9, Mar. 2014. doi:10.1186/1747-5341-9-7.

[2] "Ethical rules for the pharmaceutical industry in Sweden," Swedish Association of the Pharmaceutical Industry (LIF); 2016, https://www.lif.se/contentassets/dba784d2b9554f1eb404812ba33305a8/1er-english-20160302.pdf. Accessed on September 14, 2018.

[3] R.L. Kravitz, R.A. Bell, "Media, messages, and medication: strategies to reconcile what patients hear, what they want, and what they need from medications," *BMC Med. Inform. Decis. Mak.,* vol. 13, suppl. 3, 2013. doi: 10.1186/1472-6947-13-S3-S5.

[4] P. Salari, B. Larijani, "Ethical Issues Surrounding Personalized Medicine: A Literature Review," *Acta Med. Iran.*, vol. 55, no. 3, pp. 209-217, Mar. 2017.

[5] R. Sloane, O. Osanlou, D. Lewis, D. Bollegala, S. Maskell, M. Pirmohamed, "Social media and pharmacovigilance: A review of the opportunities and challenges," *Br. J. Clin. Pharmacol.*, vol. 80, no. 4, pp. 910-920, Oct. 2015. doi: 10.1111/bcp.12717.

[6] S. Kim, C. Mayhorn, M. Wogalter, "Use of information sources referenced in and attitudes about televised DTC prescription drug advertisements,"

Proceedings of the Human Factors and Ergonomic Society, 54th Annual Meeting, pp.1880-1884, 2010.

[7] A.Lyles, "Direct marketing of pharmaceuticals to consumers," *Annu. Rev. Public. Health*, vol. 23, pp. 73-91, 2002.

[8] American Medical Association, "Direct-to-Consumer Advertisement of Prescription Drugs. Code of Medical Ethics Opinion 9.6.7. 2018," https://www.ama-assn.org/delivering-care/direct-consumer-advertisement-prescription-drugs. (Accessed on September 15, 2018).

[9] M.B. Pinto, "On the nature and properties of appeals used in direct-to-consumer advertising of prescription drugs," *Psychol. Rep.*, vol. 86, no. 2, pp. 597-607, 2000.

[10] R.L. Norris, R.L. Bailey, P.D. Bolls, K.R. Wise, "Effects of emotional tone and visual complexity on processing health information in prescription drug advertising," *Health Commun.*, vol. 27, no. 1. pp. 42-48, 2012. doi: 10.1080/10410236.2011.567450.

[11] R.A. Bell, R.L. Kravitz, M.S. Wilkes, "Direct-to-consumer prescription drug advertising, 1989-1998. A content analysis of conditions, targets, inducements, and appeals," *J. Fam. Pract.*, vol. 49, no. 4, pp. 329-335, 2000.

[12] M. Gerteis, S. Edgman-Levitan, J.D. Walker, D.M. Stoke, P.D. Cleary, T.L. Delbanco, "What patients really want," *Health Manage Q.*, vol. 15, no. 3, pp. 2-6, 1993.

[13] R. Fadlallah, H. Nas, D. Naamani, F. El-Jardali, I. Hammoura, L. Al-Khaled, H. Brax, L. Kahale, E.A. Akl, "Knowledge, Beliefs and Attitudes of Patients and the General Public towards the Interactions of Physicians with the Pharmaceutical and the Device Industry: A Systematic Review," *PLoS One,* vol. 11, no. 8, pp. e0160540, Aug. 2016. doi: 10.1371/journal.pone.0160540.

[14] S. Green, "Ethics and the pharmaceutical industry," *Australas Psychiatry,* vol. 16, no. 3, pp. 158-165, Jun 2008. doi: 10.1080/10398560701842595.

[15] The Medical Council - Republic of Ireland, "Guide to Professional Conduct and Ethics for Registered Medical Practitioners. Relationships between doctors and industry," 2018. https://www.medicalcouncil.ie/News-and-Publications/Reports/Relationships-between-doctors-and-industry-FAQs-2018.pdf. (Accessed on September 15, 2018).

[16] P.J. Reitemeier, J.Ozuna, R.Taylor, J.Cook, E.Fox, "Gifts to Health Care Professionals from the Pharmaceutical Industry. A Report by the National Ethics Committee of the Veterans Health Administration," National Center for Ethics in Health Care Veterans Health Administration Department of Veterans Affairs, 2003. https://www.ethics.va.gov/docs/necrpts/NEC_Report_20031201_Gifts_From_Pharma_Industry.pdf. Accessed on September 15, 2018.

[17] S.L. Torvund SL, "Ethics in the international marketing of pharmaceuticals," Master's Thesis, University of Agder, 2013. https://brage.bibsys.no/xmlui/bitstream/handle/11250/135835/Torvund%2C%20Sondre%20Lislegard%20oppgave.pdf?sequence=1. (Accessed on September 15, 2018).
[18] L. Tran, "Drug Samples: Why Not?" *Virtual Mentor.*, vol. 16, no. 4, pp. 245-251, April 2014.
[19] B. Shaw, P. Whitney, "Ethics and compliance in global pharmaceutical industry marketing and promotion: The role of the IFPMA and self-regulation," *Pharmaceuticals Policy and Law*, vol. 18, pp. 199–206, 2016. doi: 10.3233/PPL-160443
[20] "Code of Ethics for Pharmaceutical Marketing Communication (A Gyógyszer-Kommunikáció Etikai Kódexe)", Budapest, 2014, http://www.egis.hu/wps/wcm/connect/egisEN/dd735640-0f0a-4074-8ec0-b8d0efa06be1/Code+of+Ethics+for+Pharmaceutical+Marketing+Communications.pdf?MOD=AJPERES; Accessed on September 14, 2018
[21] "Ethical rules for the pharmaceutical industry in Sweden," 2016, https://www.lif.se/contentassets/dba784d2b9554f1eb404812ba33305a8/1er-english-20160302.pdf, Accessed - September 14, 2018
[22] P. Manchanda, E. Honka, "The effects and role of direct-to-physician marketing in the pharmaceutical industry: an integrative review," *Yale J. Health Policy Law Ethics,* vol. 5, no. 2, pp. 785-822, 2005.
[23] "Code of Ethics for Pharmaceutical Marketing – Lithuania," Innovative Pharmaceutical Industry Association. Edition of 2015. http://ifpa.lt/wp-content/uploads/2014/09/1-HCP-Code-Lithuania-2015.pdf, Accessed: September 14, 2018
[24] "Pharma Industry Finland Code of Ethics," *Pharma Industry Finland (PIF)*, 2017. https://www.pif.fi/sites/default/files/attachments/lt_code_of_ethics_2017_netti_final_id_128083_0.pdf. Accessed on September 14, 2018.
[25] B. Mintzes, "Direct-to-Consumer Advertising of Prescription Drugs in Canada: What are the Public Health Implications?" *Health Council of Canada*, 2006. ISBN 0-9739725-6-4. http://publications.gc.ca/collections/collection_2007/hcc-ccs/H174-3-2006E.pdf. Accessed on September 15, 2018.
[26] B. Mintzes, "For and against: Direct to consumer advertising is medicalising normal human experience: For," *BMJ,* vol. 324, no. 7342, pp. 908-909, Apr. 2002.
[27] D. Egilman, N.M. Druar, "Spin your science into gold: direct to consumer marketing within social media platforms," *Work,* vol. 41, suppl. 1, pp. 4494-4502, 2012. doi: 10.3233/WOR-2012-0751-4494.

[28] W. Fleischman, S. Agrawal, M. King, A. K. Venkatesh, H. M. Krumholz, D. McKee, D. Brown, J. S. Ross, "Association between payments from manufacturers of pharmaceuticals to physicians and regional prescribing: cross sectional ecological study," *BMJ,* vol. 354, Aug. 2016. doi: 10.1136/bmj.i4189.
[29] American Medical Association, "Gifts to Physicians from Industry. Code of Medical Ethics Opinion 9.6.2," 2018. https://www.ama-assn.org/delivering-care/gifts-physicians-industry. (Accessed on September 15, 2018).
[30] IFPMA Code of Practice. International Federation of Pharmaceutical Manufacturers & Associations. 2012. https://www.ifpma.org/wp-content/uploads/2016/01/IFPMA_Code_of_Practice_2012_new_logo.pdf. (Accessed on September 16, 2018).

Disclosing Research Results to Participants: Is There a Consensus?

Adél Tóth

University of Szeged, Faculty of Medicine,
Department of Behavioural Sciences, Hungary

Abstract

Traditionally, research results have not been disclosed to participants during medical research. However, recently a debate on the ethical duty of disclosing research results to participants has emerged. In this review, three positions in the debate are outlined: first, the disclosure is deemed an ethical imperative, second, a prima facie duty, and third, a consideration only. Although the notions of respect for persons, beneficence, and reciprocity can provide strong arguments for accepting disclosure as a duty, no consensus has been concluded on this issue yet. In the lack of consensus, addressing the issue of disclosure is required from every research plan.

Keywords

Medical research; duty of disclosing research results to participants; principle of respect for persons; principle of beneficence, reciprocity.

I. Introduction

Prior to the initiation of every medical research, detailed information is provided to prospective participants about their rights, the purpose of the study, the procedures to be undergone, the potential risks and benefits of participation, the expected duration of the study, the confidentiality of personal data, the randomisation process, etc. to promote their entirely voluntary participation [1]. This informed consent process, embedded in the principles of the Nuremberg Code and The Declaration of Helsinki, is considered as one of the most important ethical requirements of biomedical research. While pre-trial information-giving primarily aims at supporting the autonomous decision-making of participants and ensuring safeguards for them during research, it also reveals important data on the research project itself.

In the light of this detailed, sometimes extremely voluminous, information-giving before research initiation, it can be surprising that after the completion of research, traditionally, researchers have not been expected to disclose information about the study results directly to the participants. The conventional ways of disseminating research results have included peer-reviewed journals, scientific meetings or even the lay media, but towards the study participants, the non-disclosure policy has been in effect for decades. Considerations about informing participants of research results have not comprised the part of ethical conduct during medical research.

However, in the recent medical ethics literature, a debate on the ethical duty of disclosing research results to participants has emerged. The need to breach the tradition of non-disclosure and to share research results with participants has been raised especially by cancer clinical trials, and by genetic experiments applying whole genome sequencing or exome sequencing technologies [2, 3]. These types of trials can generate results with direct consequences for the health status of the research participant, thus the non-disclosure of these results can hardly be justified. Over the years, the debate has extended, and nowadays the dispute focuses on the issue whether the disclosure of research results to participants should be assessed as a universal duty.

II. Conditions influencing the perspective of a debate on the duty of disclosing research results to participants

When studying the history of disclosing research results to participants, it is recognisable that the theoretical debate has partly been triggered by the problems occurring in the sporadic application of this idea. Seemingly, in the case of disclosing research results to participants, the steps of practical implementation have preceded the theoretical clarification.

In 2005, introducing the new disclosure policy was demanding for a research project named "North Central Cancer Treatment Group Study n9831," conducted in the United States, since the planned, interim analysis had revealed that the experimental treatment regimen for breast cancer was significantly more effective than the standard treatment. The research constituted a randomised, phase III clinical trial to study the effect of trastuzumab on disease-free survival of women with operable, HER-2 positive breast cancer, obtaining adjuvant chemotherapy [4]. When the interim analysis explored that the disease-free survival among women who had been randomised to chemotherapy plus trastuzumab showed statistically significant improvement compared to women randomised to chemotherapy alone, a decision was made to communicate the findings in writing to all participants. The patient letter included also the recommendation that women on the control arm should be offered trastuzumab if they had completed chemotherapy within six months [5].

In the trastuzumab trial, the decision to share research results with participants was entirely governed by the principle of beneficence, meaning that promoting the well-being of women whose treatment might change based on the research findings enjoyed preference. The disclosure decision was in accordance with the recommendation of the US National Consensus Conference on Clinical Trials in 2000, which suggested that providing the results to trial participants be considered the ethical norm [6]. As no routine procedure on disclosing results had previously been worked out, the information given to participants of trastuzumab trial focused essentially on one of the main goals of biomedicine, saving lives, thus the option of rejecting results was not offered to participants. However, a study conducted four years later to examine the effect of disclosure of results in the trastuzumab trial highlighted that not all participants wanted to get the research results. Some others mentioned that they would have valued to get the offer of disclosure first, and in this way, they could have made a decision about acceptance or rejection [5]. These findings have drawn the attention of researchers that other aspects than only the principle of beneficence should be taken into account through the disclosure process.

The perspective of the debate has further been widened by those theoretical articles that discuss the ethical problems of the disclosure of research results by using a framework different to the beneficence-based assessment. Conrad Fernandez, in his often-cited article, gives an interpretation of the issue [7], which goes far beyond the scope of the four principles of biomedical ethics (respect for autonomy, beneficence, non-maleficence and justice) [8], as he evaluates the lack of disclosure of results to research participants as "betrayal". In the hypothetical case history of Fernandez, a 33-year-old woman noticed a slowly growing mass in her breast. Medical examination confirmed she had breast cancer, which had formed metastases to lungs and bone. The patient had Hodgkin's disease at age 14 and received radiation therapy as part of medical research. After reaching her maturity, she had no further contact with the paediatric clinic where her follow-up had happened. Long-term follow-up studies showed that there was an approximately 30-fold increase in risk of breast cancer in those who had received radiation therapy in the research. The patient became upset and angry that she had never been informed of these findings and felt "betrayed" [7].

In the analysis of this case history, the author refers not only to the rational part of ethics e.g. that non-disclosure can cause harm and disseminating medical knowledge only among professionals may limit the maximisation of good produced by biomedical research, but to the emotional aspect of morality as well. By addressing non-disclosure as betrayal, the extremely vulnerable nature of trust between participants and researchers, and between patients and doctors is also highlighted. The inclusion of the aspects of emotion and social relationship into the debate is again questioned the exclusive relevance of the beneficence-based approach, which promotes the disclosure of only those clinical trial findings that can directly benefit the health of the participant. Owing to practical experiences and theoretical efforts, the scope of the debate has widened, and the question has been raised whether disclosing research results to participants is a universal duty.

III. Different positions in the debate on disclosure

The articles dealing with the problem of returning research results to participants embrace three large fields of biomedical research, such as fundamental, translational and clinical research. Researchers belonging to these divergent territories of scientific research are expected to take a stand on the issue whether similarly to informed consent prior to research, disclosure of results to participants after the completion of research should be regarded as an essential part of biomedical trials. The post-trial information can be provided in the form of aggregate research result, which is a summary of overall study findings drawn from the population of research participants, or in the form of individual research result, which is a research finding regarding the person herself. When the individual research result contains information having direct consequences on the health of the participant herself, it is required to provide it through a face-to-face communication [9, 10].

A. Disclosure as an ethical imperative

One position in the debate of disclosure contends that disclosing research results to participants should be considered as an ethical imperative, which derives from the principle of respect for persons. The most prominent representative of this view is the above-mentioned Conrad Fernandez, who believes that the disclosure of research results can uphold the requirement that participants should be treated as more than a means to a scientific end. By disclosure, it is possible to diminish the chance that participants may feel exploited during the research process [7]. As respect does not end with pre-trial informed consent, sharing results with participants comes from the commitment of showing respect for persons through the whole research process.

In this approach, the duty of disclosure is deemed universal also in the sense that not only investigators of clinical trials but all investigators are expected to offer to provide a summary of research results on completion of their study. Investigators owe a debt to research participants since, for example during clinical trials, participants are exposed to risks that they must assume; in genomic trials, participants provide tissues that possess their personal and private characteristics [11]. For their altruistic contribution, all participants deserve the possibility of getting information on research results because the disclosure can demonstrate reciprocity and effectively express appreciation [12]. Fernandez and his colleagues declare: "There are no conditions under which an offer of disclosure of research results should not be made" [7]. This opinion is objected by experts of medical genetics, who point out that fundamental research results in the field of genetics are not understandable or significant for participants since exploratory genetic factors have not yet reached the point of general clinical acceptance [3]. However, it should be mentioned that in this conception special attention is paid for those who cannot benefit directly from research (e.g. participants of fundamental research), claiming that the gesture of returning research results can help acknowledge their contribution to the development of scientific knowledge and can raise their self-esteem.

It is noteworthy that, in this concept, universal duty refers to the offer of disclosure of research results and not to the disclosure itself. Making a distinction between the offer of disclosure and the realisation of disclosure is demanded by the principle of respect for persons, which claims that the decision of participants about the rejection of getting results should be accepted. Thus, in this view, it is mandatory for the researcher to offer the disclosure of results, but it is not mandatory for the participant to accept this offer.

Besides the principled approach, utilitarian arguments can also justify the duty of disclosing research results as disclosure can provide benefit for the individual (increased self-esteem, improved health), for the community (raised public awareness of science), and for the researcher (enhanced public trust). Although the possible negative psychological, social and economic impacts of disclosure and the logistic problems (what, when, who, by whom to be informed) have been revealed, the supporters of this view assess these problems as pragmatic, which can be overcome by well-designed research projects. As David Resnik notes, the demand of respect for persons outweighs the obligations of preventing harm or saving financial resources. However, in well-designed research projects investigators are aware of the possibility of negative consequences and this awareness helps them to develop strategies to minimise risk. This precautionary approach includes counselling, psychological support and medical follow-up strategies [13].

B. Disclosure as a prima facie duty

Another approach, elucidated mainly by Annelien Bredenoord and her colleagues, accepts the disclosure of research results as a moral duty, but this duty is perceived as *prima facie*, i.e. a duty that must be fulfilled unless it conflicts with an equal or a stronger duty [14]. The idea of prima facie duty, elaborated by W. D. Ross, is not unknown in biomedicine as the four principles of biomedical ethics represent this type of obligation [8]. Prima facie duties are always binding if they constitute single duties, but in the case of their conflicts, it is possible to choose the most stringent among them. However, the infringement of a prima facie duty leaves guilty conscience. By interpreting disclosure as a prima facie duty, researchers can gain a middle position between the universal duty of disclosure and the total rejection of the duty to disclose, which makes them possible to adjust disclosure policy to the context [15].

To justify the duty of disclosure, this position alludes to the principles of respect for autonomy and beneficence. Respect for autonomy obligates researchers to return research data if participants wish to receive these. Respect for autonomy is defined in a broad sense, not only non-interference is included in its meaning but also the idea of fostering individuals' capacity of being capable of controlling their own lives and living according to their own values. It is recognised that if beneficence is completed with respect for autonomy, these principles together can give ground for the disclosure of wide range of research results, not only to data of immediate clinical relevance [14].

However, the richness of research results may cause difficulties for both participants and researchers. For example, selecting among the wide range of research findings can be too demanding for participants. They need counselling to make their decision and later they need psychological support or follow-up to cope with the awareness of research results. Result disclosure may have negative consequences and can cause harm for participants. Furthermore, the disclosure of huge amount of data can be too demanding for researchers as well since they should make efforts to communicate the findings in an understandable way, to help participants comprehend the information, and to promote the participants' adequate selection among research data. The practical problems of disclosing research results may impose unmanageable burden on researchers and on the research infrastructure. The financial, occupational, psychosocial burdens posed by the duty of disclosing wide range or the full spectrum of research results can be so enormous that it threatens the execution of the original aim, the conduction of research [14].

Perceiving disclosure of research results as a prima facie duty offers solution for the problem as the duty of disclosing results can be subordinated to the more stringent duty, to the duty of performing research, which constitutes the fundamental professional obligation of a researcher. First, it is possible to take into consideration the circumstances (capabilities of researchers, limits of financial resources, capacities of laboratories, etc.) and then make a selection about what research results to be disclosed [16]. Bredenoord notes that "In the ideal world, researchers would offer disclosure of all data if participants wish so, but in reality, individual benefits have to be balanced against societal benefits" [14]. Since offering full data disclosure is currently not realistic, in research programs where the disclosure policy is applied, generally life-saving data and data of immediate or potential clinical utility are communicated. Future contextual changes can make possible to widen the scope of result disclosure, e.g. to share data of reproductive, personal or recreational significance with participants.

C. Disclosure as a consideration

In the third type of view, the universal duty of disclosing research results to participants is questioned by researchers contending that respect for persons, beneficence, reciprocity and justice can be served by other means than disclosure. Some alternative actions are mentioned such as conducting formal ceremonies, arranging post-trial access to new medication, or planning experiments to study the special problems of the group from which participants have been recruited [17]. Based on the conviction that the above-mentioned principles cannot effectively justify disclosure as a universal duty, in this opposing view, disclosure is perceived only as a consideration [17-19].

A further element of criticism refers to the phenomenon that the ethically relevant distinction between research practice and clinical care may be blurred if the full disclosure or the individualised disclosure of research results are universally required. The disclosure practice can strengthen the therapeutic misconception, the mistaken belief that therapy and research are governed by the same primary purpose: to promote the best interest of the individual [20]. Therapeutic misconception leads to a major ethical problem in biomedical research because some participants join a research project with the false belief that it will benefit them personally. However, the goal of research is to create generalisable knowledge, and research does not focus on the optimal health care of individuals [21]. During research, participants can justifiably be exposed to risks, for example when the efficacy and side effects of an experimental therapy are not fully known, while during medical care, causing this type of risk is hardly acceptable. The therapeutic misconception can result in the overestimation of the clinical significance of research findings, which can lead to participants' disappointment and erosion of trust in research, namely, to the very contrary for which the disclosure of research results has been implemented.

Another difference between research and clinical care appears in the access of health data. While in clinical care, respect for autonomy mandates patients to have a high-level access to their own health data, in research, the participants' access to their health data explored by the researcher is not only influenced by the wishes of the participants but also by the special features of the research process itself. Since disclosure of results is regarded as a part of research, plans about what and how to disclose should be submitted to a research review board that can give an ethical approval for the outset of a research project. Thus, ethics review boards have the authority to control individuals' access to information about themselves, but this kind of authority cannot exist in relation to clinical care [22]. The divergent nature of research data, compared to health care data, has made profoundly important to determine what to disclose after the completion of research. Fiona Miller, a strong opponent of treating disclosure as universal duty, poses the more generic issue of what counts as a precise definition of research results. To eliminate the confusion about the notion of research results, she demands to make "clearer conceptual distinctions between aggregate and individual results, amongst different types of research, and across different degrees of result veracity". In her view, without a clear definition of what counts as a research result, no consensus can be reached on what should be disclosed [23].

Currently, various types of research results are communicated to participants e.g. aggregate or individual, targeted or incidental, peer-reviewed or non-peer-reviewed, clinically valid or scientifically valid, clinically actionable or clinically non-actionable, etc. The inconsistent practice and the lack of theoretical clarification have motivated some scholars to declare that the implementation of disclosing research results as a universal duty is premature [23, 24].

IV. Consequences of the lack of consensus - Concluding remarks

Although no consensus has been reached in the debate yet, owing to the rapidly increasing number of publications on the duty of disclosing research results to participants, it seems unavoidable to address the issue of disclosure in every research project. In research plans, where the disclosure policy is not accepted, good reasons should be given for its rejection.

The need to close the debate on the duty of disclosure and to begin to elaborate an adequate disclosure practice has been expressed by researchers [25]. Over the years, a strong agreement has been formed on the key elements of disclosure practice. The main points of agreement comprise that 1) the issue of disclosure should be involved in the pre-trial informed consent process; 2) disclosure should include a two-step procedure (offer plus disclosure) with the opportunity of refusal; 3) research review boards should make decisions about what to disclose; 4) sponsors and patient support groups should be involved in the implementation; 5) educational and psychosocial support may be necessary for participants [26, 27]. In sum, the avoidance of intuitive decision-making and the execution of careful planning are required from researchers through the implementation of disclosure.

The decade-long debate on disclosing research results has efficiently directed the attention of researchers to the importance of maintaining respect for participants through the whole research process. It has also helped scientists recognise the role of disclosing research results to participants in the process of democratising scientific knowledge. Although, the debate has remained unsettled, the aspects of disclosure cannot be ignored by researchers in the future.

References

[1] U.C. Gupta, "Informed Consent in Clinical Research: Revisiting Few Concepts and Areas," *Perspect. Clin. Research,* vol. 4, no. 1, pp. 26–32, Jan-Mar. 2013.

[2] C.V. Fernandez, E. Kodish, S. Shurin, C. Weijer, "Offering to Return Results to Research Participants: Attitudes and Needs of Principal Investigators in the Children's Oncology Group," *J. Pediatr. Hematol. Oncology,* vol. 25, no. 9, pp. 704-708, Sep. 2003.

[3] B.M. Knoppers, Y. Joly, Y. Simard, F. Durocher, "The Emergence of an Ethical Duty to Disclose Genetic Research Results: International Perspectives," *Eur. J. Hum. Genetics,* vol. 14, no. 11, pp. 1170–1178, Nov. 2006.

[4] E.H. Romond, E.A. Perez, J. Bryant et al, "Trastuzumab plus Adjuvant Chemotherapy for Operable HER2-Positive Breast Cancer," *N. Engl. J. Medicine,* vol. 353, no. 16, pp. 1673–1684, Oct. 2005.

[5] A.H. Partridge, A.G. Wolff, P.K. Marcom et al, "The Impact of Sharing Results of a Randomized Breast Cancer Clinical Trial with Study Participants," *Breast Cancer Res. Treatment,* vol. 115, no. 1, pp. 123–129, May 2009.

[6] A.H. Partridge, E.P. Winer, "Informing Clinical Trial Participants About Study Results," *JAMA,* vol. 288, no. 3, pp. 363-5, Jul. 2002.

[7] C.V. Fernandez, E. Kadish, C. Weiler, "Informing Study Participants of Research Results: An Ethical Imperative," *IRB Ethics Hum. Research,* vol. 25, no. 3, pp. 12-19, May-Jun. 2003.

[8] T.L. Beauchamp, J.F. Childress, *Principles of Biomedical Ethics*, 7th ed. New York: Oxford University Press; 2013

[9] D.I. Shalowitz, G. Franklin, F.G. Miller, "Communicating the Results of Clinical Research to Participants: Attitudes, Practices, and Future Directions," *PLoS Medicine*; vol. 5, no. 5, pp. 0714-0720, 2008. Available at https://journals.plos.org/plosmedicine/article?id=10.1371/journal.pmed.005009. Accessed: March 18, 2019.

[10] L.M. Beskow, W. Burke, S.M. Fullerton, R.R. Sharp, "Offering Aggregate Results to Participants in Genomic Research: Opportunities and Challenges," *Genetics in Medicine*; vol. 14, no. 4, pp, 490-496. Apr. 2012.

[11] S.N. Doernberg, D. Wendler, "Ensuring Respect for Human Research Participants. Institutional Review Boards and Sharing Results from Research," *JAMA,* vol. 316, no. 11, pp. 1149-1150, Sep. 2016.

[12] I. Petersen, R. Kollek, "The Symbolic Relevance of Feedback: Return and Disclosure of Genomic Research Results of Breast Cancer Patients in Belgium, Germany and the UK," *J. Clin. Res, Bioethics,* vol. 6, no. 4, pp. 1-7, Jul. 2015.

[13] D.B. Resnik, "Disclosure of Individualized Research Results: A Precautionary Approach," *Account Research;* vol. 18, no. 6, pp. 382–397, Nov. 2011.

[14] A.L. Bredenoord, N.C. Onland-Moret, J.J.M. Van Delden, "Feedback of Individual Genetic Results to Research Participants: In Favor of a Qualified Disclosure Policy," *Hum. Mutation,* vol. 32, no. 8, pp. 861–867, Nov. 2011.

[15] L.M. Beskow, W. Burke W, "Offering Individual Genetic Research Results: Context Matters," *Science Translat. Medicine,* vol. 2, no. 38, pp. 1-6, 2010. Available at http://stm.sciencemag.org/content/2/38/38cm20. Accessed: March 18, 2019.

[16] V. Ravitsky, B.S. Wilfond, "Disclosing Individual Genetic Results to Research Participants," *Am. J. Bioethics,* vol. 6, no. 6, pp. 8-17, Nov-Dec. 2006.

[17] P.N. Ossorio, "Letting the Genie Out of the Bottle: A Comment on Returning Individual Research Results to Participants," *Am. J. Bioethics,* vol. 6, no. 6, pp. 24–25, Nov-Dec. 2006.

[18] L. Meltzer, "Undesirable Consequences of Disclosing Individual Genetic Results to Research Participants," *Am J Bioethics,* vol. 6, no. 6, pp. 28-29, Nov-Dec. 2006.

[19] L.S. Parker, "Best Laid Plans for Offering Results Go Awry," *Am. J. Bioethics,* vol. 6, no. 6, pp. 22-23, Nov-Dec. 2006.

[20] E.W. Clayton, L.F. Ross, "Implications of Disclosing Individual Results of Clinical Research," *JAMA,* vol. 295, no. 1, p. 37, Jan. 2006.

[21] F. Lemaire, "Patient Care Versus Research: Does Clinical Research Provide Individual Benefit to Patients Enrolled in Trials?" *Curr. Opin. Crit. Care,* vol. 10, no. 6, pp. 565-569, Dec. 2004.

[22] W. Burke, B. J. Evans, G.P. Jarvik, "Return of Results: Ethical and Legal Distinctions between Research and Clinical Care," *Am. J. Med. Genet. Semin. Med. Genetics,* vol. 166C. no. 1. pp. 105–111, Mar. 2014.

[23] F.A. Miller, R. Christensen, R.M. Giacomini, S. Robert, "Duty to Disclose What? Querying the Putative Obligation to Return Research Results to Participants," *J. Med. Ethics,* vol. 34, no. 3, pp. 210–213, Mar. 2008.

[24] A.L. McGuire, J.R. Lupski, "Personal Genome Research: What Should the Participant Be Told?" *Trends in Genetics,* vol. 26, no. 5, pp. 199-201, May 2010.

[25] A.N. Bredenoord, H.Y. Kroes, E. Cuppen, M. Parker, J. J. M. Van Delden, "Disclosure of Individual Genetic Data to Research Participants: The Debate Reconsidered," *Trends in Genetics,* vol. 27, no. 2, pp. 41-47, Feb. 2011.

[26] A.H. Partridge, E.P. Winer, "Sharing Study Results with Trial Participants: Time for Action" *J. Clin. Oncology,* vol. 27, no. 6, pp. 838-839, Feb. 2009.

[27] L.G. Dressler, "Disclosure of Research Results from Cancer Genomic Studies: State of the Science," *Clin. Cancer Research,* vol. 15, no. 13, pp. 4270-4276, Jul. 2009.

FSC
www.fsc.org
MIX
Papier | Fördert gute Waldnutzung
FSC® C083411

Kolibri 360 GmbH
Ferdinand-Jühlke-Straße 7
99095 Erfurt, Deutschland
produktsicherheit@kolibri360.de